Collaborative Governance of Tropical Landscapes

*Edited by Carol J. Pierce Colfer
and Jean-Laurent Pfund*

London • Washington, DC

First published in 2011 by Earthscan

Copyright © Center for International Forestry Research (CIFOR), 2011

All rights reserved. No part of this publication may be reproduced, stored in a retrieval system, or transmitted, in any form or by any means, electronic, mechanical, photocopying, recording or otherwise, except as expressly permitted by law, without the prior, written permission of the publisher.

Earthscan Ltd, Dunstan House, 14a St Cross Street, London, EC1N 8XA, UK
Earthscan LLC, 1616 P Street, NW, Washington, DC 20036, USA

Earthscan publishes in association with the International Institute for Environment and Development

For more information on Earthscan publications, see www.earthscan.co.uk or write to earthinfo@earthscan.co.uk

ISBN: 978-1-84971-177-7

Typeset by MapSet Ltd, Gateshead, UK
Cover design by Susanne Harris

A catalogue record for this book is available from the British Library

Library of Congress Cataloging-in-Publication Data

Collaborative governance of tropical landscapes / edited by Carol J. Pierce Colfer and Jean-Laurent Pfund.
    p. cm.
  Includes bibliographical references and index.
  ISBN 978-1-84971-177-7 (hardback : alk. paper) 1. Forest management—Tropics. 2. Landscape protection—Tropics. 3. Sustainable forestry—Tropics. I. Colfer, Carol J. Pierce. II. Pfund, Jean-Laurent.
  SD247.C635 2010
  333.70913—dc22

2010028009

At Earthscan we strive to minimize our environmental impacts and carbon footprint through reducing waste, recycling and offsetting our CO$_2$ emissions, including those created through publication of this book. For more details of our environmental policy, see www.earthscan.co.uk.

This book was printed in the UK by TJ International, an ISO 14001 accredited company. The paper used is FSC certified and the inks are vegetable based.

Contents

List of Figures, Tables and Boxes   *v*
Contributors   *viii*
Acknowledgements   *x*
Abbreviations   *xiii*

1   An Introduction to Five Tropical Landscapes, their People and their Governance   1
Carol J. Pierce Colfer and Jean-Laurent Pfund, with Etienne Andriamampandry, Stella Asaha, Amandine Boucard, Manuel Boissière, Laurène Feintrenie, Verina Ingram, Emmanuel Lyimo, Endri Martini, Salla Rantala, Michelle Roberts, Terry Sunderland, Zora Lea Urech, Heini Vihemäki, Vongvilai Vongkhamsao and John Daniel Watts

2   The Governance of Tropical Forested Landscapes   35
John Daniel Watts with Carol J. Pierce Colfer

3   Role of the District Government in Directing Landscape Dynamics and People's Futures: Lessons Learnt from Bungo District, in Jambi Province   55
Laurène Feintrenie and Endri Martini

4   Information Flows, Decision-making and Social Acceptability in Displacement Processes   79
John Daniel Watts, Heini Vihemäki, Manuel Boissière and Salla Rantala

5   Changing Landscapes, Transforming Institutions: Local Management of Natural Resources in the East Usambara Mountains, Tanzania   107
Salla Rantala and Emmanuel Lyimo

6   Traditional Use of Forest Fragments in Manompana, Madagascar   133
    *Zora Lea Urech, Mihajamanana Rabenilalana, Jean-Pierre Sorg
    and Hans Rudolf Felber*

7   The Role of Wild Species in the Governance of Tropical
    Forested Landscapes   157
    *Bronwen Powell, John Daniel Watts, Stella Asaha,
    Amandine Boucard, Laurène Feintrenie, Emmanuel Lyimo,
    Jacqueline Sunderland-Groves and Zora Lea Urech*

8   Governance and NTFP Chains in the Takamanda-Mone
    Landscape, Cameroon   183
    *Verina Ingram, Stella Asaha, Terry Sunderland and
    Alexander Tajoacha*

9   A Dozen Indicators for Assessing Governance in Forested
    Landscapes   217
    Carol J. Pierce Colfer and Laurène Feintrenie

10 Minefields in Collaborative Governance   233
    *Carol J. Pierce Colfer, with Etienne Andriamampandry,
    Stella Asaha, Imam Basuki, Amandine Boucard,
    Laurène Feintrenie, Verina Ingram, Michelle Roberts,
    Terry Sunderland and Zora Lea Urech*

11 The Essential Task of 'Muddling Through' to Better
    Landscape Governance   271
    Carol J. Pierce Colfer, Jean-Laurent Pfund and Terry Sunderland

Index   279

List of Figures, Tables and Boxes

Figures

1.1	Landscape Mosaics project sites	4
1.2	Landscape Mosaics project sites in Takamanda-Mone Technical Operations Unit, Cameroon	10
1.3	Landscape Mosaics project sites in Bungo District, Jambi Province, Indonesia	12
1.4	Landscape Mosaics project sites in Laos	15
1.5	Landscape Mosaics project sites in Madagascar	18
1.6	Landscape Mosaics project sites in Tanzania	21
3.1	Land cover in Bungo District, 2005	57
3.2	Deforestation in Bungo District, 1973–2007	58
3.3	Election system of Indonesia	61
3.4	Organization of government in Indonesia	62
3.5	Bungo District forestry office activity, 2000–2008	69
3.6	Process of application for authorization to plant	72
4.1	Phadeng and Phoukhong villages, including proposed village sites	84
4.2	Location of Derema Corridor and study villages in East Usambaras	93
6.1	Sustainable livelihood approach	136
6.2	Research area, Madagascar	138
6.3	The use of Pandanaceae	146
6.4	Agreement with legal prohibition of *tavy* on natural forest	147
6.5	Problems with current forest management	148
6.6	Timber conflicts	149
7.1	Benjamin Njiku showing an immature fruit on a *msambu* tree on his farm in Shambangeda	166
7.2	Truck loaded with *eru*	172
7.3	Boys looking for fish	176
8.1	Takamanda-Mone Technical Operations Unit, Southwest Region, Cameroon	185
8.2	Flow diagrams of Takamanda-Mone NTFP market chains, values and profit margins, in CFA per kg, 2007	189
8.3	Distribution of average profit and percentage profit margins in 2007 among surveyed NTFP market chains in Takamanda-Mone	194

8.4 Production cost, selling prices and profit margins from bush mango in Takamanda-Mone, 2008 — 195
8.5 Average profit of NTFPs per trader in surveyed Takamanda-Mone markets, 2008 — 196
10.1 Years and levels of decentralization in Landscape Mosaics countries — 238
10.2 Governance indicators in five countries, 1996–2008 — 241

Tables

1.1 Landscape Mosaics sites and administrative contexts — 7
1.2 Landscape Mosaics site contexts – policy, agriculture, markets — 8
1.3 Human Development Index and rankings for Landscape Mosaics countries, 1990 and 2009 — 9
1.4 Population statistics for Landscape Mosaics sites — 9
2.1 Area of Landscape Mosaics sites — 38
3.1 Stakeholders' perceptions of governance in Bungo District — 65
4.1 Results from household perception surveys in Phadeng — 88
4.2 Land uses in East Usambaras, 1993 — 91
4.3 Communication with affected people in two displacement processes — 100
5.1 Laws and policy processes in decentralization of land, forest and tree management in mainland Tanzania — 111
5.2 Forest uses according to village bylaws in Kwatango and Misalai — 123
6.1 Characteristics of four field research villages, Madagascar — 138
8.1 Major NTFPs traded from Takamanda-Mone — 187
8.2 Legal classifications of Cameroon forests — 199
9.1 Governance assessment tool (GAT) — 222
9.2 Questionnaire for Jambi respondents on perceptions of village or region — 225
9.3 Questionnaire for Jambi respondents on rights, power and participation — 226
9.4 Synthesis of results on people's perceptions of governance (from first part of questionnaire) — 228
9.5 Synthesis results on rights (6) and access (3) of stakeholders (from second part of questionnaire) — 229
10.1 National laws relating to decentralization in five countries — 237
10.2 Governance indicators in five countries, 1996–2008 — 240
10.3 Perception of corruption rankings and indices in five countries, 2008 and 2009 — 241
11.1 Significant issues in collaborative landscape governance — 272

Boxes

6.1	The origin of an *ala fady*	141
6.2	Example of change in a *fady*	143
6.3	How the Drongo bird became a *sandrana*	144

Contributors

Etienne Andriamampandry, Association Intercooperation Madagascar, Antananarivo, Madagascar

Stella Asaha, Forests, Resources and People, Limbe, Cameroon

Imam Basuki, CIFOR, Bogor, Indonesia

Manuel Boissière, CIFOR, Bogor, Indonesia; CIRAD, Montpellier, France

Amandine Boucard, CIFOR, Viengkham, Laos

Carol J. Pierce Colfer, CIFOR, Bogor, Indonesia; Cornell University, Ithaca, New York, USA

Laurène Feintrenie, CIFOR; Institut de Recherche pour le Développement, University of Paul Valéry-Montpellier, Montpellier, France

Hans Rudolf Felber, NADEL (Nachdiplomstudium für Entwicklungsländer), ETH Zurich, Switzerland

Verina Ingram, CIFOR, Yaoundé, Cameroon

Emmanuel Lyimo, Tanzania Forest Conservation Group, East Usambaras, Tanzania

Endri Martini, World Agroforestry Center, Bogor, Indonesia; University of Hawaii, Honolulu, Hawaii, USA

Jean-Laurent Pfund, CIFOR, Bogor, Indonesia

Bronwen Powell, McGill University, Montreal, Canada; Bioversity International, East Usambaras, Tanzania

Mihajamanana Rabenilalana, Département des Eaux et Forêt, Université d'Antananarivo, Manompana, Madagascar

Salla Rantala, University of Helsinki, Finland; World Agroforestry Center, Nairobi, Kenya and East Usambaras, Tanzania

Michelle Roberts, University of Nevada, Reno, Nevada, USA and Viengkham, Laos

Jean-Pierre Sorg, Groupe de Foresterie pour le Développement, ETH, Zurich, Switzerland

Terry Sunderland, CIFOR, Bogor, Indonesia

Jacqueline L. Sunderland-Groves, Bogor, Indonesia

Alexander Tajoacha, Forests, Resources and People, Limbe, Cameroon

Zora Lea Urech, Groupe de Foresterie pour le Développement, ETH, Zurich, Switzerland and Manompana, Madagascar

Heini Vihemäki, University of Helsinki, Finland; World Agroforestry Center, Nairobi, Kenya and East Usambaras, Tanzania

Vongvilai Vongkhamsao, National Agriculture and Forestry Research Institute, Vientiane, Laos

John Daniel Watts, CIFOR, Bogor, Indonesia and Viengkham, Laos

Acknowledgements

Trying to acknowledge all the people who have contributed to this book is a mind-boggling task. However, we definitely need to make the effort, inadequate though it will inevitably be! Without our donors, we could not have begun the Landscape Mosaics (LM) project, so we give sincere thanks to the Swiss Agency for Development and Cooperation (SDC), especially to the project supervisor, Ueli Mauderli, and to the World Agroforestry Center (ICRAF). The European Union funded a crucial early international workshop in Lombok, Indonesia, in December 2007. The Japanese International Cooperation Agency (JICA) funded an equally important writing workshop in the fall of 2009, giving many of our authors an opportunity to focus on their chapters.

Within the Center for International Forestry Research (CIFOR), we are particularly thankful for the support of various supervisors, including Robert Nasi (Director, Forests and Environmental Services Programme and Leader of the Sustainable Forest Management Domain), Patrice Levang (Acting Director, Forests and Livelihoods Programme, and LM Theme Leader, Livelihoods), Elena Petkova (Director, Forests and Governance Programme), Terry Sunderland (Leader, Conservation and Development Domain, and Site Coordinator, Cameroon), Cyrie Sendashonga (Regional Coordinator, Cameroon) and of course Jean-Laurent Pfund (Coordinator of the LM project, from which this book derives).

Within ICRAF, we are grateful to Meine van Noordwijk (Researcher, ICRAF, and LM Theme Leader, Environmental Sciences), Brent Swallow (initial LM Theme Leader, Incentives), and Peter Minang (final LM Theme Leader, Incentives).

Our efforts were also supported by a number of junior professional officers (JPOs), including Nathalie van Vliet (supported by The Netherlands), who conducted much of the LM work in Cameroon. Salla Rantala worked with us as both a Finnish JPO and a doctoral student with the University of Helsinki. She was followed by Dr. Heini Vihemäki, another Finnish JPO, whose doctoral research had also taken place in our research areas. Both were financially supported by the Ministry for Foreign Affairs of Finland through ICRAF. Four other doctoral students supported our work:

- Bronwen Powell, from McGill University, Montreal, worked in cooperation with the World Vegetable Center (AVRDC), Bioversity International and the International Development Research Centre, Canada, in Tanzania.
- Mihaja Rabenilanana, at ESSA-Forêts of the University of Antananarivo, and Zora Lea Urech at the Groupe de Foresterie pour le Développement (Forestry Group for Development), ETH Zurich, contributed to our efforts in Madagascar.
- Michelle Roberts from the University of Nevada, Reno, worked with us in Laos and was funded by the U.S. National Science Foundation and Rotary International.
- Laurène Feintrenie began with us as a doctoral student in geography from Montpellier III in France, with the support of the French Institute for Development (IRD), later joining CIFOR as a consultant, primarily in Indonesia.

Amandine Boucard served as a junior advisor to the project.

Another major group we need to acknowledge and thank is our host-country partners. These include (but are not limited to) the following:

In Cameroon, we thank especially those in the communities of Assam, Okpambe and Mukonyong; also our partners in the Program for the Sustainable Management of Natural Resources–Southwest, notably the Ministry of Forestry and Wildlife for the Southwest Region, the Ministry of Scientific and Technical Research, the Wildlife Conservation Society, German Technical Cooperation and German Development Bank. We are immensely grateful for the dedication of our collaborating field team from Forests, Resources and People (FOREP), especially Stella Asaha and Alexander Atabong. Daniel Slayback of SSAI/NASA Goddard Space Flight Center also did an excellent job of the land cover change study for Cameroon, which underpins many of our findings. We also thank Jacqueline Sunderland-Groves for extensive voluntary contributions to our efforts, based on her long-term knowledge of the research area and its wildlife.

In Indonesia, we are indebted to the villagers of Desa Danau, Tebing Tinggi and Lubuk Beringin; to Bungo District's government officials, ICRAF's Bungo staff, and the Forest Governance and Learning Group (FGLG)/Bungo team. We also benefited from the contributions of the following masters students: Xavier Bonnart, from the Institut des Régions Chaudes, University of Paul Valéry-Montpellier III, Montpellier, France; Chong Wan Kian, from AgrisMundus Sustainable Development in Agriculture, Montpellier; Clara Therville and Ameline Lehébel-Peron, both from the University of Technology and Sciences in Montpellier, all supervised by Patrice Levang and Laurène Feintrenie.

In Laos, we thank first the villagers of Phadeng, Bouammi, Vangmat, Muangmuay and Phoukhong for their assistance and patience. We also thank our partners at the National Agriculture and Forestry Research Institute (NAFRI), Vongvilai Vongkhamsao, Khamphay Manivong, Phouthone Sophathilath, Vilaphong Kanyasone, Saiyasith Phonpakdy, Khamsao

Mouaxeng-Cha, and colleagues at the Northern Agriculture and Forestry Research Center (NAFReC), Manithaythip Thephavanh from the Policy Research Center (PRC), Bounthan Manivong and district-level government officials, without whom our research could not have occurred. Yulia Rahma Fitriana, a masters student at the Institut des Régions Chaudes-SupAgro, Montpellier, France, provided the project with important baseline information on the Laotian sites. Jean-Christophe Castella supported the scientific coordination.

Our central partner in Madagascar was AIM (Association Intercooperation Madagascar) with Lina Raharisoavelohanta as director, and Onésime Rapanoel and later Etienne Gershom Andriamampandry as coordinators. Other important players included the Communes Rurales de Manompana and Ambahoabe, all the staff of the EU-funded KoloAla Manompana project (Jean Romuale Randrianarisoa, Ruffin Maheva, Rivo Raminosoa and Andriantsialonina Nomenjanahary Andriamanandratra), the University of Antananarivo, ESSA-Forêts (Forest Department of the Agricultural Sciences Superior School) and finally the delegation of Intercooperation, especially Annette Kolff.

In Tanzania, we worked most closely with the people of Kwatango, Misalai and Shambangeda. We are indebted to our primary partners, the Tanzania Forest Conservation Group (TFCG), especially the dedicated field coordination by Emmanuel Lyimo and backstopping by Nike Doggart and Charles Meshack, and the district of Muheza. The WWF Tanzania Programme Office and Amani Nature Reserve provided valuable support. We would also like to recognize the contributions of Renee Bullock, Jaclyn Hall, Joel Meliyo, Abduel Kajiru, Felix Jovit and Muheza District personnel.

This listing would be even more inadequate without acknowledgement of the various people who have contributed to the production of the book itself: Sally Atwater was eternally patient, flexible and careful in her technical editing; Tim Hardwick originally negotiated the publication of the book with us and has been patient and supportive as well with our stretched deadlines; Hamish Ironside has worked hard to produce the proofs; and Rachel Butler and Anna Rice have been persistent in publicizing the work. At CIFOR, we thank the staff of the Information Services Groups, particularly Edith Johnson and Gideon Soeharyanto. We are grateful to all.

Finally, and most important, we would like to thank the individuals living in the landscapes we have analysed for their kindness and hospitality. They have accepted our questions, observations and interest, routinely helping us to accomplish what we set out to do, almost always with good humour. We hope that this book will contribute to better understanding of their situations and more effective policies that can lead to healthier environments and fuller lives for the people.

Abbreviations

ACM	Adaptive Collaborative Management
AHP	analytic hierarchy process
AIM	Association Intercooperation Madagascar
APL	*areah pemanfaatan lain* (nonforestry use areas) (Indonesia)
ASB	Alternatives to Slash and Burn
AVRDC	World Vegetable Center
BAPPEDA	Department of Regional Planning (Indonesia)
BPD	Badan Permusyawarahan Desa, village council (Indonesia)
CAPRi	Collective Action and Property Rights
CATIE	Centro Agronómico Tropical de Investigación y Enseñanza
CBFM	Community Based Forest Management
CCM	Chama Cha Mapinduzi (Party of the Revolution) (Tanzania)
CEESP	Commission on Environmental, Economic and Social Policy
CIFOR	Center for International Forestry Research
CIRAD	Centre International de Recherche en Agronomie du Développement
CMWG	Collaborative Management Working Group
COBA	Communauté de Base (Madagascar)
DPD	Dewan Perwakilan Daerah (Regional Representative Council) (Indonesia)
DPR	Dewan Perwakilan Rakyat (People's Representative Council) (Indonesia)
DPRD	Dewan Perwakilan Rakyat Daerah (Independent Representative Council) (Indonesia)
EUCAMP	East Usambara Conservation Area Management Programme (Tanzania)
EUCFP	East Usambara Catchment Forest Project
FAO	Food and Agriculture Organization of the United Nations
FBD	Forestry and Beekeeping Division (Tanzania)
FGLG	Forest Governance and Learning Group
FLEGT	Forest Law Enforcement, Governance and Trade
FOREP	Forests, Resources and People
GCF	La Gestion Contractualisée des Forêts (Contract Management of the Forests)

GELOSE	La Gestion Locale Sécurisée (Local Securised Management)
ICDP	Integrated Conservation and Development Programme
ICRAF	World Agroforestry Center
IFPRI	International Food Policy Research Institute
IIED	International Institute for Environment and Development
IPKR	Izin Pemanfaatan Kayu Rakyat (privately owned forest) (Indonesia)
IRD	Institut de Recherche pour le Développement
ISHS	International Society for Horticultural Science
IUCN	International Union for Conservation of Nature
JICA	Japanese International Cooperation Agency
JPO	junior professional officer
KENRIK	Kenya Resource Centre for Indigenous Knowledge
LFA	Land and Forest Allocation (Laos)
LM	Landscape Mosaics project
MAF	Ministry of Agriculture and Forestry (Laos)
MinFoF	Ministry of Forestry and Wildlife (Cameroon)
MNRT	Ministry of Natural Resources and Tourism (Tanzania)
MPR	Majelis Permusyawaratan Rakyat (People's Consultative Assembly) (Indonesia)
NAFReC	Northern Agriculture and Forestry Research Center (Laos)
NAFRI	National Agriculture and Forestry Research Institute (Laos)
NES	Nucleus Estates and Smallholders project (Indonesia)
NGO	non-governmental organization
NTFP	non-timber forest product
ODI	Overseas Development Institute
PAR	participatory action research
PES	payments for environmental services
PIR	Perkebunan Inti Rakyat (People's Nucleus Plantation) (Indonesia)
REDD	Reducing Emissions from Deforestation and Forest Degradation
REPOA	Research on Poverty Alleviation
RUPES	Rewarding Upland Poor for Environmental Services
SDC	Swiss Agency for Development and Cooperation
SIA	social impact assessment
SRAP	Smallholder Rubber Agroforestry Project
STEA	Science Technology and Environment Agency (Laos)
TFCG	Tanzania Forest Conservation Group
TOU	technical operations unit (Cameroon)
UNDP	United Nations Development Programme
URT	United Republic of Tanzania
USAID	United States Agency for International Development
UXO	unexploded ordinance
WARSI	Warung Konservasi (a conservation network) (Indonesia)

WGI	Worldwide Governance Indicators
WRI	World Resources Institute
WWF	World Wildlife Fund

1
An Introduction to Five Tropical Landscapes, their People and their Governance

*Carol J. Pierce Colfer and Jean-Laurent Pfund,
with Etienne Andriamampandry, Stella Asaha,
Amandine Boucard, Manuel Boissière, Laurène Feintrenie,
Verina Ingram, Emmanuel Lyimo, Endri Martini,
Salla Rantala, Michelle Roberts, Terry Sunderland,
Zora Lea Urech, Heini Vihemäki, Vongvilai Vongkhamsao
and John Daniel Watts*

As one floats down a river in the tropics, the diversity of plant and animal life assaults the senses. On the island of Borneo, in Indonesia, for instance, one encounters upland rice fields, which show up as cleared patches dotting the hillsides; in one field, a lone woman bends low to weed. There may be small patches of paddy rice along the river's edge. The forest itself is often a patchwork of varying stages of regrowth – new fallows thick with weeds, medicinal plants and perennial crops planted with upland rice the previous year, older fallows replete with fruit trees and various edible leaves, palm hearts and tubers, and still older fallows full of timber and fibres. Other patches are covered in old-growth forest, known for its lush foliage and abundant wildlife. From time to time one spots an orchard, but not a tidy, western-style orchard with its cleared underbrush and straight lines. What emerges is a chaotic profusion of greenery, identifiable as an orchard only by the comparative abundance of rubber or durian (*Durio* spp.) or duku (*Lansium domesticum*) trees.

Travelling from the highlands to the east coast of central Madagascar, one can also see a patchwork of regrowth and rice fields in the mountainous areas, blackened where fields have been recently burnt, dark green where the rice is in full bloom, or golden when it is ready for harvesting. The surrounding fallow areas also hold medicinal and other useful plants. Here, the last remnants of the forest massif can be seen, and people often use wild substitutes in degraded areas for the products they had once gathered from the forest.

After a week of collecting bush mangos in the Cameroonian 'bush' (a luxuriant forest that is a habitat for gorillas), away from the national park and the timber concession, a local woman has a long trek home. The woman may wade up to her armpits through the brown waters of Takamanda-Mone's many rivers, with a head load of kernels, a baby in her arms, a rattan rucksack full of eru vines, and the rest of her family tagging along behind. The scent of cocoa flowers provides a welcome to the village. Eventually, after she brushes past manioc and cocoyam leaves, the mud-walled huts plastered with drying bush mangos come into view.

In Borneo, as in Madagascar and Tanzania, such farming systems, which made ecological sense when population levels and external pressures on land were low, are now being transformed (see Mertz et al, 2009 on Southeast Asia). The sustainability of the country's forests and people's sources of sustenance are in question, and population growth and the expanding demand for agricultural land (for commercial and biofuel crops, as well as food) have raised concerns from the local to the global level. So far, 'solutions' have caused major problems for local peoples and for forests; the work described here has sought solutions to these, many of which are governance problems.

Observing such changes – and their variable human and ecological effects – led us to try to understand landscape dynamics more diachronically. We recognized that the relative proportion of forests, agroforests, swiddens and permanent farms depends on multiple influences, nowadays increasingly removed from local forest and human systems. In an attempt to understand the interactions among people, trees and forests over time, scientists from various backgrounds analysed five sites of high biodiversity value, situated in diverse contexts. To conduct the research, we developed a transdisciplinary project at the landscape scale to reflect variations in human use, as well as the dynamism of tropical vegetation. The landscape scale enables a holistic consideration of social–ecological systems, but also reveals pathways for local action at the landscape level, focusing on the links among decentralized governance systems,[1] actors and extant management of natural resources (Görg, 2007).

For the project as a whole (as distinct from the project's governance theme *per se*), we imagined a structured, if demanding, progression on all five sites:

- from collaborative work at community and district levels, using participatory action research (PAR), complemented by standardized ecological and socio-economic studies;

- to a holistic understanding of the local context, both human and biophysical;
- to a shared process of multistakeholder problem solving;
- concluding with plans of action and agreements among stakeholders regarding the governance and management of landscapes.[2]

The reality was, of course, less tidy. Challenges to conducting PAR in its theoretical form (Colfer et al, in press) included differing interpretations of the approach by the teams, incompatibilities with externally imposed timelines, difficulties getting permissions, logistical constraints and more (see Chapter 11). Progress occurred at different speeds from site to site; different disciplinary backgrounds compounded communication difficulties among researchers at far-flung sites representing a variety of cultures. Such problems hindered collaboration and made gathering comparative data more difficult than we had expected. Relations with local and national government and partners also differed across sites, affecting how shared problem solving and the PAR process evolved.

The results we present here differ from those we had initially envisioned: we were aiming to go beyond the planning stage and to engage in action. We expected to support the development of incentive mechanisms and landscape-level agreements that would point to the conduct and decisions of individuals and groups – those whose actions partly shape the futures of these five landscapes – with the aim of achieving greater sustainability. However, the findings we report may in fact prove to be more widely applicable because our analyses resonate far beyond the countries discussed. The relevance of the issues – swidden agriculture, resettlement, customary governance and forest use, government policies and corruption – has only increased since this work began in 2006 (see German et al, 2009; Brito et al, 2009; Ghazoul et al, 2009).

Climate change has shifted from the preoccupation of a few to a major global concern; and funding related to Reducing Emissions from Deforestation and Forest Degradation (REDD) has put forests and forest peoples at the front and centre of the debate. REDD has opened up enormous possibilities and potential risks for peoples living in landscapes such as those examined here. The Rights and Resources Initiative (2010) identified particular problems that hold relevance for our sites, including land grabs, usurpation of forest peoples' customary claims to forests by powerful parties, corruption and illegitimate claims to ownership of the new commodity, carbon.

Ultimately, for ethical and practical reasons, approaches must be found that address the well-being of the people who live in the forests and the forests themselves – these are intimately interconnected. Such genuinely 'wicked problems'[3] in tropical landscape governance form the subject matter of this book. In addressing people's and forests' well-being at a landscape scale, we raise as many questions as we answer, and we clearly identify critical issues. Many issues require action at higher levels. However, we remain convinced of the importance of mobilizing forest peoples to build on their own self-governance, cultures, environmental knowledge, self-respect and self-determination

– in collaboration with other actors – as counterweights to external power, in efforts to govern landscapes more effectively (also proposed in Locatelli et al, 2008).

Landscape Mosaics Project Design

The analyses herein are based on a five-country study, 'Integrating Livelihoods and Multiple Biodiversity Values in Landscape Mosaics' (Pfund, 2007), conducted between 2007 and 2010, in Cameroon, Indonesia, Laos, Madagascar and Tanzania (Figure 1.1).[4] The study included four research themes – livelihoods, landscape patterns, governance and (potential) rewards and incentives – with the studies reported here contributing most to the governance theme.[5] The project was designed to include PAR at both community and landscape levels (often the district level, in governance parlance), a series of quantitative cross-site studies and a component to develop a landscape-level livelihood monitoring system (particularly in Laos and Cameroon), in collaboration with another project with Royal Roads University, Canada.

Fundamentally, the research design for the overall project involved the identification of three communities and their territories in each country. These three communities were selected along a continuum of remote to comparatively accessible, with definitions varying significantly from country to country. We anticipated, and often found, a considerable degree of variation along this continuum. We imagined that the more remote communities would be in better condition, from an environmental perspective, and more dependent on their own customary governance structures and practices – in both cases, this was true to some extent. We anticipated that the more accessible communities

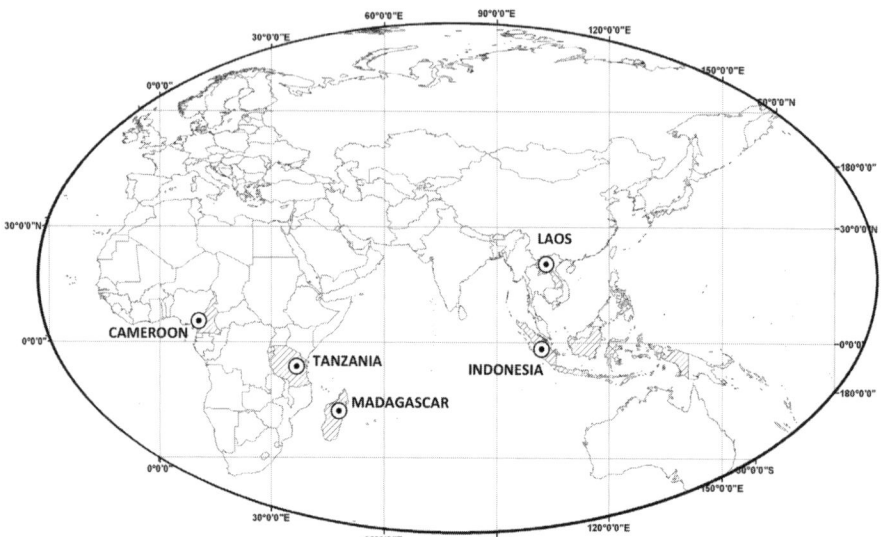

Figure 1.1 *Landscape Mosaics project sites*

would be more integrated into the broader national governance and economic systems. We also expected to find different participatory solutions – appropriate to local problems and opportunities – based on these differences.

Although further analysis remains to be done, some differences have cast doubt on the ubiquity of these patterns, so common in the past. In Madagascar, for instance, traditional methods of forest protection re-emerged in degraded landscapes, whereas significant numbers of people in remote and forested villages maintained that they could survive without forests (see Chapter 6). A similar pattern emerged in Sumatra, where the landscape is being dramatically transformed into oil palm plantations (see Feintrenie and Levang, 2009, or Chapter 3). There, customary forest management in remote areas is also being transformed (see Cramb et al, 2009; Mulyoutomi's discussion of our Indonesian landscape), whereas the larger district is striving to return to the former *Rio* system of governance (Martini and Van Noordwijk, 2009).

Until recently, communication was often so difficult that formal governments and other powerful actors had little or no access to remote areas, allowing their communities, with all their internal differentiations, the freedom to govern themselves (cf Scott, 2009). Improved roads, the spread of mobile phones and their networks, along with the growing use of the internet, have granted many actors a far greater reach, and the recent emphasis on decentralization has lengthened the arm of government (cf Edmunds and Wollenberg, 2003; Colfer and Capistrano, 2005). These new communications have connected remote communities to distant markets and previously unknown opportunities and knowledge. However, they have also made clear to communities all along the access continuum that a number of significant disadvantages exist in terms of interactions with more powerful actors, including corruption, resettlement, land seizures and ethnic marginalization (see Chapter 10).

In collaboration with government officials and partners, interdisciplinary teams selected project sites and linked our activities to ongoing policies at the national level and to the partners' work locally. Our intent was to maximize impact, minimize costs and capitalize on local expertise. In Manompana, Madagascar, for instance, our project partnered with an EU-funded scheme managed by the Association Intercooperation Madagascar (AIM). AIM planned activities in a forest corridor of the Manompana region, along the island's eastern coast, in support of the government's KoloAla programme. In this area, KoloAla was designed to pass responsibility for forest and timber management to local community groups. Team members contributed to both the AIM project and to our own Landscape Mosaics studies. Similar collaborative arrangements were made on all sites. We also relied heavily on the efforts of master's degree and doctoral students conducting related studies (many of whom are among this book's contributors).

This approach has meant that the teams represented different kinds and levels of expertise and emphases. Such differences, compounded of course by site-specific differences, complicated our quantitative cross-site studies. They proved less of a problem for the governance theme, since the uniqueness of individual sites is commonly acknowledged among students of governance (cf Swiderska et al, 2009).

Governance Theme: Research Design

In recognition of the considerable site-to-site variation in terms of team expertise and local conditions, our governance research design was simple and open-ended. We defined governance as:

> cover[ing] the ways and institutions through which individual citizens and groups express their interests, exercise their rights and obligations, and mediate their differences. Governance is thus a complex matter – the essence of which is trust and cooperation. (Carter, 2007; Robledo et al, 2008)

The governance study addressed the following six topics:[6]

- stakeholders relevant for landscape governance in each site;
- levels of formal government;
- governance of natural resources by traditional regulatory mechanisms;
- links between governing bodies at local and district or landscape (meso) levels;
- descriptions of five examples of species management; and
- descriptions of relationships between environmental and cultural services and local management.

The collection of information on these topics was designed to (1) understand the local context, (2) identify pertinent cross-site topics and insights, and (3) contribute to effective participatory action research that would take into account local structures and power dynamics.

In this book, rather than present the full studies for each site, we have selected topics that bring out each site's unique characteristics, while addressing issues with wider global applicability. Some chapters deal with a single site; others feature cross-site analyses.

Snapshots of Landscape Mosaics Sites

Here, we compare and briefly describe the five sites. Table 1.1 lists all the research villages and their administrative units in each country.

Table 1.2 provides brief synopses of major policy drivers, farming systems (including agroforestry components), market drivers and management types.

The United Nations Development Programme's Human Development Index[7] for each of our study countries provides some useful comparisons both among our countries and on a global level (Table 1.3). Four of the five countries are ranked in the bottom third of nations; all have improved since

AN INTRODUCTION TO FIVE TROPICAL LANDSCAPES | 7

Table 1.1 *Landscape Mosaics sites and administrative contexts*

Country	Province or region	District	Division	Ward, kumban, sub-district or commune	Village	Remoteness
Cameroon	Southwest	—	Manyu	Akwaya	Assam	Remote
				Akwaya	Okpambe	Intermediate
				Akwaya	Mukonyong	Accessible
Indonesia	Jambi	Bungo		Bathin Tujuh	Lubuk Beringin	Remote
				Muko-Muko	Tebing Tinggi	Intermediate
				Pelepat	Desa Danau (and Padang Palangeh)	Accessible
Laos	Luang Prabang	Viengkham		Samsun	Phadeng	Remote
				Muangmuay	Bouammi	Intermediate
				Muangmuay	Muangmuay	Accessible
Madagascar	Analanjirofo	Soanierana-Ivongo		Ambahoabe	Maromitety	Remote
				Ambahoabe	Ambofampana	Intermediate
				Manompana	Ambohimarina	Accessible
Tanzania	Tanga	Muheza	Misozwe	Misozwe	Kwatango	Remote
			Amani	Misalai	Misalai	Intermediate
			Amani	Misalai	Shambangeda	Accessible

1990. Since conditions in forests typically include less-than-average access to cash, markets and medical and educational facilities, the index probably overstates forest dwellers' welfare.

A third important factor is population, which appears to be growing on each of our sites (Table 1.4). Our Indonesian site, for example, is located in the province of Jambi, which has a population density of 61 persons per km^2 and a net reproduction rate of 1.108 per cent (www.bps.go.id/download_file/booklet_maret_2009.pdf). Tanzania's population growth rate was 1.95 per cent in 2005 (United Republic of Tanzania, 2008).

Subsequent chapters reveal the five Landscape Mosaics sites more fully, but here we provide a brief introduction, with special emphasis on the communities with which we have worked. Our desire to avoid repetition and complement data presented in the chapters requires more variability in the snapshots provided below than would otherwise be ideal.

Cameroon

Takamanda-Mone is situated at the northernmost point of Cameroon's Southwest Region, northeast of the Cross River Valley, adjacent to Nigeria (Figure 1.2). One of the wettest parts of Africa, its mean annual rainfall ranges from 2500mm to 3500mm per year (Sunderland-Groves et al, 2003). The landscape is primarily lowland closed-canopy forest, bounded by grasslands to

Table 1.2 Landscape Mosaics site contexts – policy, agriculture, markets

Site	Major policy drivers affecting conservation, development	Major elements of landscape-level farming system	Dominant agroforest components	Major market drivers	Management type
Takamanda-Mone, Cameroon	Ongoing land classification, including recent creation of national park and logging concessions	NTFPs, fishing, hunting, cassava, cocoa, timber	NTFPs, cocoa, bush mango, oil palm	NTFPs, cocoa, palm oil, timber, cassava, bush mango	Swidden, smallholder management, timber concession, protected area
Bungo District, Indonesia	Transmigration, forestry regulation, land-use planning, agro-industrial concessions	Paddy rice, oil palm, rubber, fruit	Rubber and durian agroforests	Palm oil and rubber	Smallholder management with swidden history, strong industrial trends
Viengkham District, Laos	Opium eradication, land-use planning to counter shifting cultivation, conservation and forestry policies, village resettlement	Upland and paddy rice cultivation, NTFPs, cattle, monocrop plantations	Cardamom, galangal, cassava	Rice, teak, NTFPs (broom grass, *peuak meauk*)	Swidden, tree plantations, smallholder management
Manompana Corridor, Madagascar	Conservation and forestry policies	Upland rice cultivation, clove trees, home gardens, timber, zebu, poultry	Mixed home gardens, clove and fruit trees, vanilla	Rice, cloves, coffee, vanilla	Swidden, smallholder management
East Usambaras, Tanzania	Conservation and forestry policies, land allocation, land-use planning	Maize, cassava, beans, spices, sugar cane, tea	Cardamom and other spices in uplands; teak, oranges and beans in lowlands	Spices, sugarcane, tea	Swidden, smallholder management with industrial tea plantations

Table 1.3 *Human Development Index and rankings for Landscape Mosaics countries, 1990 and 2009*

Country	1990 rank (of 134)*	1990 index	2009 rank (of 180)	2009 index
Cameroon	100 (75%)	0.519	153 (85%)	0.523
Indonesia	88 (66%)	0.623	111 (62%)	0.734
Laos	109 (81%)	0.449	133 (74%)	0.619
Madagascar	111 (83%)	0.436	145 (81%)	0.543
Tanzania	118 (88%)	0.413	151 (84%)	0.530

Note: * Percentages (in parentheses) indicate the percentage of countries in the data set with a higher HDI than the country in question.
Sources: http://hdrstats.undp.org/en/countries (accessed 14 September 2010)

the east and north, which occur on highlands at elevations of approximately 1000m to 1500m.

Besides producing a variety of natural resources and services (Comiskey et al, 2003), this rich biophysical environment is home to the rare Cross River gorilla (*Gorilla gorilla diehli*) (Sunderland-Groves and Maisels, 2003). The Takamanda-Mone Technical Operations Unit (TOU), which aims to improve coordination and balance protection and sustainable use, has assumed legal responsibility for the landscape since June 2007. The TOU encompasses the Takamanda National Park (69,599ha), the Mone Forest Reserve (55,872ha), a forest management unit (15,360ha) and the newly established Kagwene Gorilla Sanctuary (1900ha).

All villages within the TOU are governed by traditional councils, including the Anyang villages selected by Landscape Mosaics. Each council, the highest

Table 1.4 *Population statistics for Landscape Mosaics sites*

Landscape	Density (persons/km^2)	Village	Population	Households	Indicators of population growth
Takamanda-Mone, Cameroon	10 to 30	Mukonyong	804	125	50% of population <14 years old
		Okpambe	187	49	50% of population <18 years old
		Assam	180	39	—
Bungo District, Indonesia	35.9	Desa Danau	3,934	—	—
		Tebing Tinggi	1,050	270	—
		Lubuk Beringin	382	89	—
Viengkham District, Laos	20	Muangmuay	956	154	2008 population = 882
		Bouammi	349	70	—
		Phadeng	285	40	—
Manompana Corridor, Madagascar	20	Ambohimarina	640	80	34 in-migrants, 10 out-migrants
		Ambofampana	103	55	'Young'
		Maromitety	115	31	—
East Usambaras, Tanzania	96	Shambangeda	778	158	—
		Misalai	2,237	398	—
		Kwatango	1,069	182	—

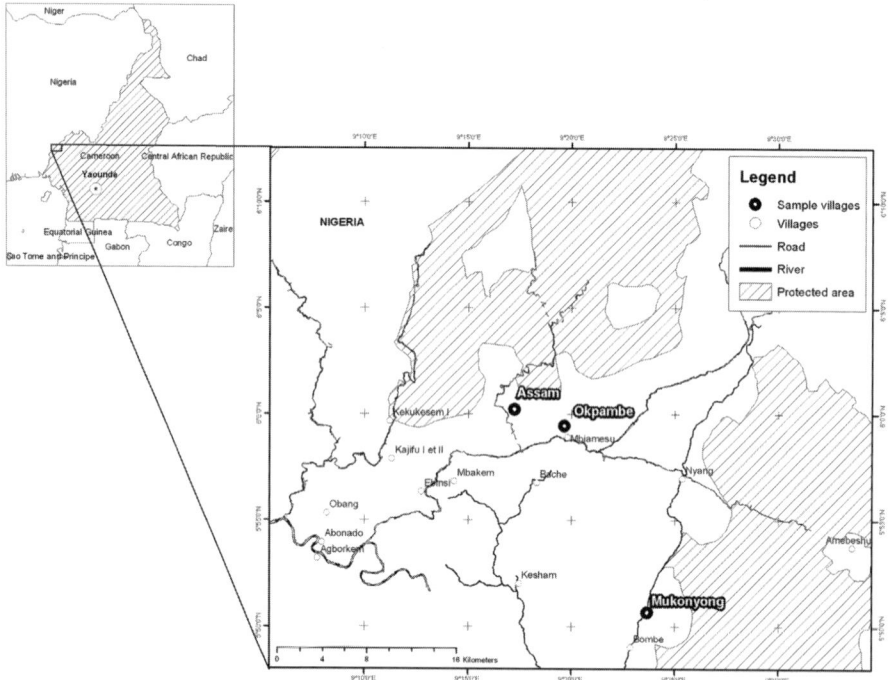

Figure 1.2 *Landscape Mosaics project sites in Takamanda-Mone Technical Operations Unit, Cameroon*

village authority, is headed by a chief whose role is to enforce customary law and to provide an administrative link to the formal government. Women's representation on councils exists but is muted.

People's livelihoods are based on subsistence swidden and cash crops (especially cocoa since road access was improved), supplemented with fishing, hunting, logging and non-timber forest product (NTFP) collection. Local people have specific and differing legal rights depending on the formal forest classification (see Chapter 8). This classification divides the forests into permanent forests (including state forest and council forests), and non-permanent forests (which can be communal, community or private), each with defining characteristics and differing management requirements (see Van Vliet et al, 2009b).

Descriptions of our research communities at Assam, Okpambe and Mukonyong follow.

Assam (remote)

This traditional Anyang village was previously quite isolated, despite its status as the seat of the local court, but in the mid-1990s a logging road to Okpambe was built. This enabled people to reach Assam in only three hours by foot (including wading waist-deep across a river in the dry season); and in March

2009, a road was built between Okpambe and the border of Takamanda National Park, providing motorbike access (one still needs to cross the Munaya River by canoe). In addition to traditional economic activities, recently people have shown considerable interest in enlarging their cocoa farms and growing other cash crops – interests that worry formal managers, given the community's proximity to Takamanda National Park. Van Vliet et al (2009a) found that only five families had cocoa farms in 2000, whereas 90 per cent of men owned such farms in 2008. Cocoa farming is less physically strenuous and generates considerably more income than hunting or trapping. Households typically have three farms, each between 0.5ha and 5ha.

Okpambe (intermediate accessibility)
Okpambe is a small, close-knit, traditional Anyang-speaking village in which the young population engages mainly in swidden farming for both subsistence and income, supplemented with NTFP collection, hunting, fishing, crafts and petty trading. The road construction that made Assam more accessible has had a more pronounced effect in Okpambe; the village can now be reached in two hours by car or motorbike from the border town of Mamfe (Van Vliet, 2009). Women farm mainly food crops such as cassava (also a source of cash income), and men grow cocoa; more than 40 per cent consider cocoa to be the main driver of 'development'. People have increased farm sizes by half a hectare per year on average and almost abandoned hunting, although fishing remains a common activity. The average farm size is 1.3ha, with two to three farms per household. Although forest products contribute to the income of 75 per cent of households, they are decreasing in importance as a cash source, in comparison with farm products such as cocoa, cassava and oil palm. Forest products remain important subsistence sources of food and medicine. Despite increased access, basic social amenities, such as a market, a health centre and an electricity supply, are conspicuously absent.

Mukonyong (accessible)
Mukonyong was settled in 1885; in the 1990s, the government began funding a road connecting Mamfe to Akwaya, reaching Mukonyong (Sunderland-Groves et al, 2003), now only an hour by car from Mamfe. The village has shops, a weekly market and a primary school, but no electricity, potable water infrastructure or health facilities. The easy accessibility of this originally Anyang village has encouraged other ethnic groups to settle within the village territory.

More than 80 per cent of adults are small-scale farmers, cultivating perennial cash crops and food crop swiddens. In contrast to the other two villages, Mukonyong households have been involved in cocoa farming for longer, thus have larger, more mature cocoa trees. Other activities involve teaching, petty trading, hunting, illegal logging and construction. Income sources have shifted from forest to farm products since 2000. Products such as oil palm and cocoa (for men), and cassava and melon (for women) are now the major cash sources, with the latter also contributing to subsistence. Collection of a few high-value NTFPs is an important, if increasingly secondary, source of income, but fishing and hunting are no longer a primary income source.

This Cameroonian site is further discussed in Chapters 7, 8 and 9.

Indonesia

Located in the lowlands and foothills of the Bukit Barisan Range in Sumatra's Jambi Province (Figure 1.3), Bungo District[8] (716,000ha) borders Kerinci Seblat National Park to the west and once provided ecological connectivity, via rubber agroforests, to Bukit Dua Belas National Park to the southeast. Conversion of agroforests to monocultures has left few such 'stepping stones' of biodiversity in place. The role of agroforests in providing corridors between protected areas has not been widely recognized in conservation planning or policy-making at various levels. Like Takamanda-Mone, the site has a significant level of rainfall (3000mm per year).

The district comprises 10 per cent protected natural forest; 34 per cent logged-over and degraded forests, 50 per cent agricultural land and 6 per cent other land uses (Bungo dalam Angka, 2002). The agricultural landscape includes remnants of traditional upland swidden rice agriculture with fallow rotations in remote areas, intensive paddy cultivation along rivers; and, in the

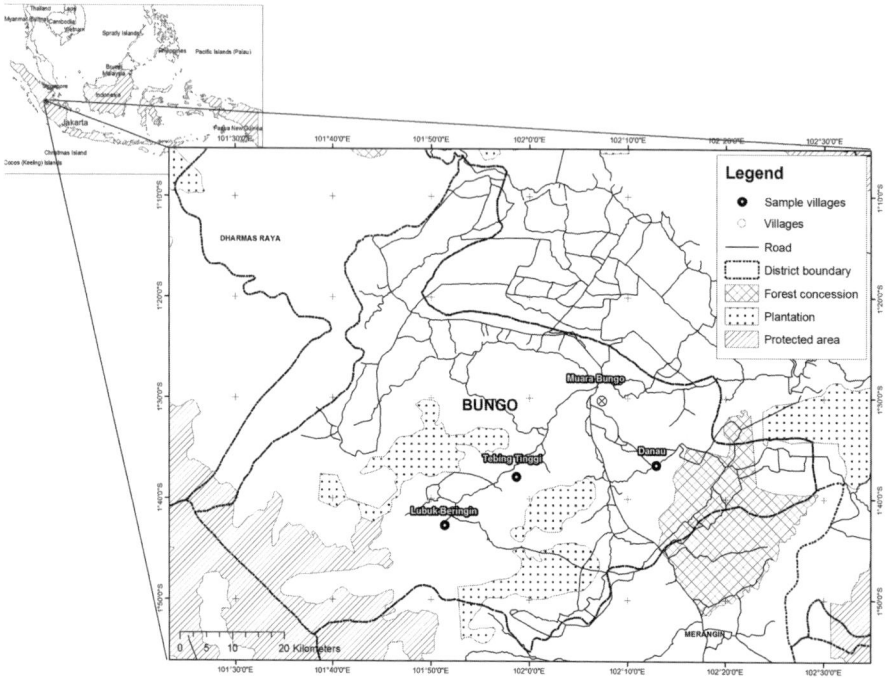

Figure 1.3 *Landscape Mosaics project sites in Bungo District, Jambi Province, Indonesia*

peneplains, complex multistrata rubber agroforests, home gardens, and increasingly, monocultural plantations of rubber and oil palm.

The early 1990s brought an end to commercial logging and a transition in many villages from food crop production to rubber and oil palm, encouraged by the development of secondary roads, as well as government policy. People have planted extensive rubber gardens, prompted by secure rice production, high rubber prices and the desire to strengthen their land claims.

Although government-managed, commercial logging stopped in 2000, the loss of natural forest cover has continued. The policy vacuum after Suharto's fall in 1998 led to a free-for-all of illegal logging, land claims and conversion (Resosudarmo and Dermawan, 2002; Barr et al, 2006). The provincial government planned and licensed oil palm concessions for virtually all logged-over forests, including smallholder-managed (and 'owned') rubber agroforests (see Chapter 3). Local elites and entrepreneurs filled the power vacuum.

Open-cast mining for coal and fluvial sand deposits for gold are major environmental threats but supplement some local incomes. Small- and medium-scale oil palm production now complement large-scale plantations and outgrower schemes. Forests and rubber agroforests are the main natural capital, but their area has been decreasing from year to year, a process that intensified in the 1980s. During this decade, transmigration programmes provided labour for rubber and oil palm plantations (36,000ha), and many forest concessions were granted (Ekadinata and Vincent, 2008).

After 1998, illegal logging and forest encroachment became more significant drivers of forest degradation. The 1998 devaluation of the Indonesian rupiah, which favoured exports, strengthened rubber and oil palm prices, thus also contributing to the conversion of forested lands. Between 2005 and 2008, rubber and oil palm price increases (as much as double the normal price) drove conversion to the more productive monocultures increasingly rapidly (Feintrenie and Levang, 2009).

Descriptions of the three Landscape Mosaics villages in Bungo District follow.

Lubuk Beringin (remote)

This village of Melayu Jambi people has a long history of NGO involvement. It is located in the Buat River Valley, comprises three hamlets and is classified as 'poor' by the Indonesian government. The people have a matrilineal system in which children, particularly women, inherit the house and paddy rice fields and agroforestry plots from their mothers, whereas men (particularly brothers) have considerable authority over decision-making about farming.

The village has a nursery school, a public primary school and several mosques, as well as a micro-hydro electricity supply from several waterwheels. Most houses are made of wood and sit on stilts. From a cross-site perspective, even this most 'remote' Indonesian site is accessible, by motorbike, and cars can approach within easy walking distance. In 2002, land uses consisted of 1400ha of protected forest, 700ha of rubber plantations, 47ha of paddy rice fields and 300ha of other land uses. In 2010, the amount of monocultural rubber has grown, although it remains secondary to the rubber agroforests, which include other species such as *Parkia speciosa* (petai) and *Durio zibethinus* (durian). The local system also has long included both paddy rice along irrigable river beds and the officially disregarded upland swiddens. Since 2000, interest in paddy rice has increased, both in response to wild pig invasion of swiddens and the support of a government programme.

Tebing Tinggi (intermediate accessibility)
Located in an area of low hills in the Bungo River Valley, at the confluence with two other rivers, Tebing Tinggi also has three hamlets. One hamlet is not accessible by car. Another is located on an asphalt road 40 minutes' drive from the district's capital city, Muara Bungo. Housing in this more accessible hamlet is of better quality (cement and bricks), and the people have electricity on the district government's grid. The people are Muslim Melayu, and their inheritance system is a mixture of the matrilineal influences from West Sumatra and the patrilineal patterns of eastern Jambi.

Agriculture here has gradually shifted from a conventional rice swidden system to a system that relies primarily on rubber agroforests to, most recently, rubber monocultures. The government has consistently encouraged the development of higher-yielding rubber monocultures, which have led to accelerated deforestation rates during times of high prices, but people have also continued to use their own diverse systems. There are fruit-based agroforests of *Durio zibethinus* (durian) and *Lansium domesticum* (duku), and some forest fallows that include various fruits (eg, *Coffea canephora*, robusta coffee, and *Archidendron pauciflorum*, Jering tree). Furthermore, some rice cultivation continues. Paddy rice has been hampered by falling water levels in the rivers, which have made waterwheels less effective for irrigation, and by low prices in the 1990s. In 2008, a government-sponsored project funded the rejuvenation of rice fields, with improvements to the irrigation system when natural rubber prices were low. In 2009, half of the households cropped rice. The local price of rice doubled from 2008 to 2009, also motivating people to continue their practice of rice cropping.

Desa Danau-Padang Pelangeh (accessible)
Desa Danau lies 20km from Muara Bungo, on an asphalt road. The village was divided into two administrative units in 2008, with four hamlets in all. Desa Danau's ethnic composition is mixed, with local Melayu people and both spontaneous and government-sponsored transmigrants who arrived, mostly from Java, to work on the plantations. Housing is of comparatively good quality, mainly brick and cement, and the area is electrified.

Land use includes some fields of rice cultivation (which were expanded under a government irrigation programme in the 1980s), but plantations dominate in the area. These include rubber agroforests (39 per cent of the area), rubber monocultures (35 per cent) and oil palm plantations (26 per cent). Nevertheless, the so-called rubber agroforests of this village are comparatively simple mixed-tree plantations, with a few fruit trees (especially durian). Oil palm, growing in importance (see Chapter 3), is planted in two systems: plasma and individual small holdings. The prevalent plasma system involves 25-year contracts with a private company, during which, in this case, farmers entrust their land to the company, which manages the plantation and pays a monthly fee to the landowner, after deducting planting and production costs.[9]

The compounding of traditional and formal regulations on natural resources results in a pluralistic legal situation that all find confusing. This is further complicated by the changes that have occurred in the last decade, relating to decentralization and capitalist expansion.

Indonesia is further discussed in Chapters 7, 8, 9 and 10.

Laos

Viengkham District[10] lies in the northeast, close to the border with Vietnam. The district is also influenced by the proximity of China (Figure 1.4). The climate in this mountainous zone is tropical and monsoonal, with the rainy season lasting from May to October. The timing and strength of the monsoon varies significantly from year to year, causing some problems for rain-fed agriculture.

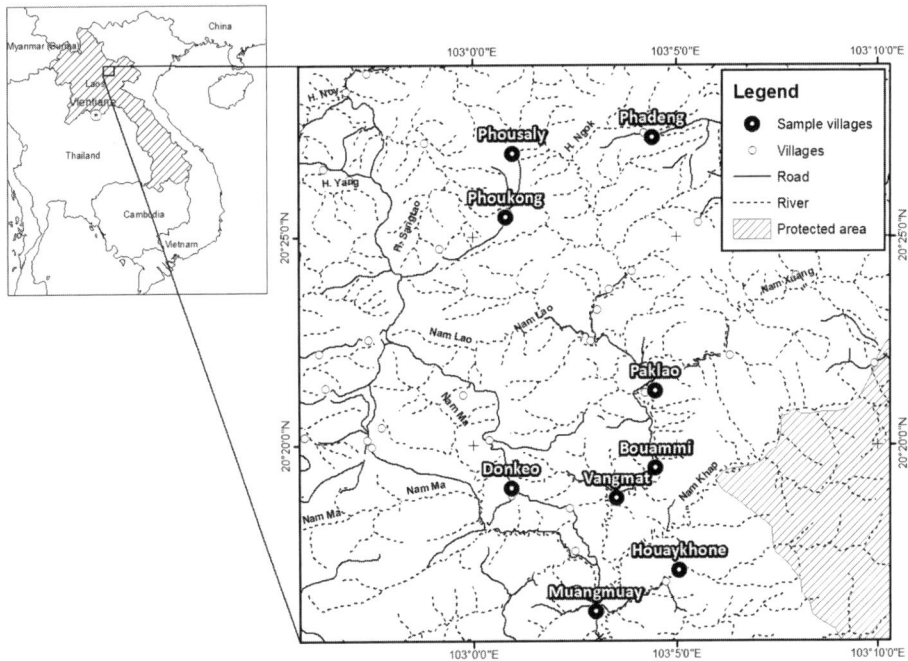

Figure 1.4 *Landscape Mosaics project sites in Laos*

The second half of the 20th century was characterized by warfare in Laos. Conflicts relating to independence from the French were followed by the establishment of the republic, and what is known by some as 'the American war'. The aftermath of this violence includes vast areas affected by unexploded ordinance (UXO – an acronym well known to our research communities). Also, as a result of this conflict, it is estimated that half of all villages moved or were moved (Rigg, 2005).

After the 1975 revolution, the position of ethnic minorities, although officially recognised in the Lao constitution, has often been one of marginalisation. Government policies aimed at eradicating opium cultivation, stabilising shifting cultivation and village relocation have disproportionately affected upland people, predominateatly from non-Tai groups (Rigg, 2005; Baird and Shoemaker, 2007).

Governmental antagonism to swidden agriculture has reinforced the informal prejudices that all ethnic groups in the local population face. Another governmental effort affecting swiddens has been the relocation of whole communities to more accessible areas (see Chapter 4). A land and forest allocation policy – which formalizes three plots of land to each family in forests – has been in place from the early 1990s, when it was considered very progressive, until recently. However, its effects have not always been salutary (Fujita and Phengsopha, 2008). Efforts to protect the neighbouring Nam Et-Phou Louey National Biodiversity Conservation Area have also affected local communities, by reducing their access to a variety of forest products and lands.

Viengkham policies on land, agriculture, infrastructure and services are characterized by contestations and consensus between the district as an implementing agent of national and provincial policy, and villages where accessibility and ethnicity are key factors. An intermediate tier, known as the *kumban*, or village cluster, presents a new forum for interaction between village and district-level actors.

In Laos, ethnic groups are commonly classified into three groups based on the altitude at which they are purported to live. According to this scheme, Lao Loum reside in the lowlands and come from the Tai-Kadai language family (ethnic Lao and other Tai-speaking groups, such as the Lue and the White Tai, are included in this category). Next are the Lao Theung, who reside in the uplands and include mainly Mon-Khmer-speaking Khmu in northern Laos. Last are the Lao Soung, who reside on the mountaintops and include the Hmong, Yao, Akha and other Tibeto-Burmese groups (Evans, 1999; Ovesen, 2004; Schliesinger, 2003; Stuart-Fox, 1986). This common tripartite classification can be misleading, since lowlander groups frequently reside in the uplands. For example, in our research villages of Muangmuay and Bouammi, Lao Loum (a lowland group) and Khmu (an upland group) live side by side in the uplands. Phadeng village, located at the top of the mountain, is composed of Hmong, a highland group. In Muangmuay, 19 per cent of the population is Lao Loum and 81 per cent is Khmu. The population of Bouammi proper is 100 per cent Lao Loum, and its separate neighbourhood of Vangmat is 46 per cent

Lao Loum and 53 per cent Khmu. Phadeng is 100 per cent Hmong (personal communication, Michelle Roberts; Fitriana, 2008).[11]

Our accessibility gradient for the three research villages corresponds to this altitudinal and 'ethnic' gradient, with the least accessible Phadeng at the highest altitude, with the heaviest rainfall and the coolest temperature. The three villages also differ in terms of their distance to the Nam Et-Phou Louey National Biodiversity Conservation Area: the villages of Phadeng and Bouammi are directly adjacent, and Muangmuay is the most removed.

Phadeng (remote)

Phadeng is a four- to five-hour trek from the nearest road, depending on the season. In late 2008, this Hmong village was designated for relocation to the Hmong village of Phukhong, which has better road access, education and a reliable water supply (see Chapter 4). The villagers still retain their original lands, however, and the travel time (three hours' walk) between Phadeng and Phukhong has brought significant changes to their farming practices.

The centrality of the resettlement issue in Laos is one of this site's distinguishing features. Like Indonesia, a swidden-agroforestry landscape appears to be evolving into a different mosaic, one marked by larger swathes of monocultures and the increasing importance of agriculture vis-à-vis forestry in people's lives (see Fox et al, 2009, for a discussion of such transformation throughout Southeast Asia). Phadeng has the oldest secondary forest of the three villages, the result of the long fallow rotations with upland rice crops possible with such a low population density. In Bouammi and Muangmuay, young secondary forest or bush predominates, reflecting the likely lower sustainability of the systems in these resettled areas.

Bouammi (intermediate accessibility)

Officially, land allocated for forest conservation and protection is most common in Bouammi. Bouammi is accessible by a semi-permanent road, which involves a two-hour walk or a one-hour motorcycle ride. Although administratively a single village, in actual fact it comprises two villages, Bouammi and Vangmat, separated by a 30-minute walk by road. In 1998, the two villages were subject to the national village consolidation policy, which required villages of less than 50 households to be merged. The settlement and land-use structures remain relatively undisturbed, however: the village head resides in Bouammi, and the sub-head in Vangmat, with each village possessing its own land and property boundaries while undertaking development projects and administrative decisions in collaboration with its neighbours.

Prior to the creation of Nam Et-Phou Louey National Biodiversity Conservation Area, the villagers of Bouammi collected NTFPs from its forests. But for the past ten years, they have refrained from entering the park and using its resources. The border area of the park is renowned for gold, which is panned by villagers between November and March.

Muangmuay (accessible)

Muangmuay is located on the national road to Vietnam and thus has comparatively good access to education, health facilities and markets. Its inhabitants were relocated to the area between 1990 and 1994. Some significant changes underway here include more collection of NTFPs from fallows than from forests, more domestication of wild products (especially broom grass), new regulations prohibiting NTFP harvests in other people's fallows, cultivation of vegetables during the dry season (because wild vegetables are locally scarce) and greater interest in cash crops (related to ease of marketing).

Laos is further discussed in Chapters 4, 7 and 10.

Madagascar

The Manompana landscape[12] is located on the eastern seaboard, between two protected areas: Mananara Nord to the north (Sodikoff, 2009) and Ambatovaky to the southwest. It includes dense, humid, evergreen primary forest, as well as other forest and non-forest land types (Figure 1.5). The area is hilly, with dramatically steep slopes, high rainfall and many rivers.

Many local threats to agriculture are related to the weather, such as recurrent cyclones and subsequent landslides, which affect the entire landscape. Politics have also played a key role in shaping the environment. Madagascar's upland group, the Merina, has held strong political positions before and after

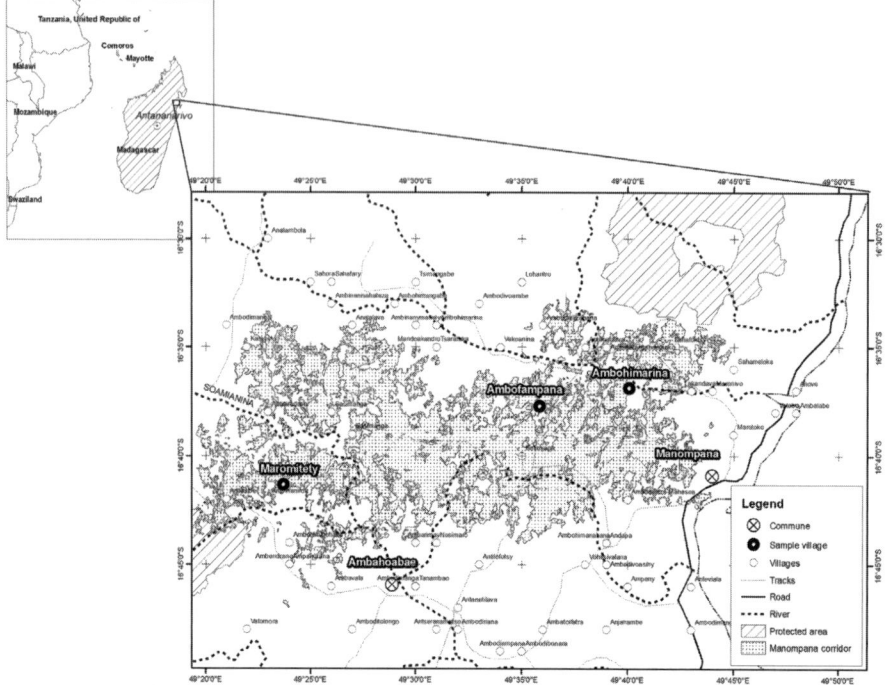

Figure 1.5 *Landscape Mosaics project sites in Madagascar*

the arrival of the French colonialists, resulting in the current political dualism between the Merina and the many coastal groups, including Manompana's Betsimisaraka. Horning (2010, p244), drawing on Harper (2002), notes that 'Since King Andrianampoinimerina united the Kingdoms and societies of Madagascar into a relatively unified polity in the late 1700s, the state had presented itself as a fierce guardian of forests.' The roles of foreign donors have also been central in recent decades.

Madagascar is divided into 22 regions. The Forest Administration, part of the Ministry of Forests, Environment and Tourism, through its deconcentrated technical services, theoretically collaborates with the decentralized authorities at different levels: regions, districts, communes and fokontany. Corruption, as in all our sites, remains a significant concern in the region (see Chapter 10).

One of the most important policy decisions affecting Madagascar's natural resource use was the 2003 decision by then-President Ravalomanana to triple the size of Madagascar's formal protected areas to 10 per cent of the island (see Horning, 2010, for a general account of conservation policies). Considerable political unrest started to occur in February 2009, when the agents of the same president opened fire on demonstrators protesting against plans to release 1.3 million hectares to a Korean company (Sodikoff, 2009). Ravalomanana was soon replaced by President Rajoelina, but the political uncertainty was still evident as 2009 drew to a close.

Of particular interest in Manompana were the presence of a forested corridor with a high management potential and a Malagasy government policy called KoloAla. This programme is designed to support sustainable forest management and allow the devolution of forest management responsibilities and rights to local communities. The emphasis clearly falls on sustainable timber production and involves the development of associations at the fokontany level. These associations are then required to develop forest management plans. The functioning of these plans will be assessed after three years, and local rights to the forest will be formalized over the longer term if the forest is well managed and the rules are followed.

Local people use the forest, young and old forest fallows, home gardens, orchards, marshes, and agricultural fields. Rice is the main crop, with taro, manioc, banana and sweet potato as supplements. Yams from the forest further supplement diets during the lean months. Swidden agriculture, although illegal, is the dominant mode of practice. Coffee, vanilla and cloves are grown as cash crops, on a small scale. A few families are involved in logging, herding, aquaculture, carpentry and basket and mat weaving. Animal husbandry includes cattle, pigs, chickens and some bees, all primarily used as 'living capital', to be sold for health care expenses or other urgent cash needs. Poverty is widespread and extreme, with rice production often lasting only six months; hunger is a reality. Outsiders, as well as some locals, consider swidden agriculture and logging to be significant environmental threats.

None of our research communities here have any formal medical facilities or personnel, although all use medicinal plants. There are no credit facilities available, either formal or informal, nor are there any formal markets. No external projects have operated in the remote areas; the two that have worked in a fourth, accessible site, had little impact (although the research design called for three project sites, we studied four in Madagascar, as discussed in Chapter 6).[13]

Maromitety (remote)
The young Betsimisaraka Avaratra village of Maromitety (ten years old) requires a two-day hike on foot to reach the nearest town (Soanierana Ivongo). It has only a primary school and no traditional healer.

Ambofampana (intermediate accessibility)
Ambofampana is situated in an enclave of the corridor. It is reached by an all-weather road from Manompana to Anove (30 minutes by car), then a river route from Anove to Lakandava (one hour by motorized canoe), with a small village path from Lakandava to Ambofampana (six hours on foot). The community has a primary school and a traditional medical practitioner.

Ambohimarina (accessible)
Ambohimarina is served by an all-weather road (30 minutes by car to Manompana), one hour in a motorized canoe along an accessible river from Anove to Lakandave, and a two-and-a-half hour walk along a footpath from Lakandava to Ambohimarina. Ethnically, this community is primarily Betsimisaraka, with a few of the nationally more dominant Betsileo. This larger community has a middle school and no traditional healer.

Bevalaina (accessible)
Bevalaina is an older and larger village, with about 110 households of five principal kin groups. Although some 85 per cent of the population is Betsimisaraka, the one Sakalava kinship group, originating in western Madagascar, exerts a strong influence in the village: the chief of the village and the *tangalamena*, a traditional leader and traditional healer, are Sakalava. This traditional healer is well known in the whole region, and people come from far away to get his advice. The village is reachable in two-and-a-half hours on foot from Manompana and has a primary school with two classes. Established in 1895, this village has an agricultural system that is much more developed and diverse than the other sites. The agroforestry systems have a great diversity of trees and are more productive. The name Bevalaina derives from *be* ('much') and *valaina* (a local species of liana).

There is widespread concern about the role of swiddens in Madagascar's deforestation. International interest in biodiversity has manifested in significant national policy and international donor contributions (with resulting effects on conservation policy).

Additional information is provided on the Madagascar landscapes in Chapters 6, 7 and 10.

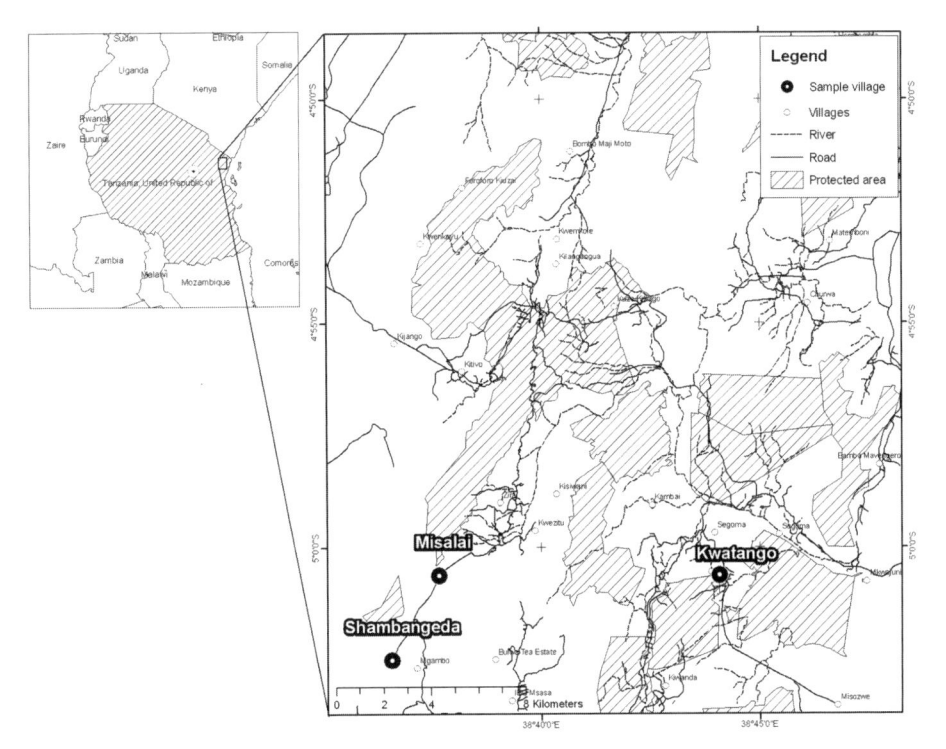

Figure 1.6 *Landscape Mosaics project sites in Tanzania*

Tanzania

The East Usambara Mountains,[14] located in northeastern Tanzania, are part of the Eastern Arc, which runs about 40km from the coast (Figure 1.6). The mountains supply water to the 220,000 inhabitants of the nearby city of Tanga (Reyes et al, 2005). The region is marked by a mosaic of agriculture and lowland forest, and sub-montane rainforest, with high levels of biodiversity in terms of number of flora and fauna species, and a high level of endemism in the forest (Rodgers and Homewood, 1982; Burgess et al, 2007).

There is archaeological evidence of human settlement in the region from ~2,000 years ago (Conte, 2004). For centuries, the Shambaa people, and other groups residing in the mountains, gained access to and controlled the forests through traditional institutions (Woodcock, 2002; Conte, 2004). In 1902, the Germans established the Amani Botanical Garden in the southern part of the East Usambaras to promote plantation agriculture (Conte, 2004). After World War I, Tanganyika became a British colony, and in the 1950s, the British closed the botanical garden, established several forest reserves and expanded the tea plantations. After Tanzanian independence in 1961, immigration from other parts of the country increased. Since the 1990s, more control over the forests has been devolved to local people, but at the same time, the size of the forest area under state control has also increased (Woodcock, 2002; Vihemäki,

2009). In 1993, the Tanzania Forest Conservation Group (TFCG, our main partner) became active in the region, encouraging the establishment of village forest reserves, tree planting and income generation projects. The Amani Nature Reserve, which covers 8380ha at an elevation of approximately 800m to 1000m above sea level, was gazetted in 1997 by the Tanzanian government. The process of establishing the Derema Corridor (970ha) was started a few years later (see Chapter 4).

The colonial administration and the socialist era in Tanzania were characterized by highly centralized control over natural resources. However, since the early 1990s, decentralized and devolved forms have gained ground, including policy reforms in forestry and wildlife management designed to benefit the rural population by reducing poverty. Despite this shift in policy, gaps remain between theory and practice in forest conservation and management (see Vihemäki, 2005; Rantala and Vihemäki, forthcoming; Blomley et al, 2010). In the East Usambaras, the involvement of both 'old' and 'new' actors, and institutions of forest control, has contributed to the fragmentation of the control of protected forests and opened up new arenas of negotiation (Vihemäki, 2009).

The Village Land Act (1999) devolves authority over land administration, land management and dispute resolution to village councils, which are issued certificates of village lands by the commissioner of lands in the name of the president (see Chapter 5). Unlike in our other landscapes, customary rights of occupancy are legally recognized in Tanzania. Local land-use planning was an important policy 'hook' for our involvement.

Current land uses include government forest reserves and nature reserves, some areas of unreserved forest, tea plantations in the uplands, agroforests (with cultivation of spices under the forest canopy), food crop farms, village forest reserves, teak plantations and wetlands (Rantala et al, 2009) and plantations of eucalyptus. In 2007, 33 per cent of the nation's entire population and 37 per cent of the rural population were estimated to be poor, in terms of their inability to meet basic needs (United Republic of Tanzania, 2008). Although the poverty rates have decreased slightly from 2000, the reduction in poverty has not kept pace with the relatively high economic growth or with population growth. Main food crops include maize, cassava, rice, yams and bananas, typically grown without fertilizers, improved seeds or pesticides.

Historically, the establishment of commercial plantations such as tea and sisal has contributed to forest degradation and loss, directly and indirectly. The expansion of small-scale agriculture and fires in the lowland areas have been significant causes of forest degradation in the recent past. The logging ban since the 1990s has resulted in a reduction in timber harvesting, which boomed in 2003 and 2004. Local tea companies log their eucalyptus plantations for firewood used in drying tea. Typically, species-rich cardamom agroforests are converted to more shade-intolerant crops when the soil fertility declines (Newmark, 2002; Reyes, 2008).

Our three villages[15] differ somewhat from other sites in that the least accessible site is located in the lowlands, whereas the two more accessible ones are situated in the uplands. All three villages have established village forest reserves, and the people generally describe the conservation functions of forests as pre-eminent over use. Agriculture is the main source of their livelihoods.

Kwatango (remote)

Kwatango, the lowland village, has established the Kwezimagati Forest Reserve (the only officially registered reserve of the three sites). But as of early 2010, Shambangeda has also been preparing to register its reserve, and has begun implementing its plan and bylaws. Most people have ready access to land that they consider to be their own.

Farms are the most important source of food and key suppliers of firewood, medicine and construction materials. We found the median farm size was 1.61ha. The most important cash crop was maize, followed by bananas, groundnuts, cassava and beans. Agroforestry efforts (orange trees, palms and teak) were in the early stages; timber species were also being grown.

This community has forest on village land and informal access to the adjacent Kwamarimba and Manga forest reserves. Important forest products include firewood, wild vegetables, fruits, mushrooms, medicine, bush meat and various timber species. Nearby are large rubber and teak plantations, providing some employment for local people.

Misalai (intermediate accessibility)

Misalai is developing a management plan for its village forest reserve. Important products from the village forest have included medicine, firewood, wild foods and marketable forest products such as *msambu* (*Allanblackia stuhlmannii*), although these uses are locally seen as likely to decrease in the future because of the recent enforcement of restrictive bylaws. However, people anticipate that forests could support livelihoods through the commercialization of timber (if allowed) and NTFPs, and they see potential in *msambu*.

Agriculture is an important source of food, medicine and timber. The median farm size was 0.80ha. Major cash crops included sugar cane, yams, beans, cassava, maize, cardamom and bananas. A large tea plantation is conspicuous in the landscape and plays an important role in local livelihoods: for one person in five, working in the factory or picking tea is the most important livelihood activity. For others, such employment provides a secondary source of income. Many people also depend on forest products from the company forest.

Agroforestry includes cardamom, cinnamon, cloves and black pepper mixed with mainly native trees and bananas, and is ranked highly for livelihood (food, medicine, construction materials, marketable NTFPs and timber) and conservation benefits. However, people say that the conversion of cardamom agroforestry systems to sugar cane (which is considered more profitable) and tree cutting without sustainable management are significant threats to the existence of agroforests.

Shambangeda (accessible)

Shambangeda lies adjacent to a major tea plantation. As of 2010, its village forest reserve has awaited district-level approval, but the community has begun implementing its plan and bylaws. Most Shambangeda respondents obtain secondary incomes from picking tea, keeping cattle or farming butterflies. For almost half the farmers, the most important cash crop was sugar cane, and for another third, it was spices grown under the tree cover in agroforestry systems. The median farm size was 1.89ha. Most people reported conserving trees on their farms to shade cardamom and other crops, but also to provide firewood and income, especially *msambu*. Community members acknowledge that forests supply many products, in addition to water. They showed great interest in improving water management through collective action. As an outcome of development projects, supplementary incomes are also obtained from *msambu* and butterfly farming.

Additional information on Tanzania is available in Chapters 5, 7 and 10.

Organization of the book

This book is organized thematically, taking into account the significant features on each site that hold relevance more widely for equitable landscape governance and management. Following a brief review of relevant literature, we provide two chapters emphasizing policy, two on the links between traditional and formal landscape management, two on wild products, one on a proposed governance assessment tool and two concluding chapters.

Chapter 2, by John Watts with Carol J. Pierce Colfer, provides a conceptual framework by pulling together some of the salient literature on landscapes, biodiversity and governance. They also clarify the relevant elements of the overall Landscape Mosaics project design.

Chapter 3, by Laurène Feintrenie and Endri Martini, takes us to Indonesia and its serious landscape dilemmas. The government and industry are providing tempting incentives to local stakeholders to adopt monocultures of oil palm, a crop that often proves profitable but reduces biodiversity. This chapter examines the factors that encourage such policies and the dangers and potential delights related to this crop in this context.

In Chapter 4, Manuel Boissière, Salla Rantala, Heini Vihemäki and John Watts address the potential and limitations of different strategies to improve communication and mitigate negative social impacts of human displacement using materials from two of our landscapes, in Laos and Tanzania. This is an important issue, since possible and sometimes actual displacement has been raised in all the Landscape Mosaics countries, among many others.[16]

Chapter 5, by Salla Rantala and Emmanuel Lyimo, supplements our understanding of pluralistic governance in Tanzania's East Usambaras. They examine the evolution of traditional landscape management in a context of multiple ethnicities residing together, and the links between formal and 'traditional' management as they function today.

Chapter 6, by Hans Rudolf Felber, Mihajamanana Rabenilalana, Jean-Pierre Sorg and Zora Urech, examines similar issues in eastern Madagascar, where there is only one main ethnic group. The authors have analysed local customary rules with regard to forested landscapes and their resources. They point out the existing differences between forest fragments and the larger forest massif. Land management is examined as it relates to forest management in this chapter.

Chapter 7, by Stella Asaha, Amandine Boucard, Laurène Feintrenie, Emmanuel Lyimo, Bronwen Powell, Jacqueline Sunderland-Groves, Zora Urech and John Watts, focuses on the products people obtain from the wild landscapes in all the sites. This cross-site comparison looks at the uses, management, synergies and conflicts that characterize the relationships between people and wild products.

Chapter 8, by Stella Asaha, Verina Ingram, Terry Sunderland and Alexander Tajoacha, takes an in-depth look at non-timber forest products, their economic value chains and their governance in Cameroon. This chapter demonstrates the economic and other links that tie even very remote areas to people and events across various scales.

Chapter 9, by Colfer and Feintrenie, proposes a governance assessment tool designed to help project leaders, certifiers, REDD implementers and others evaluate the competencies of landscape-level stakeholders to manage in an equitable and transparent manner. The authors also report their experience of testing this approach in Indonesia.

Chapter 10, by Carol Colfer with Etienne Andriamampandry, Stella Asaha, Imam Basuki, Amandine Boucard, Laurène Feintrenie, Verina Ingram, Michelle Roberts, Terry Sunderland and Zora Urech, takes a theoretical stance, questioning our own assumptions about collaborative governance, given the contexts in which we work. The chapter looks first at the formal governance of land and forests on each site, within a conventional approach to governance (including decentralization, corruption and policies relating to landownership, land-use planning, trees and landscapes). The chapter then turns to three policy issues – swidden agriculture, resettlement and participation – that are not typically considered relevant for governance but have significant effects on people and landscapes in tropical forests. These issues will need to be more seriously addressed if we are to collaboratively govern the world's forests. The chapter concludes with a discussion of equity issues on the ground.

Chapter 11, the final section, summarizes the key issues dealt with in the book (Table 11.1) and discusses our practical attempts to facilitate collaborative governance. Carol Colfer, Jean-Laurent Pfund and Terry Sunderland describe the constraints that we faced and recap the encouraging evidence that we found in this study. The final word is that there is no other viable option than collaborative governance. Progress in this difficult task is made all the more urgent by our growing understanding of climate change. We simply *have* to find ways to overcome the political, logistic, funding and capability constraints, which are so commonly encountered, to enable collaborative governance.

Notes

1. We defined *governance* as 'the ways and institutions through which individuals and groups express their interests, exercise their rights and obligations, and mediate their differences.... [It] is thus a complex matter – the essence of which is trust and cooperation' (see Carter, 2007; Robledo et al, 2008; UNDP, 1997; Appendix 1.1).
2. Differentiating governance and management is difficult and not particularly useful in our case (see Appendix 1.1). The variety of definitions of management available, for instance, on this website (www.leadership501.com/definition-of-management/21, accessed 15 March 2010) include features comparable to 'governance', as we have defined it. One could say that management is to a defined organization or sector, such as forestry, what governance is to society at large. Since our interest lies in governance of the forestry sector, the governance-management overlap is significant.
3. A 'wicked problem' in social planning describes a situation that is 'difficult or impossible to solve because of incomplete, contradictory and changing requirements that are often difficult to recognize. Moreover, because of complex interdependencies, the effort to solve one aspect of a wicked problem may reveal or create other problems' (www.en.wikipedia.org/wiki/Wicked_problem, accessed 26 October 2009).
4. This study, termed Landscape Mosaics for short, was a collaborative effort between the Center for International Forestry Research (CIFOR) and the World Agroforestry Center (ICRAF) in leadership roles, in partnership with host-country institutions and researchers. Funded by the Swiss Agency for Development and Cooperation (SDC), the research was led by Jean-Laurent Pfund at CIFOR.
5. The livelihoods theme, which also contributed directly to this book, was led first by Patricia Shanley and later by Patrice Levang (both of CIFOR); the rewards and incentives theme was first led by Brent Swallow and later by Peter Minang, both from the World Agroforestry Center (ICRAF); the landscape patterns theme was led by Meine Von Noordwijk (ICRAF); and the governance theme was headed by Carol J. Pierce Colfer (CIFOR).
6. The study design is available at www.biodiversityplatform.cgiar.org, and we hope that the full governance reports will eventually be similarly available.
7. The Human Development Index is a composite measure of a country's achievements overall, using three aspects of human development: lifespan, knowledge and standard of living. Lifespan is measured by life expectancy at birth; knowledge, by a mixture of adult literacy rate and school enrolment ratios; and standard of living, by GDP per capita.
8. Materials adapted primarily from Bonnart (2008) and Martini and Van Noordwijk (2009).
9. Contracts between plantation companies and farmers vary greatly from area to area, depending on the negotiation skills of local farmers.
10. Materials adapted primarily from Fitriana (2008) and Fitriana et al (2009).
11. As Scott (2009) predicts, there is considerable local imprecision about the exact ethnic affiliations, within the broad tripartite framework, of community sub-groups.
12. Materials adapted from Etienne Andriamampandry (2008) and Madagascar Landscape Mosaics Team (2009).

13. The difficulty of reaching these sites, combined with the team's responsibilities as part of the government's KoloAla programme, resulted in less attention being paid to the three villages selected by the project. Hence these snapshots are less complete than those for the other countries.
14. Materials derived primarily from Andel (2007) and Rantala et al (2009).
15. Materials on the three villages are from Rantala and Lyimo (2009a, b, c).
16. This issue may become even more germane as REDD researchers and implementers struggle to find ways in which to control forest use at local levels, often without significant understanding or concern for the ways local communities depend on their surroundings.

References

Andel, S.V. (2007) *Livelihood and Landscape Indicators in East Africa in Relation to Conservation and Development,* Universiteit Utrecht, Utrecht

Andriamampandry, E. (2008) 'Madagasikara Site: KoloAla Manompana', Governance Report for CIFOR, Bogor

Baird, I. G. and Shoemaker, B. (2007) 'Unsettling Experiences: Internal Resettlement and International Aid Agencies in Laos', *Development and Change,* vol 38, no 5, pp865–888

Barr, C., Resosudarmo, I. A. P., Dermawan, A., McCarthy, J. with Moeliono, M. and Setiono, B. (eds) (2006) *Decentralization of Forest Administration in Indonesia: Implications for Forest Sustainability, Economic Development and Community Livelihoods,* CIFOR, Bogor

Blomley, T., Ramadhani, H., Mkwizu, Y. and Böhringer, A. (2010) 'Hidden Harvest: Unlocking the Economic Potential of Community-based Forest Management in Tanzania', in L. German, A. Karsenty and A. M. Tiani (eds) *Governing Africa's Forests in a Globalized World,* Earthscan/CIFOR, London, pp126-143

Bonnart, X. (2008) 'How Can the Improved Livelihoods of Rural Community and Biodiversity Conservation be Integrated in the Landscape Mosaics in Bungo District?', University College of Isa, Institut des Régions Chaudes-SupAgro, Montpellier

Brito, B., Micol, L., Davis, C., Nakhooda, S., Daviet, F. and Thuault, A. (2009) *The Governance of Forests Toolkit (Version 1): A Draft Framework of Indicators for Assessing Governance of the Forest Sector,* World Resources Institute, Washington, DC

Bungo dalam Angka (2002). BAPPEDA, Muara Bungo, Jambi

Burgess, N., Butynski, T., Cordeiro, N., Doggart, N., Fjeldså, J., Howell, K., Kilahama, F., Loader, S., Lovett, J., Mbilinyi, B., Menegon, M., Moyer, D., Nashanda, E., Perkin, A., Stanley, W. and Stuart, S. (2007) 'The Biological Importance of the Eastern Arc Mountains of Tanzania and Kenya', *Biological Conservation,* vol 134, no 2, pp209–231

Carter, J. (2007) *Natural Resource Governance: 25 Years of Inspiring Change: Intercooperation 1982–2007,* Intercooperation, Berne

Chambers, R. and Conway, G. (1992) *Sustainable Rural Livelihoods: Practical Concepts for the 21st Century,* Institute of Development Studies, Brighton, available at www.eldis.org/vfile/upload/1/document/0708/DOC12443.pdf

Colfer, C. J. P. and Capistrano, D. (eds) (2005) *The Politics of Decentralization: Forests, Power and People,* Earthscan, London

Colfer, C. J. P., Andriamampandry, E., Asaha, S., Lyimo, E., Martini, E., Pfund, J.-L. and Watts, J. (in press) 'Participatory Action Research for Catalyzing Adaptive Management: Analysis of a "Fits and Starts" Process', *Journal of Environmental Science and Engineering*

Comiskey, J. A., Sunderland, T. C. H. and Sunderland-Groves, J. L. (eds) (2003) *Takamanda: The Biodiversity of an African Rainforest,* Smithsonian Institution, Washington, DC

Conte, C. (2004) *Highland Sanctuary: Environmental History in Tanzania's Usambara Mountains,* Ohio University Press, Athens, OH

Cramb, R. A., Colfer, C. J. P., Dressler, W., Laungaramsri, P., Le, Q. T., Mulyoutami, E., Peluso, N. L. and Wadley, R. L. (2009) 'Swidden Transformations and Rural Livelihoods in Southeast Asia', *Human Ecology,* vol 37, pp323–346

Edmunds, D. and Wollenberg, E. (eds) (2003) *Local Forest Management,* Earthscan, London

Ekadinata, A. and Vincent, G. (2008) 'Dinamika Tutupan Lahan Kabupaten Bungo', Jambi, in H. Adnan, D. Tadjudin, L. Yuliani, H. Komarudin, D. Lopulalan, Y. L. Siagian and D. W. Munggoro (eds) *Belajar Dari Bungo: Mengelola Sumberdaya Alam di Era Desentralisasi,* CIFOR, Bogor

Evans, G. (1999) 'Introduction: What is Lao Culture and Society?', in G. Evans (ed) *Laos Culture and Society,* Silkworm Books, Chiang Mai, Thailand

Feintrenie, L. and Levang, P. (2009) 'Sumatra's Rubber Agroforests: Advent, Rise and Fall of a Sustainable Cropping System', *Small-Scale Forestry,* vol 8, no 3, pp323–335

Fitriana, J. R. (2008) *Landscape and Farming System in Transition: Case Study in Viengkham District, Luang Prabang Province, Lao PDR,* Institut des Regions Chaudes-SupAgro, Montpellier

Fitriana, Y. R., Boucard, A., Vongkhamsao, V., Langford, K. and Watts, J. (2009) 'Landscape Mosaics Project: Laos', Governance Report for CIFOR, Bogor

Forman, R. T. T. (1995) *Land Mosaics: The Ecology of Landscapes and Regions,* Cambridge University Press, Cambridge

Fox, J., Fujita, Y., Ngidang, D., Peluso, N., Potter, L., Sakuntaladewi, N., Sturgeon, J. and Thomas, D. (2009) 'Policies, Political-Economy and Swidden in Southeast Asia', *Human Ecology,* vol 37, pp305–322

Fujita, Y. and Phengsopha, K. (2008) 'The Gap Between Policy and Practice in Lao PDR', in C. J. P. Colfer, G. R. Dahal and D. Capistrano (eds) *Lessons from Forest Decentralization: Money, Justice and the Quest for Good Governance in Asia-Pacific,* Earthscan/CIFOR, London, pp117–132

German, L., Karsenty, A. and Tiani, A.-M. (eds) (2009) *Governing Africa's Forests in a Globalized World,* Earthscan/CIFOR, London

Ghazoul, J., Garcia, C. and Kushalappa, C. G. (2009) 'Landscape Labelling: A Concept For Next-generation Payment for Ecosystem Service Schemes', *Forest Ecology and Management,* vol 258, no 9, pp1889–1895

Görg, C. (2007) 'Landscape Governance: The "Politics of Scale" and the "Natural" Conditions of Places', *Geoforum,* vol 38, no 5, pp954–966

Harper, J. (2002) *Endangered Species: Health, Illness and Death among Madagascar's People of the Forest,* Carolina Academic Press, Durham, North Carolina

Horning, N. R. (2010) 'Bridging the Gap Between Environmental Decision-makers in Madagascar', in L. German, A. Karsenty and A. M. Tiani (eds) *Governing Africa's Forests in a Globalized World,* Earthscan/CIFOR, London, pp234–257

IPCC (International Panel on Climate Change) (2007) *Fourth Assessment Report*, Intergovernmental Panel on Climate Change, Geneva, Switzerland, www.ipcc.ch

LM Team (2008) *Integrating Livelihoods and Multiple Biodiversity Values in Landscape Mosaics: Field Methods*, CIFOR, ICRAF, Joint Biodiversity Platform, Bogor, Indonesia

Locatelli, B., Kanninen, M., Brockhaus, M., Colfer, C. J. P., Murdiyarso, D. and Santoso, H. (2008) *Facing an Uncertain Future: How Forests and People Can Adapt to Climate Change*, CIFOR, Bogor, Indonesia

Madagascar Landscape Mosaics Team (2009) 'Synthèse des Données sur Les Trois Villages Landscape Mosaics [Data Synthesis for the Three Landscape Mosaics Villages]', Report for CIFOR, Bogor, Indonesia

Martini, E. and Van Noordwijk, M. (2009) 'Rubber Agroforest Potency as the Interface Between Conservation and Livelihoods in Bungo District, Jambi Province, Indonesia', Report for CIFOR, Bogor, Indonesia

McElwee, P. (2004) 'Becoming Socialist or Becoming Kinh? Government Policies for Ethnic Minorities in the Socialist Republic of Vietnam', in C. R. Duncan (ed) *Civilizing the Margins: Southeast Asian Government Policies for the Development of Minorities*, Cornell University Press, Ithaca, New York, pp182–213

Mertz, O., Leisz, S. J., Heinimann, A., Rerkasem, K., Thiha, Dressler, W., Pham, V. C., Vu, K. C., Schmidt-Vogt, D., Colfer, C. J. P. and Epprecht, M. (2009) 'Who Counts? Demography of Swidden Cultivators in Southeast Asia', *Human Ecology*, vol 37, no 3, pp281–290

Newmark, W. (2002) 'Conserving Biodiversity in East African Forests: A Study of the Eastern Arc Mountains', Report for Springer-Verlag, Berlin-Heidelberg

Ovesen, J. (2004) 'All Lao? Minorities in the Lao People's Democratic Republic', in C. R. Duncan (ed) *Civilizing the Margins: Southeast Asian Government Policies for the Development of Minorities*, Cornell University Press, Ithaca, NY, pp214–240

Petit, P. (2008) 'Rethinking Internal Migrations in Lao PDR: The Resettlement Process Under Micro-analysis', *Anthropological Forum*, vol 18, no 2, pp117–138

Pfund, J. L. (2007) 'Integrating Livelihoods and Multiple Biodiversity Values in Landscape Mosaics', proposal submitted to the Swiss Agency for Development and Cooperation, Berne, Switzerland

Rantala, S. and Lyimo, E. (2009a) 'Summary on Livelihoods and Conservation – Kwatango', Report for CIFOR, Bogor, Indonesia

Rantala, S. and Lyimo, E. (2009b) 'Village Feedback Workshop Report – Misalai Village', Report for CIFOR, Bogor

Rantala, S. and Lyimo, E. (2009c) 'Village Feedback Workshop Report – Shambangeda Village', Report for CIFOR, Bogor

Rantala, S. and Vihemäki, H. (forthcoming) 'Forest Conservation and Human Displacement: Lessons from the Establishment of the Derema Corridor, Tanzania', in I. Mustalahti (ed) *Features of Forestry Aid: Analyses of Finnish Forestry Assistance*, Institute of Development Studies, University of Helsinki, Helsinki

Rantala, S., Lyimo, E., Powell, B., Kitalyi, A. and Vihemäki, H. (2009) 'Natural Resource Governance and Stakeholders in the East Usambara Mountains: Governance Report', Report for CIFOR/ICRAF, Bogor

Reid, H., Cannon, T., Berger, R., Alam, M. and Milligan, A. (eds) (2009) 'Community-based Adaptation to Climate Change', *Participatory Learning and Action*, vol 60, IIED, London

Resosudarmo, I. A. P. and Dermawan, A. (2002) 'Forests and Regional Autonomy: The Challenge of Sharing the Profits and Pains', in C. J. P. Colfer and I. A. P. Resosudarmo (eds) *Which Way Forward? People, Forests and Policy-making in Indonesia*, Resources for the Future/CIFOR, Washington, DC, pp325–357

Reyes, T. (2008) 'Agroforestry Systems for Sustainable Livelihoods and Improved Land Management in the East Usambaras, Tanzania', Report for University of Helsinki, Helsinki

Reyes, T., Quiros, R. and Msikula, S. (2005) 'Socio-economic Comparison Between Traditional and Improved Cultivation Methods in Agroforestry Systems, East Usambara Mountains, Tanzania', *Environmental Management*, vol 36, no 5, pp682–690

Rigg, J. (2005) *Living with Transition in Laos: Market Integration in Southeast Asia*, Routledge, Oxon

Rights and Resources Initiative (2010) *The End of the Hinterland: Forests, Conflict and Climate Change*, Rights and Resources Initiative, Washington, DC

Robledo, C., Blaser, J., Byrne, S. and Schmidt, K. (2008) 'Climate Change and Governance in the Forest Sector: An Overview of the Issues on Forests and Climate Change with Specific Consideration of Sector Governance, Tenure and Access for Local Stakeholders', Report for Rights and Resources Initiative, Washington, DC

Rodgers, W. A. and Homewood, K. M. (1982) 'Species Richness and Endemism in the Usambara Mountains Forests, Tanzania', *Biological Journal of the Linnean Society* vol 18, pp197–242

Schliesinger, J. (2003) *Ethnic Groups of Laos*, White Lotus Press, Bangkok, Thailand

Scott, J. C. (2009) *The Art of Not Being Governed: An Anarchist History of Upland Southeast Asia*, Yale University Press, New Haven, CT

Sodikoff, G. (2009) 'The Low-wage Conservationist: Biodiversity and Perversities of Value in Madagascar', *American Anthropologist*, vol 111, no 4, pp443–455

Stuart-Fox, M. (1986) *Laos*, Frances Pinter Publishers, London

Sunderland-Groves, J. L. and Maisels, F. (2003) 'Large Mammals of Takamanda Forest Reserve, Cameroon', in J. A. Comiskey, T. C. H. Sunderland and J. L. Sunderland-Groves (eds) *Takamanda: The Biodiversity of an African Rainforest*, Smithsonian Institution, Washington, DC, pp111–127

Sunderland-Groves, J. L., Sunderland, T. C. H., Comiskey, J. A., Ayeni, J. S. O. and Mdaihli, M. (2003) 'The Takamanda Forest Reserve, Cameroon', in J. A. Comiskey, T. C. H. Sunderland and J. L. Sunderland-Groves (eds) *Takamanda: The Biodiversity of an African Rainforest*, vol 8, Monitoring and Assessment of Biodiversity Program, Smithsonian Institution, Washington, DC, pp1–7

Swiderska, K., with Roe, D., Siegele, L. and Grieg-Gran, M. (2009) *The Governance of Nature and the Nature of Governance: Policy That Works for Biodiversity and Livelihoods*, IIED, London

Tress, B. and Tress, G. (2000) 'The Landscape – From Vision to Definition,' in J. Brandt, B. Tress and G. Tress (eds) *Multifunctional Landscapes: Interdisciplinary Approaches to Landscape Research and Management*, material for international conference on 'Multifunctional Landscapes', Centre for Landscape Research, University of Roskilde, Denmark, 18–21 October 2000

UNDP (United Nations Development Programme) (1997) 'Governance for Sustainable Human Development', UNDP Governance Policy Paper, vol 97, United Nations Development Programme, New York, NY

UNISDR (United Nations International Strategy for Disaster Reduction) (2009) UNISDR Terminology on Disaster Risk Reduction, available at www.undp.org.ge/new/files/24_619_762164_UNISDRterminology-2009-eng.pdf

United Republic of Tanzania (2008) 'Mkukuta Annual Implementation Report 2007/08'

Van Vliet, N. (2009) 'Participatory Action Research in Takamanda-Mone Technical Operations Unit in Cameroon', Report for CIFOR, Bogor

Van Vliet, N., Asaha, S., Assembe, S., Tajoacha, A., Sunderland, T., Slayback, D., Sunderland-Groves, J. and Nasi, R. (2009a) 'From Small-scale Forestry-based Activities to Cash Crop Production: Finding the Balance in the Context of Conservation and Development Projects', Unpublished Paper

Van Vliet, N., Asaha, S. and Ingram, V. (2009b) 'Governance Report, Cameroon Site: Takamanda-Mone Technical Operations Unit', Report for CIFOR, Bogor

Vihemäki, H. (2005) 'Politics of Participatory Forest Conservation: Cases from the East Usambara Mountains, Tanzania', *Journal of Transdisciplinary Environmental Studies*, vol 4, no 2, 1–16

Vihemäki, H. (2009) 'Participation or Further Exclusion? Contestations over Forest Conservation and Control in the East Usambara Mountains, Tanzania', doctoral dissertation, University of Helsinki

Wollenberg, E. and Springate-Baginski, O. (2010) 'Introduction', in O. Springate-Baginski and E. Wollenberg (eds) *REDD, Forest Governance and Rural Livelihoods: The Emerging Agenda*, Center for International Forestry Research, Bogor, Indonesia, pp1–18

Woodcock, K. (2002) *Changing Roles in Natural Forest Management: Stakeholders' Roles in the Eastern Arc Mountains, Tanzania*, Ashgate, Burlington, VT

Appendix 1.1: Glossary of Terms

Term	Definition	Source
Access	Freedom or ability to obtain or make use of something.	www.merriam-webster.com/dictionary/access [accessed 11 March 2010]
Access, formal	Legal forms (either state or customary) that specify rights of access.	authors' use
Access, real	Actual availability of the resources in question to the particular stakeholder or group in question.	authors' use
Equity (social)	Fairness in the processes and outcomes related to social justice and how costs and benefits are distributed.	Wollenberg and Springate-Baginski, 2010
Governance	The ways and institutions through which individuals and groups express their interests, exercise their rights and obligations, and mediate their differences… [It] is thus a complex matter – the essence of which is trust and cooperation.	Robledo et al 2008; Carter 2007; UNDP 1997
Governance (local)	We focus on the landscape level, including local villages and the districts or other relevant administrative units that feature in the landscape – in an effort to complement the work of others typically at higher or lower levels.	authors' use
Institution	Institutions are structures and mechanisms of social order and cooperation governing the behavior of a set of individuals within a given human collectivity. Institutions are identified with a social purpose and permanence, transcending individual human lives and intentions, and with the making and enforcing of rules governing cooperative human behaviour.	Authors use; see also http://en.wikipedia.org/wiki/Institution
Landscape	Encompassing multiple dimensions, including the ecological, economic, cultural, historical and aesthetic, a landscape is a mosaic, where the mix of local ecosystems or land uses is repeated in similar form over a kilometres-wide area, and characterized by a repeated cluster of spatial elements. In LM, it is an area appropriate for intervention and including a gradient of environmental conditions.	Forman 1995; LM Team 2008; Tress and Tress 2000

Term	Definition	Source
Livelihood	A livelihood comprises the capabilities, assets (stores, resources, claims and access), and activities required for a means of living. A livelihood is sustainable when it can cope with and recover from stress and shocks, maintain or enhance its capabilities and assets, and provide sustainable livelihood opportunities for the next generation; and which contributes net benefits to other livelihoods at the local and global levels, in the long and short term.	Chambers and Conway 1992
Marginalized group	A category or group of people who have been disenfranchised, who may have less money, power, education, health than do those in more powerful positions. Typical examples include women, lower classes or castes, despised ethnic groups or occupations.	authors' use
Resources	Information, funds, land, products	authors' use
Resilience	Resilience is the ability of a system, community or society exposed to hazards to resist, absorb, accommodate, and recover from the effects of a hazard in a timely and efficient manner, including through the preservation and restoration of its essential basic structures and functions.	UNISDR 2009
Risk	Risk is expected damage or loss due to the combination of vulnerability and hazards.	Reid et al. 2009
Scenario	A plausible and often simplified description of how the future may develop, based on a coherent and internally consistent set of assumptions about driving forces and key relationships. Scenarios may be derived from projections, but are often based on additional information from other sources, sometimes combined with a narrative storyline.	IPCC 2007
Security	Certainty that one's perceived rights are not in danger of being abridged.	authors' use

Term	Definition	Source
Territory	A local territory, in our usage, corresponds to a local community-managed area within the landscape. It contains various land uses or land cover types.	LM Team 2008
Triangulation	The verification of information gained from one source or methodology with that gained from one or more other sources of methodologies.	Reid et al. 2009
Trust	Reliance on another person or entity; the degree to which one party trusts another is a measure of belief in the honesty, fairness, or benevolence of another party.	Authors' use, see also http://en.wikipedia.org/wiki/Trust {accessed 15 September 2010}
Voice	Capacity to bring one's knowledge, interests, and goals to bear in shared decision-making.	authors' use

2

The Governance of Tropical Forested Landscapes

John Daniel Watts with Carol J. Pierce Colfer

Multifunctional forested landscapes of the tropics contain rich biodiversity and sustain rural livelihoods. The use of different land types is determined by their historical, natural and cultural significance, in addition to their future potential. Different people and organizations, ranging from the village to the global level, recognize different values of these landscape mosaics (cf Boissière et al, 2010). The management objectives and processes of multifunctional landscapes are shaped by these competing values and power imbalances among people and organizations. The people who live in these landscapes are often the most dependent on them for their livelihoods, yet also the poorest and the least powerful among the different levels of governance. Landscape governance, it is proposed, should identify and reconcile the competing values of multiple levels of governance in a way that contributes to the realization of the visions of villagers and sustainable management of the landscape.

This chapter is divided into three parts. The first explores the development of the concept of a multifunctional landscape and its usefulness in contributing to adaptive management. The second discusses natural resource governance from the village to the global level and addresses the issue of multilevel governance. The third part explores how landscape-level institutions can be created for the adaptive and equitable management of multifunctional tropical landscapes.

Rural Landscape Governance in the Tropics

The rural inhabitants of tropical countries often face land management choices that can harm, sustain or improve their livelihoods. In forested landscapes, the conversion to intensive or more extensive agricultural systems typically comes at the expense of the remaining forest areas. The resulting loss of forest biodiversity affects the livelihoods of peoples living in forests, and related ecosystem services, as well as the valuable natural resources. Mitigating the loss of forest biodiversity for the purposes of conservation and sustainable management while still addressing the aspirations and poverty of villagers has emerged as a significant challenge. In this context, action at the landscape scale has been argued to be the best option for mitigating the trade-offs between conservation and development.

If the landscape scale is the best scale at which to tackle issues of human development and biodiversity loss, we must consider what is meant by the landscape. How can landscape governance, as a model of governance, contribute to the sustainable management and conservation of biodiversity while addressing the needs of rural livelihoods? The concept that the Landscape Mosaics project used was that of a multifunctional landscape encompassing multiple dimensions, including the ecological, economic, cultural, historical and aesthetic (Tress and Tress, 2000). Governance, in turn, is understood as being 'about who decides and how, and encompasses policies, institutions, processes and power' (Swiderska et al, 2009, p1; see Appendix 1.1, this volume, for another useful definition). Landscape governance examines the links between the socially constructed, multiple levels of governance and the actual conditions of landscapes (Görg, 2007). Through understanding these linkages, pathways can emerge for the development of landscape-level institutions for the more sustainable and equitable management of the environment.

To explain why the project adopted that approach, in this chapter we will briefly explore the development of landscape approaches and theories on multilevel governance regimes. We will then go on to explore some successful examples of landscape-level management, concluding with a discussion of a number of methods for facilitating the emergence of landscape-level institutions for the more sustainable and equitable management of the environment.

Multifunctional Landscapes

The concept of landscapes that emerged from the discipline of landscape ecology provides a starting point for discussion. The term landscape ecology was coined by Carl Troll in 1939 (Liu and Taylor, 2002), but only in the 1980s did the field begin to gain currency outside Europe. The landscape, in the context of this discipline, has been defined as 'a mosaic, where the mix of local ecosystems or land uses is repeated in similar form over a kilometres-wide area, thus characterized by a repeated cluster of spatial elements' (Forman, 1995).

The most basic unit in this landscape is a patch, which is a relatively homogeneous area that differs from its surroundings (Liu and Taylor, 2002; Forman, 1995). Additionally, the landscape is composed of corridors and a background matrix (Forman and Godron, 1986), which is an 'open system with flows across landscape boundaries and interactions with other landscapes'. Although emphasizing the heterogeneity of the landscape, Forman and Godron (1986) classified landscapes according to the dominant ecological or social system – natural landscapes, landscapes with forestry, agricultural and built landscapes.

The landscape-scale analysis of landscape ecology has proven effective in understanding patterns and processes of the natural world. In the context of tropical forested landscapes, we have examined how the landscape scale of analysis can be used to explain land use and land cover change. We have considered how the reciprocal impacts of social and ecological systems (Berkes and Folke, 1998) can be incorporated into such an analysis. The concept of a multifunctional landscape helps us tackle these challenges. By expanding the landscape to look at not only the ecological dimensions, but also the economic, cultural and historical contexts (Tress and Tress, 2000), we can achieve a better understanding of the actual dynamics of the landscape.

In practice, however, what does this look like? From a static perspective, an aerial photograph or satellite image will reveal mosaics of different land covers and land uses. The sites of the Landscape Mosaics project (see Chapter 1) included primary forests; agroforests for rubber, cocoa and shifting cultivation; intensive agriculture, such as sugar cane; and settlements and infrastructure, such as roads. Examining the development of these landscape mosaics, we take into account the cultural, historical and economic choices of the people who live there and others who have a role in the management of the landscape. The dynamics of the multifunctional landscape are more complex than the resulting landscape patterns, but explaining these patterns helps to reveal the social-ecological processes at play.

The landscape, as defined in landscape ecology, focuses on a dominant ecological or social system, simplifying the boundary definition. In the case of multifunctional landscapes, which contain a mosaic of land uses and land cover types, where should the boundary be drawn? Like Levin (1992), we have concluded that there is no single, correct scale at which the dynamics of natural systems should be studied. In our work, multifunctional landscapes were selected that represented a gradient from dense forest and protected areas to intensive agriculture. The rationale behind this selection was to understand the drivers of land-use and land-cover change or variation. The scale of each of the sites, however, varies greatly (Table 2.1). The reason for the variation in scale relates to three factors: the drivers of land-use change being studied, the target area for conservation or sustainable management of forests and the linkages to administrative units. So, although the analysis of the landscape scale is argued to be the best for understanding land-use change, the study of administrative units is viewed as the more appropriate for facilitating better management.

Table 2.1 *Area of Landscape Mosaics sites*

Site	Area (km²)	Administrative unit
Bungo District, Indonesia	4679	Bungo District
Manompana Corridor, Madagascar	1750	Manompana Corridor
Takamanda-Mone, Cameroon	896	Takamanda-Mone Technical Operations Unit
East Usambara Mountains, Tanzania	823	Derema Corridor
Viengkham District, Laos	620	Viengkham District, Nam Et-Phou Loey National Protected Area

Problems of scale are further complicated by disciplinary discrepancies in the use of units of analysis. Gibson et al (2000) explore these in relation to the different scales employed by the social and natural sciences. The authors argue that there is a need to harmonize notions of scale between the various disciplines in order to address issues of global environmental change. In our work, whether one focuses on governance or protected areas affects the selection of the broader landscape. There is a further need to adopt approaches that address diachronic dimensions of ecological and social systems to effectively tackle issues of environmental degradation (MacMynowski, 2007). These inconsistencies in spatial and temporal scales correctly highlight the need to acknowledge and harmonize these multiple scales. However, questions remain over the terms on which and to what ends such harmonization should occur.

In this context, we ask, does the multifunctional landscape as an analytical construct have value not only for analysing social and ecological processes but also for facilitating the emergence of adaptive management of the environment? Or are we superimposing an arbitrary boundary on the people within these landscapes? Are we imposing boundaries for the purposes of the natural sciences and achieving globally defined conservation and development goals? The value of applying the concept of landscape governance, especially to the study of tropical forested landscapes, lies in its contribution to identifying and reconciling multiple interests, and navigating the path towards negotiating sustainable development.

Multilevel Governance

The governance of tropical, rural and multifunctional landscapes is a multilevel concept. Levels of governance comprise the different tiers of government, as well as local, frequently informal and often customary governance mechanisms, and their corresponding extents of influence and levels of power. At the top of this political hierarchy is the international or global level. This level is governed by treaties between nation-states, as well as affected by transnational actors, such as corporations and international non-governmental organizations (NGOs). The international or global level has the greatest spatial coverage, but not necessarily the greatest power. The national level, in contrast, refers to the power and territory of the nation-state, which still possesses the highest level of formal sovereignty within the international system. The spatial extent of the nation-state is determined by historically defined and often contested boundaries. Typically, the constitutional powers of the nation-state can formally override all lower levels of government.

The sub-levels of national governance, their powers and spatial extents are relative to the type of national government and unique historical circumstances. All the countries discussed in this book feature a centralized system of government, with respect to national, provincial, district (sometimes sub-district) and village levels of formal government. Four of the countries included are democratic; one, Laos, is based on single-party rule. Typically, the roles of the national and provincial levels of government are seen to incorporate the design of policies and allocation of powers to implement them at lower levels. Contestation may occur regarding whether the district or sub-district is the lowest arm of government, or instead the village or hamlet. This chapter will briefly explore these levels and how they affect landscape governance.

Community and village-level governance

The village level is the essential building block for the governance of tropical multifunctional landscapes, for at least three reasons. The first is that people in villages typically manage and use their village territory's resources to sustain their livelihoods. Decisions are made at the individual and household level, of course, but at the village level we begin to see more coordinated management of the multifunctional landscape. Through these decisions, the trade-offs among conservation, sustainable management of forests and the conversion to more intensive agriculture or other more profitable land uses often occur. The second reason is that the linkages among traditional management or governance institutions and official government institutions of land and natural resource management, including tenure, are most obvious at the village level. In addition, the dynamics of how villagers adopt management strategies that respond to socio-economic drivers are visible at this level. The third reason relates to equity. Villagers in tropical landscapes are typically poor, dependent on the forests and landscapes for their lives, and have significantly less voice than people representing higher levels of governance. Local people and their aspirations must be included in any management or governance institution if landscape governance is to be equitable (and, in most cases, effective).

Researchers looking at common pool resource management have argued that under certain conditions, natural resources can be sustainably managed at the community level (see Agrawal and Gibson, 1999). Ostrom (1990), after studying many cases of community-level natural resource management, proposed a set of criteria for success, focusing on several dimensions. The first is spatial – that the resource managed has clearly defined boundaries. The subsequent criteria focus on village-level institutions, in terms of rules and processes for managing and monitoring the natural resource. The final criteria involve horizontal and vertical linkages with higher-level authorities, such as the right to organize, which often requires agreement from those in authority, and the need for nested enterprises if the resources belong to larger systems. The criteria Ostrom has proposed address the spatial and multilevel dimensions of governance and natural resources, but her central focus is typically on a single resource at the village level.

Community-based natural resource management has emerged as a distinct body of scholarship and concurrently has become widely used in conservation and development projects. Notions of equity have played a significant role in its adoption. In its 'ideal form', community-based management repositions the community from being a passive recipient of central government policies, which often lead to natural resource degradation, to a more active participant with community rights and agency for the management of local natural resources. The success of these management regimes is also contingent on communities' ability to negotiate management agreements with governments. These co-management agreements typically represent a compromise between the traditional knowledge and practices of communities and conventional scientific approaches to natural resource management and tenure (Borrini-Feyerabend et al, 2004).

Despite the popularity of community-based approaches, problems with targeting interventions at the community level have emerged. Agrawal and Gibson (1999) examine the concept of community as it has been applied in natural resource conservation. The authors note that the community is often unquestionably assumed to be in its ideal form, or as they describe it, the 'mythic community' – a spatially small, homogeneous social structure defined by shared norms. Researchers and managers often assume that the traditional systems of natural resource management, which were sustainable under previous conditions, will continue to function in similar ways in the future. Agrawal and Gibson recommend instead an approach based on the institutional analysis of natural resource management – an approach that identifies the rules affecting natural resource management, the actors affected by such rules and the processes by which such rules change in a given situation. Such an approach would differentiate community groups by their size, composition, norms and resource dependence.

Despite these critiques, the community, local or village level is still argued to be the critical level for the management of social–ecological systems, the main difference being that it must be viewed within the context of its linkages with other levels (Fabricius et al, 2007; Berkes, 2006). The major challenges to these local regimes come from the national and global levels of governance (Armitage, 2008), briefly discussed below.

National and global-level governance

Until recently, most formal, legal governance of natural resources has taken place at the national level. In most tropical countries, land is owned and formally managed by the national government – although this is changing (see Sunderlin, 2008; Larson et al, 2010). The typical remoteness of tropical forest areas has often meant that plans and policies bear little relation to forest peoples' day-to-day lives and governance (see Scott, 2009).

Global-level processes can be understood through the chains of actors that connect the global processes to the landscape. Global processes can also be addressed through the spaces – economic and political – where decisions are made by virtual communities that materially affect people in the landscape (Gezon and Paulson, 2004). Such global processes and institutions include the following:

- market forces deriving from economic globalization;
- international treaties, from trade to environmental;
- flows of goods, ranging from organisms to minerals;
- flows of information, from financial to the media; and
- transnational actors, from corporations to NGOs (Lemos and Agrawal, 2006).

In our work, particularly potent global processes have included the recent emphasis on designating and managing protected areas, sustainable forest management (for purposes of timber production), forestry decentralization and planning related to climate change mitigation and adaptation. One of our hopes for this volume is to contribute to more effective planning and implementation of such policies.

Multilevel governance

The interdependencies among levels of governance have been increasingly recognized. We understand the following:

- that policies designed at the national level are altered and reinterpreted on the basis of local-level realities;
- that local communities can benefit from the involvement of benign higher-level government actors (and that they can be adversely affected by corrupt practices of actors at all levels);
- that conflicts at one level may be resolved peacefully by facilitation from a higher level; and
- that governance in general can be improved by monitoring from any level.

The utility of Elinor Ostrom's early work (1990), which emphasized the nested nature of good local governance, has been continually reinforced and expanded over the past two decades, both by herself and by her students. In addition to nesting in multiple layers of governance, good governance of natural resources requires institutions to have the capacity for analytic deliberation and include a variety of institutional types (Dietz et al, 2003). Fikret

Berkes has contributed significantly to our understanding of such interconnections. One of his most recent papers focuses on the significance of bridging organizations and their networks. He outlines some of their major functions, such as information sharing, joint social learning and power sharing, with examples and techniques for strengthening ties among levels (Berkes, 2009).

Looking 'upwards', many authors have demonstrated the importance of multiscale links: Krishna's (2002) study of social capital in India identifies the critical role played by bridging social capital as wealthier, younger and better educated villagers serve as effective links to resources outside their villages. Hlambela and Kozanayi (2005) emphasize the importance of higher-level governmental support for the success of villagers' collective efforts in rural Zimbabwe. Fabricius et al (2007) provide a three-part typology of communities, with an excellent discussion of the characteristics and skills that facilitate multiscale action and enable 'adaptive co-managers' to adapt, manage and govern effectively. These authors contrast adaptive co-managers with the less successful 'powerless spectators' and 'coping actors'.

Two forms of multilevel governance arrangements – decentralization and co-management – have been tested globally.

Decentralization

Decentralization as both a national process and policy prescription gained increasing prominence in developing countries throughout the late 1990s. Decentralization aims to transfer powers from the central apparatus of government to lower levels of the political hierarchy or territorial units (Larson, 2005). The process is based on the concept of subsidiarity, whereby decision-making should be at the lowest appropriate level of governance (Meinzen-Dick et al, 2008). Decentralization has both directly and indirectly affected the management of natural resources in developing countries.

Effective decentralization requires the commitment of central authorities to devolve powers and authority to local-level institutions, the provision of adequate resources to these institutions to fulfil their mandates, and their commitment to ensuring that the new institutions are locally relevant – requirements rarely met to date (see Ribot and Larson, 2005; Colfer and Capistrano, 2005; Colfer et al, 2008; German et al, 2009). Ribot (2003) notes that the creation of decentralized institutions can reinforce existing undemocratic, often customary, institutions if additional mechanisms for ensuring downward accountability are not implemented as part of the process. According to Ribot, such mechanisms should have the following aims:

- to democratize local government first;
- to apply multiple accountability measures;
- to establish a domain of local autonomy within a measure of oversight;
- to transfer power before capacity is established;
- to transfer rights rather than privileges; and
- to use minimum environmental standards rather than centralized planning.

Diaw (2005, 2009) sees these links differently, emphasizing that formal governments have so far failed either to see or understand the local-level complexity of social systems. He notes, for instance,

> *major discrepancies between the dominant values and structures promoted by decentralization and conservation programmes and the still functioning, mostly invisible infrastructure of embedded institutions and local legitimating networks. We are thus far more interested in local actors as active democratic subjects and in the diverse manifestations of local agency than in the nascent controversy about what qualifies as democratic decentralization.*
> (Diaw, 2009, p57)

Several authors have noted the plurality of overlapping methods of governance in tropical forests (see Marfo et al, 2010; Wollenberg et al, 2005; Chapters 5 and 6). Some authors have shown how policies designed to decentralize and empower local communities have actually reduced the autonomy and rights of forest peoples (see Contreras, 2003; Edmunds and Wollenberg, 2003; Shackleton et al, 2001, whose findings in southern Africa are mixed). Others have emphasized the interdependence of local communities and higher levels of government (see Makapukaw and Mirasol, 2005, on the Philippines; Hlambela and Kozanayi, 2005, on Zimbabwe). This view is implicit in the analysis of Tanzania in Chapter 5.

Co-management

There are a variety of approaches to co-management; see Carter (2000) or Hobley (1996) for early analyses of global efforts to collaborate across levels. The Center for International Forestry Research's (CIFOR's) Adaptive Collaborative Management (ACM) work was initially described as 'co-management' (Buck et al, 2001). However, the ACM team was convinced early in the process that the prefix of 'co' implied the equality of rights among stakeholders – something that did not exist in reality. Collaborative management, on the other hand, clarified the need for negotiation and variable rights.

The collection by Bernardo and Snelder (1999) documents a variety of approaches, mainly in the Philippines, which can be viewed in terms of co-management – reflecting an ongoing body of work on this topic, coordinated and facilitated by Wageningen University in The Netherlands. Another locus of endeavour on studying co-management or collaborative management comes from Canada, where Armitage, Berkes and others have written extensively about such processes (see Armitage et al, 2007; Berkes, 2009). Canada's Model Forest programme, which now works in partnership with governments in several countries, represents a collaborative approach to landscape management, elements of which bear similarities to our own.

We have focused on analyses of co-management, but extensive literature is available on the implementation of government efforts to manage collaboratively in Nepal, India and Philippines and, more recently, in Tanzania and Zimbabwe, among a number of other countries.

Landscape Governance for Adaptive and Equitable Management

Comparison with other integrated approaches

Landscape governance assumes that multilevel institutions are useful in managing the multifunctional aspects of tropical, rural landscapes, addressing both livelihoods and ecological sustainability. The effort to encourage such governance assumes that landscape managers have the capacity to plan, manage and monitor multifunctional landscapes and understand the consequences of the trade-offs and integration of the different land types of the mosaic. It builds upon previous integrated approaches to natural resource management but places the trade-offs and complementarities among the aspirations of villagers and sustainable management plans at the centre of the model. Integrated water resource management, integrated natural resource management and landscape approaches to conservation are discussed briefly below.

Water resource managers have been among the proponents of integrated approaches. Even more than forests, water is a resource of common interest; its use frequently requires considerable collaboration and compromise among users. Norman Uphoff's (1996) work in Sri Lanka was a landmark study, showing the importance of working with communities in solving water management issues. His students (John Ambler, Ruth Meinzen-Dick, Ujjwal Pradhan and others) have gone on to demonstrate the significance of attending to local realities in trying to solve natural resource dilemmas. The actual level of integration that these initiatives have achieved has been critiqued, as well as their potential for success in future (Biswas, 2008).

Integrated natural resource management, a concept with a long history, made a return in the late 1990s and early 2000s (see Campbell and Sayer, 2003), stressing the interrelationships among parts of systems. The ecosystem approach, as adopted by the Convention on Biological Diversity (CBD), promotes a similar level of integration, with an emphasis on equity (UNEP/CBD, 2000). The notion of 'panarchy', which draws on chaos and systems theories, set forth in a collection by Gunderson and Holling (2002), also places emphasis on interactions among parts of systems in managing natural resources, as well as the importance of temporal and spatial change. The authors in this collection see panarchies as cross-scale, nested sets of adaptive cycles – a concept that holds relevance for landscape-level governance.

Landscape approaches to conservation have been proposed as a broader solution for reconciling the trade-offs between development and conservation in developing countries. The limitations of these approaches for achieving both conservation and development goals have been highlighted by Sayer (2009), in particular the low capacity of local institutions for managing complex multifunctional landscapes.

Land-use planning as policy process holds the potential to address both the multifunctional and the institutional dimensions of landscape management. Through land-use planning, issues of tenure and trade-offs between competing land uses can potentially be resolved and complementarities can be identified and built on. The challenges of engaging with land-use planning processes include integrating local knowledge and aspirations into land-use plans, ensuring that institutions are adaptive and accountable, addressing drivers of land-use change and linking analysis and intervention from the local user to the district and national level (Wollenberg et al, 2008). We saw land-use planning as the most relevant policy domain for tackling issues of landscape governance. However, as discussed below, our success was mixed at best.

Harmonizing values at landscape level

A significant challenge to working at multiple levels with different actors is finding a harmonized vision for the planning and management of the landscape. An early effort to harmonize values can be seen in the work of Mendoza and his team (1999) in the Philippines and Indonesia. Confronted with a variety of assessments of criteria and indicators about the same landscapes being managed for timber production, these authors developed and used 'multicriteria analysis' to come up with fair decisions. This approach involved the ranking, rating and comparison of indicators that form part of an analytic hierarchy process (AHP).

A more recent effort to harmonize values has been CIFOR's multidisciplinary landscape assessments (Sheil et al, 2003), which have systematically studied people's perceptions about their own landscapes in numerous locales (ten are reported in a recent study by Boissière et al, 2010). Besides raising the visibility of local systems and perceptions, such assessments have proven useful in making other stakeholders aware of local needs and interests. Furthermore, in some cases, they have offered negotiating tools (such as maps and species lists) to disempowered local communities. Brian Belcher, Manuel Boissière, Amandine Boucard and Eloise Pulos are implementing a local-level monitoring project in Laos and Cameroon, which complements our own efforts described herein.

Developing landscape-level institutions

Although some landscape-level institutions may already exist, we sought to see whether it was possible to contribute to their emergence. The background to this approach came from CIFOR's ACM programme (1998–2006), which initially focused primarily on the village level and a comparatively small landscape level. ACM field teams used participatory action research (PAR) to improve management and local-level governance, including the institutionalization of social learning (Colfer, 2005). As that multicountry programme evolved, the relevance of the larger landscape scale became increasingly obvious, as did the importance of working at multiple scales to address local-level problems.

Building on these earlier efforts, Komarudin and his team (2008, in press) conducted landscape-level PAR in Bungo District, Jambi (see Chapters 3 and 9). This two-year Collective Action and Property Rights (CAPRi) project[1] involved simultaneous PAR work at district and village levels, with the intention to bring shared concerns together – a partial precursor to the design of the Landscape Mosaics project. Land tenure and land-use planning evolved as a shared concern, as did dealing with elite capture, resolving conflicts among stakeholders and developing effective mechanisms to strengthen local people's participation in government. Considerable progress was made in terms of mutual analysis and understanding, building trust, information exchange and improving understanding of constraints to effective citizen participation (with some success in alleviating them). But the typical short-term funding cycle constituted a significant constraint to follow-through.

McDougall et al (2009, 2010) followed up on ACM work in Nepal, bringing together communities and district (and higher-level) government personnel in structured social learning processes. Nepal, a pioneer in the collaborative management of forests, was dealing with second-generation issues such as equity. It has since reported progress in enhancing equity and mutual understanding, and the institutionalization of multilevel, collaborative social learning processes.

Experiences at Landscape Mosaics sites

The design of the Landscape Mosaics project was complex. Field teams began building on the earlier ACM and post-ACM efforts by conducting PAR at both village and district or landscape levels – as had been done in the earlier, simpler CAPRi project described by Komarudin and colleagues. Project leaders knew that both the CAPRi and the Landscape Mosaics projects were too short in duration (two years). However, the significant progress made under CAPRi encouraged the Landscape Mosaics team to try to make further progress with this evolving approach. In hindsight, a significant difference between the CAPRi effort and our own can be seen in the contextual knowledge, experience and contacts built up during the Jambi-based ACM project. The team knew the area and its people well, and the project's formal leader was an ACM expert, also familiar with the region. In the Landscape Mosaics project, however, we were beginning anew in four of the five sites, and although the same ACM expert served as a support person on the new project, her level of engagement was far lower than in the previous work.

Despite the virtually unanimous commitment of team members to stimulating constructive action on the sites, very little in the ideal form of PAR developed until late in the project, although less formal action research took place everywhere. Indeed, in Cameroon, Indonesia[2] and Madagascar, although visioning workshops were facilitated, there was no action group formation, no routine facilitation, no iterative cycles of social learning, and of course no self-monitoring of the learning process itself. Elements of PAR did begin to emerge in Laos and Tanzania during the second year of the project, but a genuine, iterative PAR process was only beginning as the project drew to a close (see Colfer et al, in press for more detailed analysis).

Project designers were also influenced by the sustainable livelihoods approach (Carney, 1998; Farrington et al, 1999), which provided holistic grist for the livelihoods theme of the project. A variety of monitoring approaches informed the analysis, including Aldrich and Sayer's (2007) landscape outcomes assessment methodology, which built on the five capitals – natural, human, physical, social and financial – to assess outcomes, CIFOR's multidisciplinary assessments (Sheil et al, 2003) and CIFOR's work on criteria and indicators (e.g. CIFOR, 1999).

Linkages to policy processes – land-use planning and land management

A final design element in the Landscape Mosaics project was a link on each site to a national-level policy related to the landscape. In Cameroon (see Chapter 8), this was established through the ongoing process of developing a management plan for the Takamanda technical operations unit, including forest reserves for timber, a national park and small-scale agricultural lands. Other important national policies there included community and council forests, and community hunting zones. Indonesian policies of interest (see Chapter 3) included the new village forests (*hutan desa*) and informal government–industry coalitions, resulting in the transformation of the landscape, from primarily forests and agroforests to monocultures of tree crops and oil palm.

In Laos (see Chapter 4), the emphasis rested on participatory agriculture and forest land-use planning, a policy designed to strengthen security of tenure and protect forest resources, and on the new administrative structure, the village development cluster (*kumban pattana*). Madagascar (see Chapter 6) implemented KoloAla, a policy designed to transfer timber management from the government to community associations, for both environmental and social betterment. Tanzanian teams (see Chapter 5) focused on the policy encouraging local-level, participatory land-use planning, also designed to strengthen security of tenure.

On all sites, the project teams planned to negotiate agreements between local communities and actors at the landscape scale. We intended (1) to draw on the understandings and enhanced local capabilities developed in the PAR process at both village and landscape levels; and (2) to mesh with the site-specific policies and incentives outlined above. As should become clear in the analyses that follow, this step was perhaps overly ambitious for the time allotted.

Critical Perspectives

Much of what we report in this book takes a constructive and positivist view of governance. However, there is a significant body of scholarship that remains sceptical, particularly of governments per se. One such analyst to whom we devote a fair amount of attention (see Chapter 10) is James C. Scott, who has written three important books with governance implications (Scott, 1985, 1998, 2009). *Weapons of the Weak* documents the strategies available to communities to resist the formally powerful, including governments; *Seeing Like a State* explores the need of governments to simplify and make complex realities 'legible' for purposes such as taxation and military service; and *The Art of Not Being Governed* posits a different historical view of the relationship between states and hinterland peoples (such as those residing in the Landscape Mosaics sites) – a more antagonistic relationship than that typically assumed.

The work of Roe (1991, 1994) makes clear the role of 'narratives' or simplified versions of reality (needed by us all) in the making of policy. Narratives such as the destructive role of swidden farmers in deforestation or the essential role of protected areas in maintaining biodiversity – take on lives of their own. The result is the tendency to ignore any evidence that counters the narrative, such as the co-evolution over the millennia of today's tropical forests and swidden farmers or the different, but often dramatic biodiversity characterizing many agricultural landscapes. The works of Leach, Fairhead, Mearns and Scoones (see Fairhead and Leach, 1996; Leach and Fairhead, 2001; Leach and Mearns, 1996; Scoones, 2001) and their colleagues provide convincing evidence of the power (and inevitable inaccuracy) of such narratives, which are powerful engines in policy-making.

In a brief survey of critical governance scholars, one cannot ignore Ferguson (1994), whose careful analysis of the politics of development in Lesotho gives pause to anyone who has worked on either conservation or development projects. He convincingly shows how the functioning of projects was systematically depoliticized and rendered ineffective, using as evidence glaring examples of bureaucratic foot-dragging; the co-opting of development workers willing to set aside their native intelligence in interpreting statistics; the warping of priorities based on the availability of funds, personnel and personal interest; and the persistent official blindness to the overwhelmingly important economic role of South Africa in Lesotho's economy.

Finally, we mention the works of Anna Lowenhaupt Tsing (2005) and Celia Lowe (2006), both of whom write about Indonesia, its people and its biodiversity, in Kalimantan and Sulawesi, respectively. In different ways, they stress the links between local communities, on the one hand, and other actors and contexts at divergent scales, on the other. They look at the malleability or responsiveness of the world views of humans to others and to changing contexts. Tsing's concept of 'friction', analogous to what happens when 'the rubber hits the road' (or when policies come into contact with local realities), is a powerful image that reflects the interdependencies and lack of direct, predictable power of broader-scale policies as they are implemented.

Conclusion

The sustainable management of tropical, rural multifunctional landscapes requires multilevel governance arrangements that integrate the aspirations of villagers and multiple functions of land types into land-use planning and management strategies. The local level – community or village – is the essential building block for such arrangements. The success of these management regimes is contingent on the type and strength of horizontal and vertical linkages with other levels of and actors in governance. Critical to these linkages is the negotiation of divergent visions and management institutions for these landscapes. One of the main challenges that multilevel landscape governance arrangements face in rural, tropical landscapes is how well they can respond to national and global economic and policy drivers of land-use change.

The following chapters provide examples, from the five sites of the Landscape Mosaics project, of how multiple levels of governance affect local livelihoods and the management of natural resources. The authors illuminate the multiple dimensions of landscape governance through specific examples. Chapters 3 and 4 address issues relating to the influence of national policies on practices of land conversion in Indonesia and displacement in Laos and Tanzania, respectively. The other chapters focus more on the linkages between local- and higher-level management regimes, in particular the management of environmental services, both official and unofficial (see Chapter 5), the use and management of forest fragments (see Chapter 6), and the management and trade of non-timber forest products or NTFPs (see Chapters 7 and 8). The following chapter uses the Landscape Mosaics project experience, as well as other material, to develop a governance assessment tool for use at the landscape level (see Chapter 9). Chapter 10 provides more conventional governance comparisons and, building on Scott (2009), a critical assessment of our global efforts to govern collaboratively. The final chapter summarizes the main governance issues addressed in the book and assesses the 'good news' and the 'bad' about such multiscale efforts – concluding that there is in fact no viable alternative to continuing our efforts to 'make it happen'.

Through diverse examples, the complexity of landscape governance becomes clear, as well as the challenges involved in facilitating the emergence of multilevel governance arrangements for these landscapes. The most notable challenges come from global market pressures, as well as policy processes that cannot be fully addressed at the landscape scale. Another challenge that recurs is the unwillingness of higher-level actors to work with village-level actors towards negotiated agreements for the management of the landscape. Redressing these imbalances of power in a way that is beneficial for both local people and their environments is another major challenge – increasingly important as we contemplate the potential of REDD, REDD+, and other climate change-related initiatives to help or harm – that must be addressed to achieve this model of landscape governance.

Acknowledgements

We would like to thank the anonymous reviewers who have provided valuable recommendations for revising the chapter.

Notes

1. CAPRi is a system-wide initiative of the Consultative Group on International Agricultural Research. The Center for International Forestry Research (CIFOR) and World Agroforestry Center (ICRAF) are part of this global group.
2. The Jambi Landscape Mosaics team was under the direction of ICRAF, and its members included individuals who were unfamiliar with the PAR approach and were not particularly attracted to it. Thus the advantage of building on prior work was lost to this team.

References

Agrawal, A. and Gibson, C. C. (1999) 'Enchantment and Disenchantment: The Role of Community in Natural Resource Conservation', *World Development*, vol 27, no 4, pp629–649

Aldrich, M. and Sayer, J. (2007) 'In Practice: Landscape Outcomes Assessment Methodology (LOAM)', Report for WWF Forests for Life Programme, Geneva

Armitage, D. (2008) 'Governance and the Commons in a Multi-level World', *International Journal of the Commons*, vol 2, no 1, pp7–32

Armitage, D., Berkes, F. and Doubleday, N. (eds) (2007) *Adaptive Co-Management*, University of British Columbia Press, Vancouver

Berkes, F. (2006) 'From Community-based Resource Management to Complex Systems', *Ecology and Society*, vol 11, no 1, pp 45ff, www.ecologyandsociety.org/vol11/iss1/art45

Berkes, F. (2009) 'Evolution of Co-management: Role of Knowledge Generation, Bridging Organizations and Social Learning', *Journal of Environmental Management*, vol 90, no 5, pp1692–1702

Berkes, F. and Folke, C. (eds). (1998) *Linking Social and Ecological Systems: Management Practices and Social Mechanisms for Building Resilience*, Cambridge University Press, Cambridge, UK

Bernardo, E. C. and Snelder, D. J. (1999) 'Co-managing the Environment: the Natural Resources of the Sierra Madre Mountain Range', Paper Presented at International Work Conference, College of Forestry and Environmental Management and Plan International, Garita, Philippines

Biswas, A. K. (2008) 'Integrated Water Resources Management: Is it Working?' *International Journal of Water Resources Development*, vol 24, no 1, pp5–22 (March)

Boissière, M., Sassen, M., Sheil, D., Heist, M. V., Jong, W. D., Cunliffe, R., Wan, M., Padmanaba, M., Liswanti, N., Basuki, I., Evans, K., Cronkleton, P., Lynam, T.,

Koponen, P. and Bairaktari, C. (2010) 'Researching Local Perspectives on Biodiversity in Tropical Landscapes: Lessons from Ten Case Studies', in A. Lawrence (ed) *Taking Stock of Nature: Participatory Biodiversity Assessment for Policy, Planning and Practice*, Cambridge University Press, Cambridge, pp113–141

Borrini-Feyerabend, G., Pimbert, M., Farvar, M. T., Kothari, A. and Renard, Y. (2004) *Sharing Power: Learning by Doing in Co-management of Natural Resources Throughout the World*, IIED and IUCN/CEESP/CMWG, Cenesta, Tehran

Buck, L., Geisler, C. C., Schelhas, J. and Wollenberg, E. W. (eds) (2001) *Biological Diversity: Balancing Interests through Adaptive Collaborative Management*, CRC Press, Boca Raton, Florida

Campbell, B. M. and Sayer, J. A. (2003) *Integrated Natural Resource Management: Linking Productivity, the Environment and Development*, CABI Publishing, Cambridge, Massachusetts

Carney, D. (1998) 'Implementing the Sustainable Rural Livelihoods Approach', in *Sustainable Rural Livelihoods: What Contribution Can We Make?*, Policy Statement, Department for International Development, London

Carter, J. (2000) 'Recent Experience in Collaborative Forest Management Approaches: A Review Paper', Report for Intercooperation, Bern, Switzerland

CIFOR (Center for International Forestry Research) (1999) *C&I Toolbox*, CIFOR, Bogor

Colfer, C. J. P. (2005) *The Complex Forest: Communities, Uncertainty, and Adaptive Collaborative Management*, Resources for the Future/CIFOR, Washington, DC

Colfer, C. J. P. and Capistrano, D. (eds) (2005) *The Politics of Decentralization: Forests, Power and People*, Earthscan, London

Colfer, C. J. P., Andriamampandry, E., Asaha, S., Lyimo, E., Martini, E., Pfund, J.-L. and Watts, J. D. (in press) 'Participatory Action Research for Catalyzing Adaptive Management: Analysis of a "Fits and Starts" Process', *Journal of Environmental Science and Engineering*

Colfer, C. J. P., Dahal, G. R. and Capistrano, D. (eds) (2008) *Lessons from Forest Decentralization: Money, Justice and the Quest for Good Governance in Asia-Pacific*, Earthscan/CIFOR, London

Contreras, A. P. (ed) (2003) *Creating Space for Local Forest Management in the Philippines*, LaSalle Institute of Governance and Antonio Contreras, Manila

Diaw, M. C. (2005) 'Modern Economic Theory and the Challenge of Embedded Tenure Institutions: African Attempts to Reform Local Forest Policies', in S. Kant and R. A. Berry (eds) *Sustainability, Institutions and Natural Resources: Institutions for Sustainable Forest Management*, Springer, Netherlands, pp44–82

Diaw, M. C. (2009) 'Elusive Meanings: Decentralization, Conservation and Local Democracy', in L. German, A. Karsenty and A. M. Tiani (eds) *Governing Africa's Forests in a Globalized World*, Earthscan/CIFOR, London

Dietz, T., Ostrom, E. and Stern, P. C. (2003) 'The Struggle to Govern the Commons', *Science*, vol 302, no 5652, pp1907–1912.

Edmunds, D. and Wollenberg, E. (eds) (2003) *Local Forest Management*, Earthscan, London

Fabricius, C., Folke, C., Cundill, G. and Schultz, L. (2007) 'Powerless Spectators, Coping Actors and Adaptive Co-managers: A Synthesis of the Role of Communities in Ecosystem Management', *Ecology and Society*, vol 12, no 1, p29ff, www.ecologyandsociety.org/vol12/iss1/art29

Fairhead, J. and Leach, M. (1996) *Misreading the African Landscape: Society and Ecology in a Forest-Savanna Mosaic,* Cambridge University Press, New York

Farrington, J., Carney, D., Ashley, C. and Turton, C. (1999) 'Sustainable Livelihoods in Practice: Early Applications of Concepts in Rural Areas', *Natural Resource Perspectives,* vol 42, pp2005–2006

Ferguson, J. (1994) *The Anti-Politics Machine: 'Development,' Depoliticization, and Bureaucratic Power in Lesotho,* University of Minnesota Press, Minneapolis

Forman, R. T. T. (1995) *Land Mosaics: The Ecology of Landscapes and Regions,* Cambridge University Press, Cambridge, UK

Forman, R. T. T. and Godron, M. (1986) *Landscape Ecology,* John Wiley, New York

German, L., Karsenty, A. and Tiani, A.-M. (eds) (2009) *Governing Africa's Forests in a Globalized World,* Earthscan/CIFOR, London

Gezon, L. L. and Paulson, S. (2004). 'Place, Power, Difference: Multiscale Research at the Dawn of the Twenty-first Century', in *Political Ecology Across Spaces, Scales, and Social Groups,* Rutgers University Press, New Brunswick, New Jersey, pp1–16

Gibson, C. C., Ostrom, E., Ahn, T. K. (2000) 'The Concept of Scale and the Human Dimensions of Global Change: A Survey', *Ecological Economics,* vol 32, pp217–239

Görg, C. (2007) 'Landscape Governance: The "Politics of Scale" and the "Natural" Conditions of Places', *Geoforum,* vol 38, no 5, September, pp954–966

Gunderson, L. H. and Holling, C. S. (eds) (2002) *Panarchy: Understanding Transformations in Human and Natural Systems,* Island Press, Washington, DC

Hlambela, S. and Kozanayi, W. (2005) 'Decentralized Natural Resources Management in the Chiredzi District of Zimbabwe: Voices from the Ground', in C. J. P. Colfer and D. Capistrano (eds) *The Politics of Decentralization: Forests, Power and People,* Earthscan/ CIFOR, London, pp255–268

Hobley, M. (1996) *Participatory Forestry: The Process of Change in India and Nepal,* ODI, London

Komarudin, H., Siagian, Y. and Colfer, C. J. P. (in press) 'The Role of Collective Action in Securing Property Rights for the Poor: A Case Study in Jambi Province, Indonesia', in E. Mwangi, H. Markelova and R. Meinzen-Dick (eds) *Collective Action and Property Rights for Poverty Reduction,* University of Pennsylvania Press, Philadelphia

Komarudin, H., Siagian, Y. L., Colfer, C. J. P., with Neldysavrino, Yentirizal, Syamsuddin and Irawan, D. (2008) *Collective Action to Secure Property Rights for the Poor: A Case Study in Jambi Province, Indonesia,* Collective Action and Property Rights System-wide Initiative, Washington, DC

Krishna, A. (2002) *Active Social Capital,* Columbia University Press, New York

Larson, A. M. (2005) 'Democratic Decentralization in the Forestry Sector: Lessons Learned from Africa, Asia and Latin America', in C. J. P. Colfer and D. Capistrano (eds) *The Politics of Decentralization: Forests, People and Power,* Earthscan/CIFOR, London

Larson, A. M., Barry, D., Dahal, G. R. and Colfer, C. J. P. (eds) (2010) *Forests for People: Community Rights and Forest Tenure Reform,* Earthscan/CIFOR, London

Leach, M. and Fairhead, J. (2001) 'Plural Perspectives and Institutional Dynamics: Challenges for Local Forest Management', *International Journal of Agricultural Resources, Governance and Ecology,* vol 1, no 3/4, pp223–242

Leach, M. and Mearns, R. (eds) (1996) *Lie of the Land: Challenging Received Wisdom on the African Environment,* International African Institute, Oxford

Lemos, M. C. and Agrawal, A. (2006) 'Environmental Governance', *Annual Review of Environment and Resources,* vol 31, pp297–325

Levin, S. A. (1992) 'The Problem of Pattern and Scale in Ecology,' *Ecology,* vol 73, pp1943–1967

Liu, J. and Taylor, W. W. (eds) (2002) *Integrating Landscape Ecology into Natural Resource Management,* Cambridge University Press, Cambridge

Lowe, C. (2006) *Wild Profusion: Biodiversity Conservation in an Indonesian Archipelago,* Princeton University Press, Princeton, New Jersey

MacMynowski, D. P. (2007) 'Across Space and Time: Social Responses to Large-scale Biophysical Systems', *Environmental Management,* vol 39, pp831–842

Makapukaw, A. L. S. A. D. and Mirasol, F. S. (2005) 'Decentralizing Protected Area Management at Mount Kitanglad', in C. J. P. Colfer and D. Capistrano (eds) *The Politics of Decentralization: Forests, Power and People.* Earthscan, London, pp269–281

Marfo, E., Colfer, C. J. P., Kante, B. and Elias, S. (2010) 'From Discourse to Policy: The Practical Interface of Statutory and Customary Land and Forest Rights', in A. Larson, D. Barry, G. R. Dahal and C. J. P. Colfer (eds) *Forests for People: Community Rights and Forest Tenure Reform,* Earthscan/CIFOR, London

McDougall, C., Pandit, B. H., Banjade, M. R., Paudel, K. P., Ojha, H., Maharjan, M., Rana, S., Bhattarai, T. and Dangol, S. (2009) *Facilitating Forests of Learning: Enabling an Adaptive Collaborative Approach in Community Forest User Groups, a Guidebook,* CIFOR, Bogor

McDougall, C., Ojha, H., Banjade, M., Pandit, B. H., Bhattarai, T., Maharjan, M. and Rana, S. (2010) *Forests of Learning: Experiences from Research on an Adaptive Collaborative Approach to Community Forestry in Nepal,* CIFOR, Bogor

Meinzen-Dick, R., Gregorio, M. Di and Dohrn, S. (2008) 'Decentralization, Pro-poor Land Policies and Democratic Governance', CAPRi Working Paper 80, IFPRI, Washington, DC

Mendoza, G. A., Macoun, P., Prabhu, R., Sukadri, D., Purnomo, H. and Hartanto, H. (1999) *Guidelines for Applying Multi-Criteria Analysis to the Assessment of Criteria and Indicators,* CIFOR, Bogor

Ostrom, E. (1990) *Governing the Commons: The Evolution of Institutions for Collective Action,* Cambridge University Press, Cambridge

Ribot, J. C. (2003) 'Democratic Decentralization of Natural Resources: Institutional Choice and Discretionary Power Transfers in Sub-Saharan Africa', *Public Administration and Development,* vol 23, no 1, pp53–65

Ribot, J. C. and Larson, A. M. (eds) (2005) *Democratic Decentralisation through a Natural Resource Lens,* Routledge, London

Roe, E. M. (1991) 'Development Narratives, or Making the Best of Blueprint Development', *World Development,* vol 19, no 4, pp287–300

Roe, E. M. (1994) *Narrative Policy Analysis: Theory and Practice,* Duke University Press, Durham, North Carolina

Sayer, J. (2009) 'Reconciling Conservation and Development: Are Landscapes the Answer?' *Biotropica,* vol 41, no 6, pp649–652

Scoones, I. (2001) '*Dynamics & Diversity: Soil Fertility and Farming Livelihoods in Africa*, Earthscan, London

Scott, J. C. (1985) *Weapons of the Weak,* Yale University Press, New Haven, Connecticut

Scott, J. C. (1998) *Seeing Like a State: How Certain Schemes to Improve the Human Condition Have Failed,* Yale University Press, New Haven, Connecticut

Scott, J. C. (2009) *The Art of Not Being Governed: An Anarchist History of Upland Southeast Asia,* Yale University Press, New Haven, Connecticut

Shackleton, S., Campbell, B. M., Cocks, W. M., Kajembe, G., Kapungwe, E., Kayambazinthu, D., Jones, B., Matela, S., Monela, G., Mosimane, A., Nemarundwe, N., Ntale, N., Rozemeijer, N., Steenkamp, C., Sithole, B., Urh, J. and Jagt, C. V. D. (2001) *Devolution in Natural Resource Management: Institutional Arrangements and Power Shifts (a Synthesis of Case Studies from Southern Africa),* USAID SADC NRM, Project No 690–0251.12 through WWF-SARPO, EU's 'Action in Favour of Tropical Forests' through CIFOR and the Common Property STEP Project, CSIR, Bogor

Sheil, D., Puri, R. K., Basuki, I., Van Heist, M., Wan, M., Liswanti, N., Rukmiyati, Sardjono, M. A., Samsoedin, I., Sidiyasa, K., Chrisandini, Permana, E., Angi, E. M., Gatzweiler, F., Johnson, B. and Wijaya, A. (2003) *Exploring Biological Diversity, Environment and Local People's Perspectives in Forest Landscapes: Methods for a Multidisciplinary Landscape Assessment,* CIFOR, Bogor

Sunderlin, W. D. (2008) *From Exclusion to Ownership? Challenges and Opportunities in Advancing Forest Tenure Reform,* Report for Rights and Resources Initiative, Washington, DC

Swiderska, K., Roe, D., Siegele, L. and Grieg-Gran, M. (2009) *The Governance of Nature and the Nature of Governance: Policy That Works for Biodiversity and Livelihoods,* IIED, London

Tress, B. and Tress, G. (2000) 'The Landscape – From Vision to Definition,' in J. Brandt, B. Tress and G. Tress (eds) *Multifunctional Landscapes: Interdisciplinary Approaches to Landscape Research and Management,* material for international conference on 'Multifunctional Landscapes', Centre for Landscape Research, University of Roskilde, Denmark, 18–21 October 2000

Tsing, A. L. (2005) *Friction: An Ethnography of Global Connection,* Princeton University Press, Princeton, New Jersey

UNEP/CBD (eds) (2000) 'Ecosystem Approach,' in *Decisions Adopted by the Conference of the Parties to the Convention on Biological Diversity at its Fifth Meeting,* pp103–109

Uphoff, N. (1996) *Learning from Gal Oya: Possibilities for Participatory Development and Post-Newtonian Social Science,* Intermediate Technology Publications, London

Wollenberg, E., Anderson, J. and López, C. (2005) *Though All Things Differ: Pluralism as a Basis for Cooperation in Forests,* CIFOR, Bogor

Wollenberg, E., Campbell, B., Dounias, E., Gunarso, P., Moeliono, M., and Sheil, D. (2008) 'Interactive Land-use Planning in Indonesian Rainforest Landscapes: Reconnecting Plans to Practice', *Ecology and Society,* vol 14, no 1, pp35, www.ecologyandsociety.org/vol14/iss1/art35

3

Role of the District Government in Directing Landscape Dynamics and People's Futures: Lessons Learnt from Bungo District, in Jambi Province

Laurène Feintrenie and Endri Martini

The environment and sustainable management of natural resources have become topics for international meetings on global issues, such as the mitigation of climate change and adaptation to its consequences. But regardless of international agreements, in the field, only national or decentralized governments have the authority to make decisions on their natural resources. They have to deal with local and regional constraints, people's needs and demands, and the interests of groups, external or internal, which have claims on the resources or their management. Governments may affect natural resources with different tools, such as legislation and regulations, land-use planning and public programmes of rural development or of environmental conservation.

In 2005, the Indonesian forest was estimated to cover an area of 94 million hectares by the Food and Agriculture Organization of the United Nations (FAO, 2010a, 2010b). This indicated that it is the developing world's largest tropical closed forest. But the country has been criticized by the international community for its high levels of CO_2 emissions from deforestation and fires. Indonesia is an emergent country whose economic growth is bringing about rapid changes both in the landscape and in the society. One of the major changes in Indonesia in the past decades has been political decentralization, which began in the late 1990s with the end of President Soeharto's regime. Three levels of government now share authority over natural resources in a complex legislative frame. There is a constant debate over who has the authority and responsibility to manage natural resources, especially forested land and forest products. However, other powerful stakeholders challenge the government's authority or plans by leading their own agendas. The private sector appeals to people and governments with promises of economic growth and wealth; local leaders set their authority on local debates and client-oriented, paternalistic relations; environmentalist NGOs defend the preservation of biodiversity, sometimes not looking too closely at the trade-offs between conservation and development.

In this complex framework of power relations and divergent interests, what is the actual role of the district government in directing landscape dynamics and the evolution of people's livelihoods? This chapter considers the role of the district government in the economic development of the district and in changes within the environment. Bungo District, in the province of Jambi, in the centre of Sumatra, was chosen as the case study. The authors and the research institutions to which they are affiliated – Institut de Recherche pour le Développement (IRD), Centre International de Recherche en Agronomie du Développement (CIRAD), Center for International Forestry Research (CIFOR) and World Agroforestry Center (ICRAF) – have been conducting research in the district since the beginning of the 1990s. Bungo's economy has developed during the past 30 years, often as a result of forest resources and with the stimulus of a growing population. Observation of the district over a long period enables a better understanding of the district government's policies and actions and how they interact with other stakeholders' agendas.

The role of the district government in directing landscape dynamics and people's livelihoods is explored through three research questions: Who has authority over land and forest? How does the district government manage forest and land? What are the main causes of conflict in the district?

Study Site

Bungo District lies in Jambi Province, in the lowland area at the centre of Sumatra Island (Figure 3.1). The district was formed in 1999 when Bungo-Tebo District was divided into two administrative units. Several national parks of high biodiversity conservation value surround the district: Kerinci Seblat National Park in the south, Bukit Duabelas National Park in the southeast and Bukit Tigapuluh National Park in the north. Rubber agroforests and secondary forests along the riparian zone offer a potential connection between protected areas in the region, but forest and rubber agroforest are not the most profitable land covers and are threatened as people seek better livelihoods (Feintrenie and Levang, 2009).

Until the end of the 19th century, primary forests covered nearly all the island. The first valorization of this natural resource was hunting and gather-

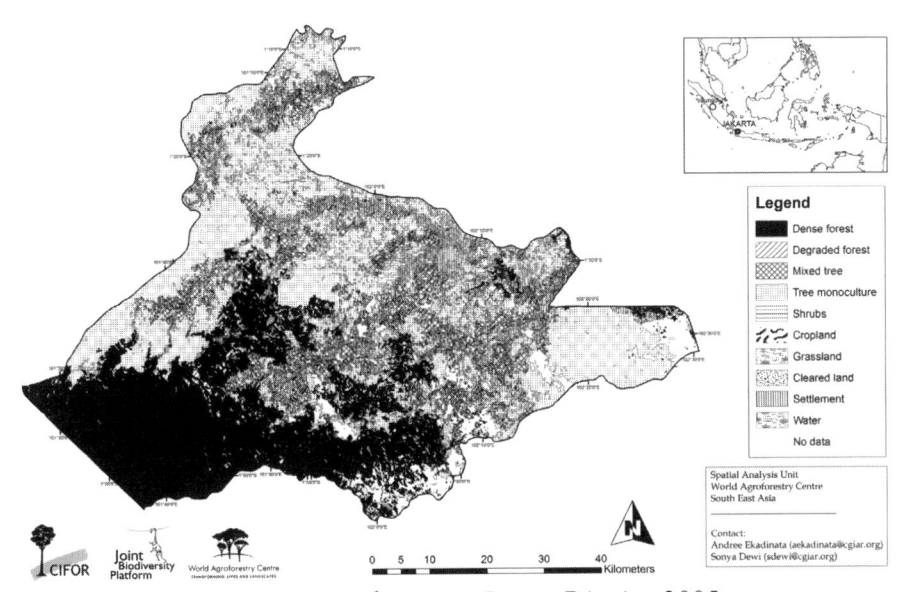

Figure 3.1 *Land cover in Bungo District, 2005*

ing, followed by swidden cultivation of upland rice. Small patches of the forest were converted into rice swiddens, alternating with bush fallows. The industrial revolution in Europe and North America at the beginning of the 19th century created a demand for rubber. In response to this new market opportunity, farmers introduced rubber seedlings in their swiddens, amid the upland rice. In doing so, they invented a new cropping system, rubber agroforests. From the beginning of the 20th century, farmers progressively converted their swiddens into rubber agroforests at the expense of neighbouring forests. From the 1980s, increased demand for rubber and reduced access to forests incited the smallholders to intensify their practices, and to convert their agroforests into monospecific rubber plantations. Public development programmes, such as the distribution of clonal rubber seedlings and improved rice varieties, have favoured intensification.

Thanks to the continually increasing demand for rubber, agroforests spread over Sumatra's eastern peneplains until the 1990s. But with growing demographic pressure, market integration and household monetary needs, agroforests have become increasingly endangered. New cropping systems have appeared, challenging agroforests' dominance in the landscape. Since the mid-20th century, monospecific rubber plantations have been competing for land and have become undoubtedly more profitable than agroforests. More recently, the national and regional governments have sponsored estates development or rubber and oil palm production through transmigration projects. And still more recently, oil palm plantations have spread over the island, quickly threatening to supersede rubber agroforestry (Feintrenie and Levang, 2009). The forested area has depleted rapidly since the 1980s, when large rubber and oil palm plantations were developed under the Nucleus Estates and

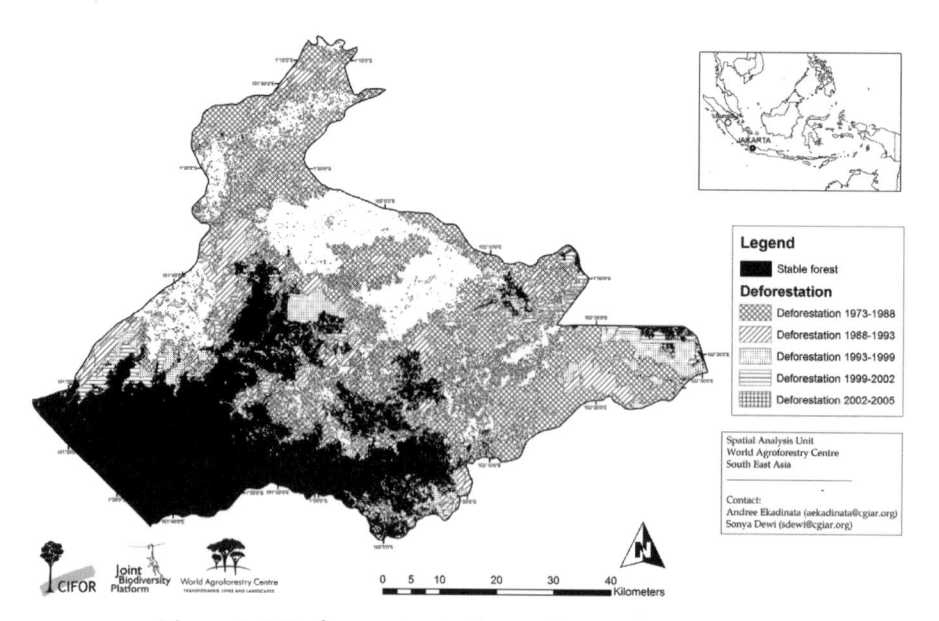

Figure 3.2 *Deforestation in Bungo District, 1973–2007*

Smallholders transmigration programme (Perkebunan Inti Rakyat)[1] and when the government authorized logging in forest concessions (Suherman and Taher, 2008; Ekadinata and Vincent, 2008) (Figure 3.2). As in many other districts in Indonesia, the transition from a centralized government to a decentralized one in 2001 also affected natural resource management. During the past 30 years, the forest cover in Bungo has decreased from 70 per cent to less than 30 per cent of the district's surface area.

Agriculture is the main source of local community livelihoods, with the main agricultural products being natural rubber and oil palm. In terms of latex sourced from rubber trees, 64 per cent was produced from rubber agroforest gardens owned by smallholder farmers (Joshi et al, 2003). Oil palm is produced mainly by private and public companies in large estates. Paddy fields are still maintained in some villages for family consumption, as well as upland rice in the most remote areas.

The majority of the remaining forests, as well as some cultivated land, is state owned. The ministry of forestry holds authority over all forest land and is willing to convert former timber concessions into timber plantations, thus retaining control over the land and its resources. Because the district government reaps no benefit from national forests, it favours the conversion of forested land into agricultural plantations, which generate income for the district. Local communities support the district's strategy. Many are willing to convert large portions of their landscape into oil palm plantations in order to improve their livelihoods. One regional economic development project is opening new access to cultivated or forested lands by building new roads, another is distributing seedlings of clonal rubber, a third one by attracting oil palm companies in the district. These programmes directly affect the landscape. Indeed the presence of a road, even unpaved, is a precondition for oil palm development. The proximity of agro-industrial companies is a determining factor on infrastructure, seedling availability, tenant contracts, technical assistance, credit schemes and processing plants. Another group of programmes aims to rejuvenate the paddy fields. These farming development schemes have had a net impact on the landscape of the district.

Methods

Analysis of governance

Governance of forests and natural resources in Indonesia is the subject of numerous studies. We reviewed publications on this issue, with a focus on transmigration and decentralization, and their consequences for rural and forest-based societies. In addition, we surveyed the literature on forests and the Indonesian legislation on natural resource management, land tenure and forests. The issue of benefits sharing between the levels of government was specifically targeted. At the district level we also consulted documents specific to Bungo District, especially the land-planning reports and district regulations on agriculture, forests, transmigration and agro-industry.

Interviews with civil servants working in the district offices (*dinas*) helped us to understand government actions in the area and the relations among public sector actors. Five departments were assessed: the Department of Forestry and Plantations (Dinas Perkebunan-Kehutanan),[2] the Department of Agriculture, Fisheries and Animal Breeding (Dinas Pertanian/Perikanan/Peternakan), the Department of Transmigration (Dinas Transmigrasi), the Department of Regional Planning (BAPPEDA), and the National Land Office (Badan Pertanahan Nasional) (Two to four interviews were conducted with 17 respondents). These civil servants also kindly provided documents in response to our questions or to illustrate their answers.

In Chapter 9, Colfer and Feintrenie describe an assessment tool that lists indicators of the quality of governance. This tool was used as a questionnaire to assess respondents' perceptions about regional and national governance, with a specific focus on forest management. We asked representatives of the main groups to give us their own evaluations of the governance system, its structure, its implementation and its efficiency. The respondents were six civil servants working in the district (BAPPEDA, Department of Forestry and Plantations, Department of Transmigration, Department of Agriculture, Fisheries and Animal Breeding, and the National Land Office), five people working for research centres and NGOs (ICRAF and CIFOR, the Indonesian conservationist association Warung Konservasi, WARSI, and an independent consultant), and about 130 villagers from 12 villages (categorized by gender, and in each village a specific interview with the head of the village). Table 3.1 shows selected results.

Survey of conflicts involving communities

Three cases of conflicts within or between villages were discovered through the literature review, interviews or observations. To understand the causes and consequences of the conflicts and the way in which the government deals with them, we interviewed all the categories of actors who were involved, as well as people who had observed the conflicts. Civil servants and representatives of private companies involved were interviewed, villages were visited and semi-structured interviews were conducted with village leaders (especially current and former heads of villages, and heads of agriculture cooperatives or farmers' groups) and farmers involved in the conflicts, as well as with villagers not involved.

Each of the three cases related to the transmigration programme. The oldest conflict involved a community in opposition to an oil palm company. The second case was related to land tenure conflict between an indigenous village and a village of transmigrants. The last case was caused by the absence of a company that was promised in the contract people had signed with the government.

Results

Authority over land and forest

Since the enactment of Indonesia's decentralization laws, 22/1999 and 25/1999, administration has been divided into a central government (*pemerintah pusat*) and regional governments (*pemerintah daerah*, i.e. provinces and districts). The president is the head of the executive branch of the central government and gives direct orders to the ministers. Provinces and districts have similar government structures, but with different types of authority.

Both men and women can vote and elect their legislative and executive leaders directly (Figure 3.3), except for the heads of ministries, nominated by the president, and the heads of subdistricts (*camat*), nominated by heads of districts (*bupati*). The legislative power is held by three levels of parliament. At the national level, the People's Consultative Assembly (Majelis Permusyawaratan Rakyat, MPR) comprises two bodies: the People's Representative Council (Dewan Perwakilan Rakyat, DPR) and the Regional Representative Council (Dewan Perwakilan Daerah, DPD). The People's Representative Council has three main functions: legislation, budgeting and oversight. It draws up and passes laws of its own and also discusses and approves government regulations in lieu of laws and proposals from the Regional Representative Council related to regional issues. Together with the president, it produces the annual budget, taking into consideration the views of the Regional Representative Council. It also has the right to question the president and other government officials. At the province and district levels, two independent representative councils (Dewan Perwakilan Rakyat Daerah Propinsi, DPRD *propinsi* and DPRD *kabupaten*) enact provincial and district

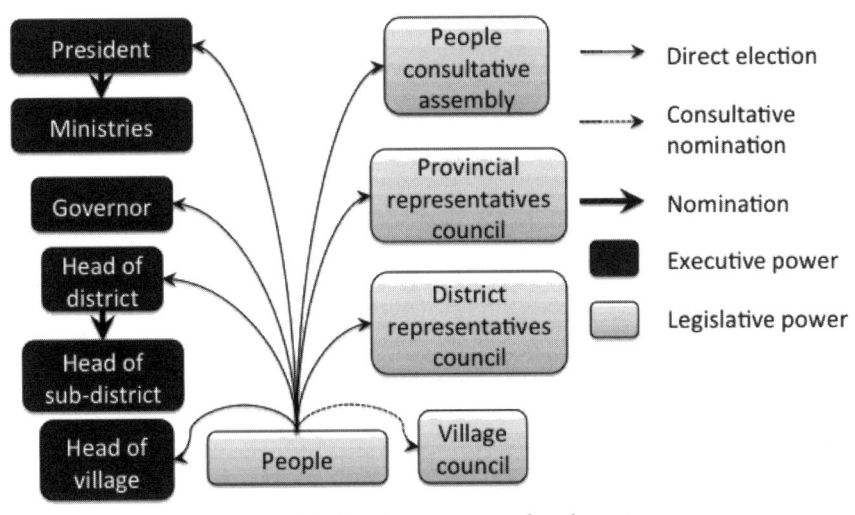

Figure 3.3 *Election system of Indonesia*

Source: Feintrenie (forthcoming)

laws. The village council (*badan permusyawarahan desa*, BPD) plays a similar role at the village level, with the difference that its members are chosen by mutual consent of the village head and villagers rather than through elections. The function of a village council, as stated in the law, is to implement village regulations together with village heads, in order to address and support the community's aspirations.

Although the Indonesian decentralization process enables the transfer of authority, certain domains remain under the central government's control. For example, the National Land Office is under the direct command of the central government and is not decentralized. In Figure 3.4, the three boxes labelled 'National Land Office' represent the three levels of management and show the hierarchy from the national office to the regional offices. One task of this office is to issue landownership certificates and land-use permits (*hak guna usaha*). Without its approval, the district government and the provincial government are not authorized to issue permits for developing agricultural estates. According to local officers from the head of district's office and the National Land Office, this permit is not required for developing oil palm plantations, but is required by banks to get a loan and by the provincial and district governments for authorizing large estates. The permit is proof that the plantation company has legitimately acquired the permitted area of land. Indeed, cultivation rights on land are defined as a 'right given to individuals or legal bodies for cultivating state land for agricultural and farming enterprises, and such land remains under the direct control of the government' (*Pengaturan Republik Indonesia* No 40/1996).

Bungo District is divided into 17 sub-districts and 124 villages. According to Law 32, Article 40, the district council (DPRD *kabupaten*) holds legislative,

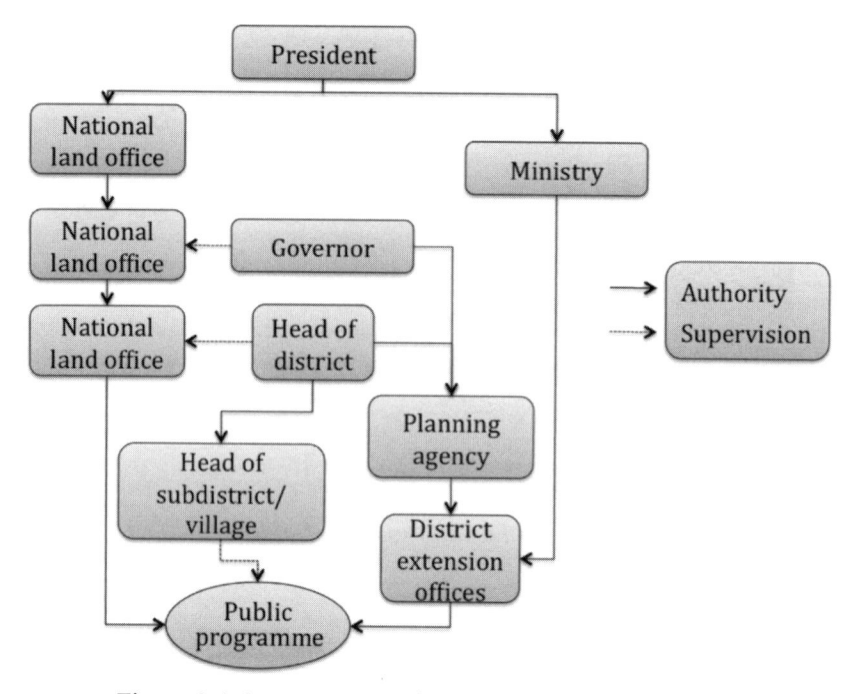

Figure 3.4 *Organization of government in Indonesia*

planning and supervisory roles; for example, if agribusiness companies are in conflict with villagers, the villagers might turn to district council members for assistance. Thus, companies seek to develop good relationships with members of the district council before conflicts arise. Land-use planning is carried out by BAPPEDA at a district level. BAPPEDA is also in charge of the coordination between district agencies (*dinas*) to achieve the goals defined by the governor and the head of the district (Figure 3.4). The *dinas* directly implements public programmes under the supervision of BAPPEDA and the heads of the ministries. Each district has a land-use planning committee, responsible for preparing or subcontracting the plan, overseeing its implementation and monitoring the use of allocated lands. This committee consists of representatives from BAPPEDA, the district office of the National Land Office, the district forest office and the head of the district. The body also plays a vital role in mediating land conflicts.

The formal administrative unit of a local community in Indonesia is a village, whose organization was made uniform throughout Indonesia in 1979 by a national law, UU No 5/1979. The implementation of this regulation resulted in the separation of some communities, named '*Bathin*' in our research area, which have a common history and customary attributes (rules, relations and territory). UU No 5/1979 was revised to UU No 32/2004, which states that each village, district and province has authority over its internal affairs. Regulation UU No 32/2004 reinforced the law it replaced, by allowing regional governments to restructure their formal administrative area, as long as it did not inhibit economic development. In early 2008, the villages' boundaries were redefined in Jambi Province, including in Bungo District, to a structure that refers to the former customary territory organization. *Kampung*, the former word for a hamlet, has replaced sub-villages (*dusun*). Several *kampung* are under the territory of a common *rio*, formerly 'village'. This reform, which aims at reinforcing customary institutions, induced very few changes in Bungo, where the administrative units of villages (formerly *desa*) were defined along the customary spatial divisions. Villagers often interpret this reform as a change of titles, but nothing more. The village restructuring has been organized by BAPPEDA since 2005 (Hasan et al, 2008), and it was still in process in late 2009. Customary leaders are included in the formal village government through heads of customs (*kepala adat*) and of religion (*kepala agama*) in the village council. This integration of the two authorities illustrates the common will to respect customary authority. This trend is also shown by the fact that the *bupati* of Bungo also governs customary practices, as the (district) head of *adat*, or custom (*kepala adat kabupaten*), as is the case in all the districts in the province of Jambi. As the district head of customs, the bupati has the highest authority in the decision-making process related to the implementation of the customary regulation at district level.

The customary rules recognized throughout Bungo are compiled in the Bungo District Customary Guide Book (Pemerintah Daerah Kabupaten Bungo, 2000). Customary regulations are still respected and used by local communities in Bungo, particularly to solve internal village conflicts. Sanctions are flexible, usually not very severe, but not necessarily equitable or fair. The weight of customary rule varies among the villages. In remote villages, which have poor access to external networks, such rule holds more importance in daily life than where a high proportion of the villagers are already working outside the agricultural sector (Feintrenie, forthcoming). Additionally, customary rules and sanctions have a weak level of power when the conflict involves outsiders.

The Indonesian Forestry Basic Law, No 41/1999, gave more authority to the regional governments in managing natural resources, including the right to issue logging permits for small-scale forestry management and harvesting. But in 2002, a new law transformed this right into an obligation to assist the implementation of regulations decided and approved by the central government. Thereafter, a Ministry of Forestry decree, No 382/kpts-II/2004, enabled regional governments to issue small-scale logging permits in non-forest state lands, 'to use timber from privately owned forests' (*izin pemanfaatan kayu rakyat*, IPKR). IPKR are to be granted to cooperatives, farmer groups or foundations that operate in regions outside state forest, covering less than 100ha of non-forestry use areas (*areah pemanfaatan lain*, APL), on privately owned land. The same decree obliges the loggers to use sustainable forest management practices. But foresters who do not respect this constraint are scarcely sanctioned, as law enforcement remains poor in the country (Obidzinski, 2005a, 2005b).

Law No 10/2004 gave authority to the districts and provinces to edit regional regulations (*peraturan daerah* or *perda*) on natural resources management, and to villages to edit local regulations (*peraturan desa* or *perdes*). These regional and local regulations must respect laws of higher levels. *Perda* are voted on by the district council (DPRD *kabupaten*) and applied by the head of the province, the governor, the head of the district or the head of the city (*walikota*). *Perdes* are voted on by the village council and applied by the head of the village. Between 2000 and 2003, 17 *perda* related to natural resources management were issued in Bungo District, including five *perda* applicable to mining (gold, coal and sand), four to logging concessions and one to plantation forest (Suherman and Taher, 2008). *Perdes* related to natural resources management are still scarce. *Perdes* should be discussed with all the villagers before being issued, but in reality participation is difficult to organize and to manage. Usually the village leaders neither possess the skills nor the interest to engage in participatory decision-making processes. Many village heads have only primary or secondary school education and little training in administration, decision-making and communication methods. Therefore, the decision-making process can easily lose transparency and end in the local elites capturing power and resources.

In Bungo, as in many places in Indonesia, most forested lands are state-owned, despite the long-term historical use of land by people and traditional customary tenure rights. According to the customary rights, forests are under common property tenure. Cultivated lands, including agroforests and plantations, have private status in the customary law. The customary law on land tenure states that ownership of a plot belongs to the first person who has cut and planted this land with trees or cultivated it with annual crops. If the planted land is on a riverside, then all the land from the river to the summit of the hill above the river is reserved for the owner, although others can harvest forest products until the land is planted, but can no longer slash and plant it. Results from the perception survey conducted in 12 villages of Bungo in 2009 show that customary plot boundaries are well respected; few conflicts arise, and when any do, they are quickly resolved through the customary regulation system (Feintrenie and Levang, 2010; see also Chapter 9 of this book). The harvesting of timber and non-timber forest products (NTFPs) is regulated by customary rules. There is little logging in the area, simply because hardly any valuable timber remains in the accessible secondary forests. Conflicts may arise if villagers claim ownership or assume rights over a forest that the government also claims and wants to allocate for a logging or industrial concession (e.g. timber, coal mining, oil palm and rubber).

Coordination between customary leaders and formal governments appears clearly in the results of the perception survey, summarized in Table 3.1 (Feintrenie and Levang, 2010). On a range of scores from 1 to 5, respondents gave a positive score of 4 to the level of agreement between the two sources of authority (for more details on the indicators, see Chapter 9). The efficiency of local and regional governments is also apparent in the scores given to efficiency

Table 3.1 *Stakeholders' perceptions of governance in Bungo District*

Indicators of governance (scores from 1 to 5)	All respondents	Villagers	Male villagers	Female villagers	Civil servants	NGOs
Formal land categories' conformity to actual land use	3.1	3.1	3.1	3.1	2.3	3.8
Rights to access forest and use forest land and products	3.4	3.5	3.6	3.3	3.1	2.5
Security of rights to access forest and use forest products	4.1	4.2	4.4	4.0	3.6	2.4
Efficiency of mechanisms of participation in decision-making	3.5	3.7	3.9	3.5	3.4	3.3
Actual voice of each interest group in decision-making	4.2	4.3	4.4	4.1	4.0	2.2
Efficiency of formal and customary governance	4.0	4.2	4.3	4.0	3.6	3.8
Enforcement of rules and regulations	4.0	4.2	4.4	4.0	3.7	4.0
Acceptability of the level of conflict	4.0	4.2	4.5	3.9	3.9	3.3
Level of trust among stakeholders	3.7	3.9	4.0	3.8	3.8	3.2
People's access to external networks	3.2	3.4	3.5	3.2	3.6	2.4

Source: Feintrenie (forthcoming)

of rules enforcement (4), low level of conflicts (4) and participation of all the actors in the decision-making process (3.5). Villagers are generally more positive in their evaluation than civil servants or conservationists. Among the villagers, women are generally more negative and less happy with the governance than men. In some villages, the interviews clearly showed that women were not used to being asked for a personal opinion on such a subject. Especially in the most remote and isolated villages, they did not feel comfortable in discussing governance matters or the decision-making process.

Average scores of 3.7 were given to the level of trust among stakeholders and 4.2 to the actual voice of each interest group in decision-making. These two scores express a common feeling of confidence among the actors living or working in the district. People are confident in their government and civil servants, believe that they participate in the decisions of their villages, and that anyone can express an opinion without fear. The villagers regularly compared the situation to circumstances under the previous Soeharto regime, when one could not complain about any political orientation or official decision, and when participation was almost nonexistent.

Does this perception of participatory governance reflect actual decision-making power for the communities? Their actual power rests on their votes in the different elections, on their capacity to protest and organize protests through strikes, demonstrations or blockades of supply to a palm oil refinery or a rubber mill. But they are interdependent with the other actors of the regional economy. They can negotiate and are in a good position for that, but they cannot improve their livelihoods without working with the private sector. The private sector is well organized, and prices are dependent on the international market. At the national and regional scales, agribusiness companies discuss the prices of natural rubber and oil palm among themselves, agree on prices and then negotiate every two weeks with the head of the district. If farmers can press the companies on the price and quality of their products, they only miss out on the discussion of in-between price boundaries. The farmers' choice of crop also depends on what they can sell. In Bungo, they have access to natural rubber and oil palm fruit buyers, not to cocoa or coffee buyers. Choosing crops without taking into account the presence of a local market would not be in the producers' interest, as it would add to the transportation costs for their products. Thus, farmers are dependent on the private sector and on the political decisions about which companies can develop activities in the district.

The authority over land and forests mainly belongs to the central government, as owner of state forests, which account for most of the forested lands in the country, and some non-forested lands. The provincial and district governments have authority only over non-state forest land and the agricultural and 'other' land-use areas. In Jambi Province, and specifically in Bungo District, the customary governance system is well represented in the formal government. Conflicts over land are generally solved under the customary system, in agreement with the formal authority. However, if people feel secure in their access to land and forest, they still depend on the central government to recognize their rights through formal authorization of land use or certificates of landownership.

Forest and land management

The main activities of the district government that influence economic development and landscape dynamics are conducted under public-sponsored development programmes. These programmes are implemented by the district offices of each department, under the coordination of BAPPEDA.

BAPPEDA publishes a land-use and economic development plan *(rencana tata ruang wilayah kabupaten Bungo)* every five years. This plan can be revised after two years. The first Bungo plan was issued in 2000 (BAPPEDA-Bungo, 2000), after the creation of Bungo District (as a result of the division of Bungo-Tebo District), and was revised in 2005 (BAPPEDA-Bungo, 2005). The 2000 plan deals with the division of Bungo-Tebo; the 2005 revision added the investment interests of oil palm plantations and coal mining to the spatial planning. The plan is based on the *bupati*'s strategy of development. It also builds on the department offices' annual reports on their activities, programmes and budgets. The presence of NGOs and research centres[3] since 1994 has influenced the land-use plan. ICRAF and WARSI, an Indonesian conservationist association, worked with the district officials and BAPPEDA on the preparation of the plan and its revisions to include some issues of forest and biodiversity conservation. But the main environmental concern addressed in the plan is the protection of water quality, especially focusing on water pollution due to gold- and coal-mining activities.

The results of this long involvement of conservation NGOs in the area can be seen in WARSI's recent success in the creation of village forest or *hutan desa* status, which authorizes villagers to manage the protected forest around their village, under constraints to preserve the land's sustainability. WARSI and the forestry office of Bungo hope to extend this status to surrounding villages to create a parcel big enough to compete for a project that can be classified in terms of a Reducing Emissions from Deforestation and Forest Degradation (REDD) project (personal communication and district forestry office data). The province of Jambi has asked the districts to submit areas of village forest for consideration. The agreement successfully reached in Lubuk Beringin has inspired all district governments in Jambi Province to urge their villages to propose their village forest.

The district agency and BAPPEDA have two sources of income, one from the district and the province and the other from the central government. In 2006, the total budget of the district was about Rp. 406 million (US$42,735), 7 per cent of which came from the district's own income sources, 3.9 per cent from Jambi Province and 89 per cent from the central government (BPS-Bungo, 2007). Each provincial government office received a part of the central government's income directly from its central ministry, and another part of it and full regional income share from BAPPEDA. The district's and province's incomes are generated by taxes and other fees received by such provincial government offices during the processes of certifications or issuance of authorizations, permits and concessions.

The main sources of income for the district are coal mining and agriculture, with natural rubber and oil palm as the main agricultural commodities, also generating income from the industrial sector. Between 2006 and 2009, public aid was given to the local community mainly through the plantations revitalization programme (Revitalisasi Perkebunan Karet, Department of Plantations, funded by the province), the forest and land rehabilitation programme (Rehabilitasi Hutan dan Lahan, funded by the Department of Forestry and Plantations and the Department of Agriculture), the programme to increase rice production (IP 4 and IP 3 programme from Agricultural Department), the land certification programme funded by the National Land Office (Badan Pertanahan Nasional) and the programme of access to small credits (Program Nasional Pemberdayaan Masyarakat).

The process of application to a national programme begins when the district office of a department proposes some locations to implement the programme to the ministry. After agreement, the ministry sends funding to the district office, which organizes a meeting with heads of sub-districts (*camat*) and heads of villages, to present the programme information, which they go on to share with villagers. To participate in the programme, farmers must form a group (*kelompok petani*) and address a letter to the district office, giving their names, the locations of plots, the surface concerned and their needs. The programme of plantation revitalization is a national programme; it was launched in Bungo in 2003 and aims at increasing the productivity of rubber smallholdings. A team comprising civil servants from the National Land Office, the Department of Forestry and Plantations, along with farmers, will check the location and issue landownership certificates (if the land is not situated in a forbidden area, such as state forest or land concession). Then the plantations office distributes rubber seedlings, fertilizers and access to bank credit to help the farmers during the crop's immature period. The land certificate is given to the bank as a guarantee until the credit has been fully re-paid by the farmer. Participants are allowed to plant other trees or annual crops in their rubber plots. The plots submitted by villagers to this programme should be old rubber plantations (*sesap*) or unproductive plantations. This programme has been a success; the forestry office receives more demands than it can accommodate (the target for 2009 was 700ha). But the results are compromised by the low quality of the seedlings received from the province. One unusual and positive characteristic of this programme is that it is advisory, thereby allowing farmers to apply their own cropping practices, rather than imposing intensive techniques and monocultures.

The programme of rehabilitation of forest and land is an older national programme that began in the 1950s, before decentralization (Nawir et al, 2008). It aims to improve the production of forests and agroforests, mainly through the distribution of seedlings of different species (e.g. timber, fruit trees and rattan). The programme focuses on degraded lands and forests. The main areas of degraded land in the district lie in areas of former gold mines, on poor and often unstable sandy soils. This programme is also a success; villagers regularly ask for seedlings to enrich their agroforests or surrounding forests. Both programmes have been implemented in numerous villages (Figure 3.5).

The programme to increase rice production was launched by the central government in 2007, when the Ministry of Agriculture asked the provincial agriculture department to promote the production of rice four times a year. Volunteer farmers receive aid from the government in the form of workers to rejuvenate paddy fields in fallows, seeds for selected varieties and money to buy fertilizers. In Bungo, the programme was introduced in the beginning of 2008. The farmers who submitted their application at the beginning of the year received the aid in October, and thus the farmers who engaged in the programme got help in time to cope with the drop in the natural rubber price induced by the world economic crisis of 2008. As a consequence of the economic downturn, more and more people applied to this programme, whereas in the previous years, participants were scarce, preferring rubber production to rice cropping. Food security through diversification is well established as a valuable strategy to cope with a crisis (Feintrenie et al, 2010).

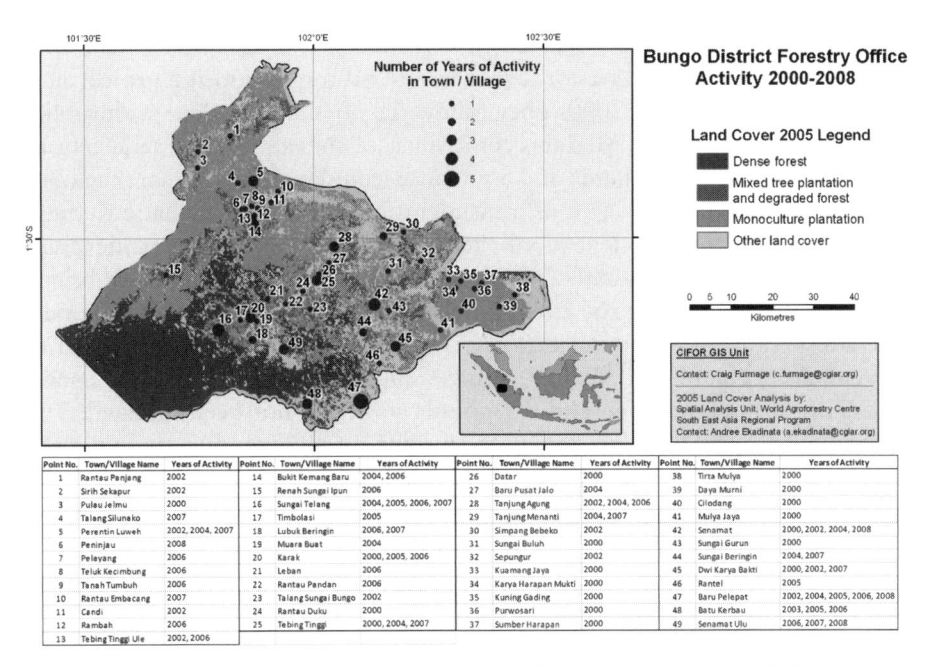

Figure 3.5 *Bungo District forestry office activity, 2000–2008*

Bungo District is among the areas chosen to host a transmigration programme. This programme began in 1905, when Indonesia was still a Dutch colony, and mixed agrarian reform, which aimed at giving land to landless farmers, with economic development and a policy to secure the country's borders by colonizing forests at the boundaries (Levang, 1997). Volunteers from the densely populated islands of Java, Bali and Madura were (in most cases) invited to move to the 'empty' islands of the periphery, especially Kalimantan and Sumatra. As a province still rich in uncultivated land, Jambi was recently selected to be part of the programme. The district's first transmigration project was implemented in the area of Kuamang Kuning, in the eastern part of the district, in 1983. It was a Nucleus Estates and Smallholders project (NES), also known as a Perkebunan Inti Rakyat project (PIR); a category of transmigration project that involves a partnership among the government, farmers, an agribusiness company and a bank. The government identifies an area, negotiates with local communities, selects volunteer participants, issues land certificates to the transmigrants and provides them with houses and facilities. The bank funds the plantation costs by giving credits to each transmigrant, under the guarantee of the land certificate. The company plants an estate that it will own (*inti*) and smallholdings that belong to the migrants (*plasma*) (Levang, 1997). The Kuamang Kuning area was the first of the district to be designated for palm oil production.

The national transmigration programme was officially closed in 1999, with the end of the Soeharto regime. But some areas, including Jambi Province and especially Bungo District, decided to continue this programme under new regional management. As a consequence, several transmigration project sites have been selected since 2000. They follow the NES model, some with rubber and others with oil palm. Migrants come alone, or sometimes in an arrangement between two provinces (Jambi and a province from Java), which share the costs of their travel. The main area of transmigration in Bungo, Kuamang Kuning, has seen rapid improvement of villagers' livelihoods and economic development. But some cases of conflict have also arisen; examples are discussed below.

Interestingly, despite the effects on land-use change and local livelihoods, the latest transmigration projects are not clearly part of the current district development planning (BAPPEDA-Bungo, 2005). This implies that the consequences on the landscape and the population have not been planned. The district transmigration department is working on its own, directly under the instructions of its central ministry. The recent transmigration projects in the sub-districts of Rantau Pandan and Bathin III Ulu are anticipated to be the last ones in the district (personal communication and transmigration office data) because there is no more land available under the 'other' land-use status that is not already designated as a plantation to be given to a company.

The district government influences people's livelihoods and landscape dynamics by implementing public-sponsored programmes and issuing agricultural plantation concessions. These programmes have hastened land-use change and produced some positive economic developments, but also caused conflicts among communities or between communities and the government or the agribusiness companies.

Causes of conflict

The oldest conflict in the district took place between 1994 and 1998 and involved a group of villagers opposed to an oil palm company (Suyanto, 2007). People from four villages united to claim rights to the concession awarded by the Ministry of Forestry. They disagreed on the compensations proposed by the company, and the conflict became more serious by mid-1998, with the Indonesian political reform (*reformasi*), and the resulting, short-term loss of governmental control and widespread unrest. After farmers burned the company's base camp and oil palm nursery, the company returned 1000ha of oil palm to the villagers. Unclear land tenure and a lack of consultation with the local communities were the sources of this conflict. Encouraged by the *reformasi* dynamics, people were not afraid to claim their land rights. Consultation with local communities is now an absolute prerequisite for any transmigration project. Although this conflict was resolved in favour of the farmers, it hasn't completely ended their protests. Indeed, they turned against heads of the *plasma* cooperatives, accusing them of stealing fertilizers and pesticides from the cooperative's stocks. To quell this new tension, the company took over the spreading of fertilizer and pesticides, which raised costs for the farmers (field survey data).

The second case of conflict related to an oil palm transmigration project took place in 2001–2002 (Chong, 2008); it was caused by unclear boundaries between villages and a non-transparent process. Land to be allocated for the transmigration project was believed to belong to one local village, with which negotiations were conducted. But when the transmigrants' plantations were installed, a second local village claimed the land as part of its customary territory and took it back from the transmigrants by force. Neither the district government nor the company interfered; transmigrants were left alone to solve the problem. It ended when some transmigrants sold their land to the locals; others lost land without any compensation. A majority of them had planted rubber on the land dedicated for food crops and then looked for a second source of income, generally off-farm (field survey data).

The problem did not directly involve the company and was mainly the result of a conflict between two villages, with land tenure uncertainty the main reason for the dispute. The lack of an official and reliable cadastre allows for often-unverifiable land claims by individuals, as well as by groups. A similar conflict arose in another area of the district, between a village and a logging company that received a timber concession from the government (involving provincial and district offices of the Forestry Department), on land claimed by villagers as a customary forest (Yasmi, 2008). The company had maps signed by the *bupati* and the district forestry agency, showing the logging area, but the community had no proof of its customary rights. The community's weak position meant that its people received no support or intervention from the district government.

Adnan and Yentirizal (2007) report another case of conflict that started in 2004 with the launch of a new oil palm transmigration project. The villagers agreed to participate in the project and gave up about 1000ha of land for its implementation, in exchange for a promise that as many local households as transmigrants would be included (Adnan and Yentirizal, 2007). The transmigrants, recruited in Jakarta, arrived in 2004 and 2005, but the oil palm company never arrived: the hilly land needed to be terraced, and potential companies were reluctant to incur the costs. Tensions over land increased between local people and migrants. In 2008, some migrants planted oil palm to get some cash income. The villagers wanted their land back, since they had not received the compensation promised by the government. Some acts of violence against the migrants were reported. The main cause of this conflict was bad governance in the organization of the project. Javanese migrants were settled on village lands before the district government had secured the participation of an oil palm company. In 2009, a company began to plant in neighbouring villages, but did not venture into this particular one. The areas' atmosphere of conflict may slow the planting process, since it erodes confidence among the partners.

The formal process of application for a planting authorization (see Figure 3.6) should, in theory, prevent disagreements among the partners of NES deals and secure the willing participation of villagers. However, the steps required to

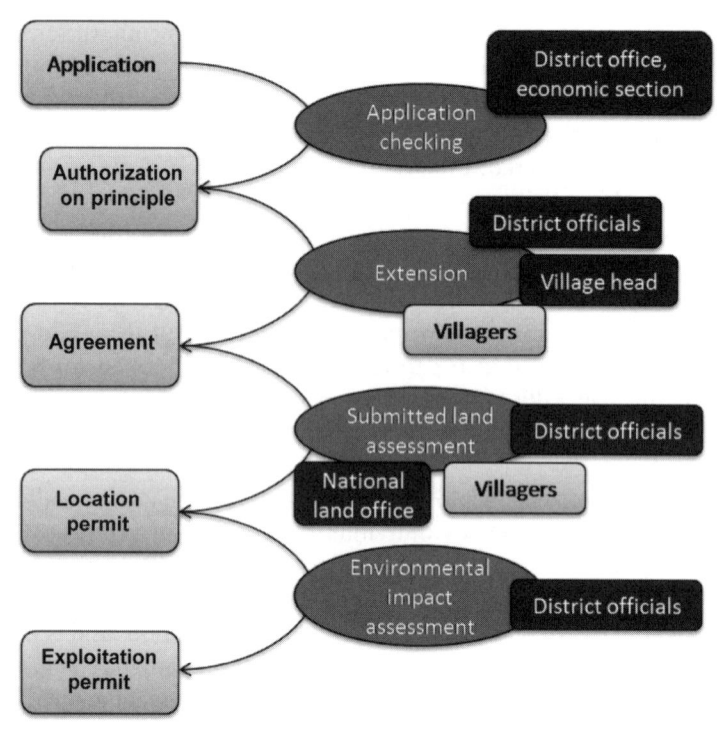

Figure 3.6 *Process of application for authorization to plant*
Source: Adapted from Chong (2008)

obtain the planting authorization can be circumvented. The agreement of the villagers, which is required for the location permit, may be obtained by negotiating only with the head of the village, and not with all the villagers. This creates the opportunity for corruption of local elites who wish to bypass difficult discussions with villagers.

The application for a planting authorization goes through several steps. An agribusiness company first submits a request to the district government, which issues an authorization on principle (*izin prinsip*) if the plan respects the legislation and regulations, and is located on lands that can legally be planted. Then representatives of the company, accompanied by sub-district officers or civil servants from the forestry and plantations office, go to the village territories covered by the plantation project to propose a deal to the heads of villages. The local leaders discuss the proposal with the villagers and respond. If an agreement (*kesepakatan*) is reached, a location survey is organized. This survey is conducted by district government representatives, civil servants from the National Land Office and the Department of Forestry and Plantations, villagers (including village leaders or heads of the group that applied for the project), and representatives of the company. The team ensures that the plots submitted for planting are not protected forest or state forest, part of another concession or owned by another party. Once the survey is finished, the district delivers a location permit (*izin lokasi*). This survey will also be used to produce the landownership certificates. District officials then conduct an environmental impact assessment, including the evaluation of the land's potential for agriculture, the design of any necessary grading (terraces) and an infrastructure development plan, and the company's commitment to follow environmentally sound practices. All these details are stated in the exploitation permit (*izin usaha perkebunan*) that awards the planting authorization to the company.

In the context of a transmigration project, this process of planting authorization is preceded by the Department of Transmigration's selection of the site and meetings with villagers to obtain their agreement on the project and define the conditions under which they agree to welcome transmigrants. These include the number of households, location of the new houses (generally in a new hamlet), the territory given to transmigrants and the compensation for local people. This procedure is intended to protect local communities and allow transmigrants to settle in secure conditions. Nevertheless, the most recent conflict (described above) began after the decentralization reforms, in 2004. Issues of bad governance, weak planning of projects, local elite capturing of power and unclear land tenure may still cause conflicts. Moreover, two of the four conflicts are still unresolved, and the district authorities refuse to get involved in them. In the best outcome, the district offices favoured participation of misled villagers in development projects, but only on land with clear formal, uncontested ownership. But when there are conflicting claims on land, the authorities wait for the villagers to find an agreement, eventually through customary judgement, and do not intervene.

These conflicts reflect bad governance related to oil palm development, but they also illustrate the willingness of the district government to support oil palm expansion. Local people as well as migrants have been deceived or even mistreated and have sometimes vented their anger at innocent third parties (either transmigrants or the company). Nevertheless, in this case, none refused oil palm development and, in fact, asked for greater participation. Such an attitude is common in the district. A perception survey of people's opinion about land uses, landscape and forest conservation, conducted in 2007, clearly showed that all surveyed villages were willing to accommodate an oil palm company on their premises (Therville et al, 2010). Many believe that their future lies in oil palm and rubber, and do not imagine alternative livelihoods.

Conclusion

Indonesian legislation gives authority over unforested land to local people, but grants the majority of forests to the central government, under the status of state forest. On these state forests, the central government is undeniably the most powerful stakeholder, with full authority on their use or conservation. On the other forests, privately owned or commonly owned by local communities, authority depends on the land status. If the land is privately owned, with a landownership certificate, then the right of its owner is secure. If the owner possesses only customary recognition, then his or her right is secured as long as it does not conflict with a governmental project, such as an industrial concession. Nevertheless, our results showed that people feel secure in their rights of access to land, forest and forest products. Few conflicts arise in the district, and trust exists between the population and the government, largely because of the good coordination and mutual respect between formal and customary authorities, and the rapid economic development of the area. The result is an optimistic and positive atmosphere – even in 2008–2009, when the world economic crisis led to a decline in natural rubber and palm oil prices, slowing down the local economy.

People believe they are actively participating in decision-making, at least at the local level. Their power lies, first, in their right to vote for the legislative and executive representatives, at all levels of governance (village, district, province and state). Second, they have informal power through their capacity to organize collective action, such as strikes, demonstrations and blockades of supply to industry mills. But agribusiness companies are also organized and make far-reaching decisions, such as fixing the price of agricultural commodities or land. Because such decisions are taken out of their reach, people are dependent on market conditions, the buyers and the government's strategy of economic development for the district. Agribusiness companies and political authorities work hand-in-hand to maintain the economic development of the area, which also creates room for corruption (Dudley, 2000, 2002; Komarudin et al, 2008; Wells, 2008).

However, the district government is responsible not only for promoting economic development, but also for managing natural resources, creating livelihood opportunities for people and dealing with external pressures, such as conservation NGOs and the central government's requirements. The district government develops a spatial plan every five years, which is revised after two years, that details its strategy for meeting the political targets defined by the head of district. Bungo District's plan was written with the help of several NGOs that have been working in the district for years, and it thus took into account some ecological matters, especially water-quality protection. Under the coordination of BAPPEDA, the public agencies conduct several programmes of development, sponsored or co-funded by the district, the province or the central government. One major development scheme in the district is the transmigration programme, which has brought economic advances and conflicts.

The district manages forest, land and other natural resources through these development programmes, via which it issues authorizations for plantations, mines or industry and enforces district and local (village) regulations to complete national legislation and adapt to the regional and local context. But another opportunity for natural resource management has appeared with the emergence of the carbon market and REDD approach, a backdrop against which Bungo District is in line to submit projects, with the support of NGOs.

The conflicts recorded in the district can generally be explained by lack of prior consent of communities in old transmigration projects, bad governance or bad planning of project implementation and little organization among villagers for collective action (such as managing an oil palm cooperative and negotiating with the industry). But one of the main causes of conflict is unclear land tenure, and especially unclear boundaries between village commons, which can lead to violent intercommunity conflicts.

Acknowledgements

We would like to acknowledge the villagers who welcomed us in their villages and families. We also would like to thank all the kind respondents to our surveys: villagers, civil servants, officials and conservationists, who kindly agreed to answer our questions. We also thank the Landscape Mosaics project and specifically the Jambi team: Joshi Laxman, Ratna and Jasnari, Meine Van Noordwijk, Patrice Levang, and the students who conducted their research and fieldwork in our site: Clara Therville, Ameline Lehébel-Péron, Xavier Bonnart and Wan Kian Chong, and also the field assistants Asep Ayat and Asep Wahyu Suherman. Last but not least, we thank Sonya Dewi, Andree Ekadinata, Gen Takao, Wim Nursal and Craig Furmage for the spatial analysis.

Notes

1. The first such transmigration project was launched in 1976 in Kalimantan. It was sponsored by the World Bank and marked an important change in the transmigration policy: before this project, only food crops were included in transmigration programmes, but this new project included rubber plantations as cash crops. A transmigration centre associated a private or public agribusiness company with migrants selected and sponsored by the government. The company paid for the infrastructure (roads, houses and school), and planted and managed a cash crop plantation (most often rubber or oil palm) until the first harvest. Then the migrants, who owned private plots, were grouped in cooperatives and asked to followed the technical instructions of the company; all their production had to be sold to the company. Planting and management costs were deducted by the company from the monthly payment for production. This type of transmigration project was profitable for the government and for the companies, which benefited from a cheap source of labour that was easy to control. The company was both buyer and employer. Although they had no influence on the cost inputs nor the price of the product sold to the company, migrants gained access to landownership and technical training through this system (Levang, 1997).
2. In Bungo District, some department offices are combined, such as forestry and plantations, or agriculture, fisheries and animal breeding, to reduce administration and infrastructure costs.
3. ICRAF and WARSI have conducted various projects on biodiversity conservation and agriculture improvement in the district. WARSI projects: Community Based Forest Management (CBFM); Integrated Conservation and Development Programme-Kerinci Seblat National Park (ICDP-KSNP). CIFOR-ICRAF projects:

Alternatives to Slash and Burn (ASB); Landscape Mosaics project; REDD-Alert project. ICRAF project: Smallholder Rubber Agroforestry Project (SRAP). ICRAF-WARSI project: Rewarding Upland Poor for Environmental Services (RUPES). CIFOR has also conducted several projects aimed at enhancing collective action through capacity building at the community level: Adaptive Collaborative Management (ACM); Collective Action and Property Rights (CAPRi). The involvement and success of these two projects, operating at both village and district level, provided important input in the planning of the Landscape Mosaics project. In the villages that have benefited from these projects, people now speak easily about their needs, difficulties and desires. Women are less shy and voice their opinions more quickly than in other villages. Some of these communities were able to negotiate more equitable and beneficial agreements with industrial companies.

References

Adnan, H. and Yentirizal (2007) 'Blessing or Misfortune? Locals, Transmigrants and Collective Action', CIFOR Governance Brief, June (vol 36), pp1–11

BAPPEDA-Bungo (2000) *Revisi rencana tata ruang wilayah kabupaten Bungo* [Revision of the Land Use Plan for Bungo District's Area], Pemerintah kabupaten Bungo, Muara Bungo

BAPPEDA-Bungo (2005) *Revisi rencana tata ruang wilayah kabupaten Bungo*, Pemerintah kabupaten Bungo, Muara Bungo

BPS-Bungo (2007) *Bungo dalam angka, 2006* [Bungo in Figures, 2006], Badan Pusat Statistik Kabupaten Bungo, Muara Bungo, Indonesia

Chong, W. K. (2008) 'Oil Palm Development and Land Management in Bungo District, Indonesia', IRC/SupAgro, Montpellier

Dudley, R. G. (2000) 'The Rotten Mango: The Effect of Corruption on International Development Projects' *Part 1: Building a System Dynamics Basis for Examining Corruption*, System Dynamics Society, Bergen, Norway

Dudley, R. G. (2002) 'Dynamics of Illegal Logging in Indonesia', in C. J. P. Colfer and I. A. P. Resosudarmo (eds) *Which Way Forward? People, Forests and Policymaking in Indonesia*, Resources for the Future/CIFOR, Washington, DC

Ekadinata, A. and Vincent, G. (2008) 'Dinamika Tutupan Lahan Kabupaten Bungo, Jambi,' [Land Cover Dynamics in Bungo District] in H. Adnan, D. Tadjudin, L. Yuliani, H. Komarudin, D. Lopulalan, Y. L. Siagian and D. W. Munggoro (eds) *Belajar Dari Bungo: Mengelola Sumberdaya Alam di Era Desentralisasi* [Learning from Bungo: Managing Natural Resources in an Era of Decentralization], CIFOR, Bogor, Indonesia

FAO (Food and Agriculture Organization of the United Nations) (2010a) 'Global Forest Resources Assessment 2010', Country Report, Indonesia, FRA 2010/095, FAO, Rome

FAO (2010b) 'Faits et Chiffres, les 10 Pays dont la Surface Forestière est la Plus Étendue' [Facts and Figures, the 10 Countries with the Most Reduced Forest Area], www.fao.org/forestry/41775@80687/fr, accessed 8 April 2010

Feintrenie, L. and Levang, P. (2010) 'Local voices call for economic development over forest conservation: trade-offs and policy in Bungo, Sumatra', *Forests, Trees and Livelihoods*

Feintrenie, L., Chong, W. K. and Levang, P. (2010) 'Why do farmers prefer palm oil? Lessons learnt from Bungo district, Indonesia', *Small Scale Forestry*, vol 9, no 3, pp379–396

Feintrenie, L. and Levang, P. (2009) 'Sumatra's Rubber Agroforests: Advent, Rise and Fall of a Sustainable Cropping System', *Small-Scale Forestry*, vol 8, no 3, pp323–335

Hasan, U., Irawan, D. and Komarudin, H. (2008) 'Rio: Modal Sosial Sistem Pemerintah Desa' [Rio: Social Capital in the Village Governance System], in H. Adnan, D. Tadjudin, L. Yuliani, H. Komarudin, D. Lopulalan, Y. L. Siagian and D. W. Munggoro (eds) *Belajar Dari Bungo: Mengelola Sumberdaya Alam di Era Desentralisasi*, CIFOR, Bogor

Joshi, L., Wibawa, G., Beukema, H., Williams, S. and van Noordwijk, M. (2003) 'Technological Change and Biodiversity in the Rubber Agroecosystem of Sumatra', in Vandermeer, J. (ed) *Tropical Agroecosystems*, CRC Press, Florida, pp133–157

Komarudin, H., Siagian, Y. L. and Colfer, C. J. P. with Neldysavrino, Yentirizal, Syamsuddin and Irawan, D. (2008) 'Collective Action to Secure Property Rights for the Poor: A Case Study in Jambi Province, Indonesia,' vol 90, Working Paper, CAPRi, Washington, DC

Levang, P. (1997) 'La Terre d'en Face, la Transmigration en Indonésie' [The Land Beyond Transmigration in Indonesia], ORSTOM, Montpellier

Nawir, A., Murniati, A. and Rumboko, L. (2008) 'Rehabilitasi hutan di Indonesia: Akan kemanakah arahnya setelah lebih dari tiga dasawarsa?' [Forest Rehabilitation in Indonesia: What Future Direction After More than Three Decades?], CIFOR, Bogor, Indonesia

Obidzinski, K. (2005a) 'Illegal Logging in Indonesia: Myth and Reality', in B. P. Resosudarmo (ed), *The Politics and Economics of Indonesia's Natural Resources*, Institute of Southeast Asian Studies, Singapore, pp193–205

Obidzinski, K. (2005b) 'Illegal Forest Activities in Indonesia', *Indonesian Quarterly*, vol 33, no 2, pp100–104

Pemerintah Daerah Kabupaten Bungo (2000) *Buku Panduan Adat Kabupaten Bungo* (Bungo District Customary Guide Book), Muara Bungo

Suherman, K. and Taher, M. (2008) 'Potret Perubahan Tutupan Hutan di Kabupaten Bungo 1990–2002' [A Change Portrait of Land Cover in Bungo District, 1990-2002], in H. Adnan, D. Tadjudin, L. Yuliani, H. Komarudin, D. Lopulalan, Y. L. Siagian and D. W. Munggoro (eds) *Belajar Dari Bungo: Mengelola Sumberdaya Alam di Era Desentralisasi*, CIFOR, Bogor

Suyanto, S. (2007) 'Underlying Cause of Fire: Different Form of Land Tenure Conflicts in Sumatra', *Mitigation and Adaptation Strategy to Global Change*, vol 12, pp67–74

Therville, C., Feintrenie, L. and Levang, P. (2010) 'Farmer's perspectives about agroforests conversion to plantations in Sumatra', Lessons Learnt from Bungo District (Jambi, Indonesia)', *Forests, Trees and Livelihoods*, vol 20, no 1, in press

Wells, A. (2008) 'Verification of Legal Compliance in Indonesia', in D. Brown, K. Schreckenberg, N. Bird, P. Cerutti, F. D. Gatto, C. Diaw, T. Fomété, C. Luttrell, G. Navarro, R. Oberndorf, H. Thiel and A. Wells (eds) *Legal Timber: Verification and Governance in the Forest Sector*, CATIE, RECOFTC, CIFOR, ODI, London

Yasmi, Y. (2008) 'Peningkatan Konflik dalam Pengelolaan Sumber Daya Hutan' [The Rise in Conflict over Management of Forest Natural Resources], in H. Adnan, D.

Tadjudin, L. Yuliani, H. Komarudin, D. Lopulalan, Y. L. Siagian and D. W. Munggoro (eds) *Belajar Dari Bungo: Mengelola Sumberdaya Alam di Era Desentralisasi*, CIFOR, Bogor

Indonesian laws

Undang-undang Republik Indonesia nomor 5 tahun 1979 tentang Pemerintahan Desa (Law on rural governance)

Undang-undang Republik Indonesia nomor 40 tahun 1996 tentang Hak Guna Usaha, Hak Guna Bangunan dan Hak Pakai Atas Tanah (Law on land-use authorization)

Undang-undang Republik Indonesia nomor 22 tahun 1999 tentang pemerintahan daerah (Law on regional governance)

Undang-undang Republik Indonesia nomor 25 tahun 1999 tentang perimbangan kuangan antara pemerintah pusat dan daearah (Law on fiscal balancing between the central government and the regions)

Undang-undang Republik Indonesia nomor 41 tahun 1999 tentang kehutanan (Law on forestry)

Undang-undang Republik Indonesia nomor 10 tahun 2004 tentang Pemerintahan Desa (Law on rural governance)

Undang-undang Republik Indonesia nomor 32 tahun 2004 tentang Pemerintahan Daerah (Law on the regional government)

4

Information Flows, Decision-making and Social Acceptability in Displacement Processes

*John Daniel Watts, Heini Vihemäki,
Manuel Boissière and Salla Rantala*

The displacement of rural people, whether for conservation or for development, demonstrates how policy processes and decision-makers operating beyond the boundaries of a landscape can dramatically affect the lives of its residents. Displacement occurs not only when people are forced to move but also when people are excluded from their previous economic activities (Brockington and Igoe, 2006), such as farming. Displacement can also be a side effect of landscape management, when new protected areas are established or old ones extended. In addition, interventions justified by development goals, such as improved access to infrastructure, condition the operations and choices available to those seeking to improve natural resource management at landscape level. Displacement is essentially a disproportionate exercise of power that overrides local systems, tenure and aspirations in favour of the objectives of those at higher levels of government and sometimes beyond, as in the case of biodiversity conservation projects.

Although displacement brings villagers and implementing agencies into contact, there is often little scope for the overarching goals to be modified. The best possible outcome is likely to be a process of negotiation. Information on the potential livelihood and environmental effects, which guides the selection of areas for displacement and the implementation process, should be comprehensive but often is not. Furthermore, ongoing monitoring, assessment and adaptive mechanisms are required to ensure that any negative impacts can be addressed (see Lasgorceix and Kothari, 2009). In the absence of adaptive management, how can the livelihood needs and concerns of the displaced people be met?

In this chapter, we present two case studies of displacement, in Laos and Tanzania. The first case explores a village relocation process in Phadeng, in Viengkham District, Luang Prabang Province, Laos. The village was relocated under the dual rhetoric of improving both conservation and development by moving the village farther away from the Nam Et-Phou Louey National Protected Area and closer to services. The second case study describes conservation-related displacement in the East Usambara Mountains of Tanzania, where a large tract of land previously used for farming was appropriated in the establishment of the Derema Corridor.

In both cases, efforts were made to mitigate negative effects by improving flow of information, anticipating the outcomes for the affected people and, to varying degrees, involving them in related planning and decision-making. The establishment of the Derema Corridor was at least outwardly participatory, in the sense that local people were consulted, and did not involve resettlement. But from the outset, the process in Phadeng was not participatory, and it included resettlement. In Derema, the owners of the farms were compensated for their lost access to land, whereas in Phadeng, no compensation was paid. In the study of Derema, the researchers' role was to document and analyse an ongoing process; in Phadeng, the researchers became actively involved in facilitating communication between parties. The cases are also described at different stages of the displacement process: the Derema case describes a nine-year process, whereas the Phadeng case is focused on the initial phase. By combining and comparing the approaches and experiences from the two cases, we explore a broader set of mechanisms and pathways for mitigating negative social effects and communicating livelihood information to decision-makers. Here we consider the stages of the displacement process and the conditions under which these mechanisms can improve social acceptability and mitigate harm. To begin with, we offer a short conceptualization of conservation- and development-related human displacement. Then we contextualize the two cases and give an account of each displacement process. In so doing, we explore the disparities between the goals and expectations of the villagers and the implementing agencies regarding the outcomes. Furthermore, we describe what strategies were used by the project implementers (in Tanzania) and researchers (in Laos) to address social impacts, enhance communication and negotiate the conditions of displacement, and how these efforts influenced the displacement processes. We then address the responses of the affected people to the interventions and their scope for influencing the process through facilitation organized by researchers (in Laos) or consultants and conservationists (in Tanzania). Finally, we reflect on how and under what sets of conditions, communication tools and other mechanisms can be used to improve the acceptability of displacement.

Displacement: By-product of Conservation and Development

Displacement can be considered in terms of the eviction of human populations to make way for development or conservation interventions. Some researchers define displacement narrowly, restricting it to the physical removal of people from their home areas (see Agrawal and Redford, 2009), but others have broadened the concept to include restricted access to resources, such as land[1] (see the World Bank's Operational Policy 4.12 on resettlement; Cernea, 2005). Furthermore, the resettlement policy of the African Development Bank covers 'loss of assets or involuntary restriction of access to assets including national parks, protected areas or national resources; or loss of income sources or means of livelihood as a result of projects, whether or not the affected persons are required to move' (Cernea, 2005, p49). We consider lost access to land to be a form of displacement because it typically damages people's livelihoods, especially in rural landscapes in developing countries.

Most international development funding agencies have policies on restoring the lives of the displaced (Cernea, 2008). The policy terms, such as whether the affected people are to be compensated financially or given access to alternative resources, can be negotiated and agreed upon with the 'target population'. For instance, mainstream conservation policy and thinking have begun to favour socially just conservation in recent years, calling for a shift away from forced to voluntary displacement, in the establishment of protected areas (Brockington and Igoe, 2006; Schmidt-Soltau and Brockington, 2007). Nevertheless, the line between voluntary and involuntary resettlement, or other forms of displacement, is not clear-cut (see Baird and Shoemaker 2007; Schmidt-Soltau and Brockington, 2007; High, 2008). It is not necessarily easy to identify the level of legitimacy of agreements or other formally binding decisions about displacement when the parties have unequal negotiating positions and access to information. As Freeman et al (2008) note, the understanding of what constitutes consent also varies among different groups. In addition, it proves challenging to implement displacement policies: alternative land may not be offered, compensation may not be paid and other measures of mitigation may be absent (see Cernea 1993, 2005). Finally, High (2008) refers to the 'experimental consensus' among poor Laotians, who seek or agree to resettlement with aspirations of modernity, a perspective that others fear could lend support to policies that may have adverse effects (High et al, 2009). While acknowledging many adverse consequences of resettlement, Petit (2008) likewise emphasizes the agency of individuals and reliance on kinship relations and other social relations in creating different lives for themselves.

Diverse cultural, social and ecological changes may be triggered by displacement. Natural ecosystems are often destroyed or degraded at the sites to which the displaced people are relocated (Lasgorceix and Kothari, 2009). In addition, displacement from forested areas may alter local livelihoods by restricting access to forest-based resources, especially if access to similar resources elsewhere is not provided. Displacement usually harms the displaced people (Cernea, 1993; Lasgorceix and Kothari, 2009). Schmidt-Soltau and Brockington (2007) argue that in the establishment of protected areas, the people who previously settled or used the area to be protected carry the largest share of social costs resulting from the land-use change. Yet this does not necessarily apply to all cases, and the impact may be mitigated through careful and inclusive planning. Overall, information on the ecological and socio-economic consequences of displacement is still scarce and partial, and often based on assumptions rather than systematic assessments (Brockington and Igoe, 2006; Lasgorceix and Kothari, 2009; High, 2008).

Approaches and Methods

In the Lao site, data was collected between 2008 and 2010, using a combination of empirical and action research methods. Focus group discussions and household interviews with district representatives, village officials and other villagers were conducted. Participatory and 3D-mapping workshops were held to facilitate discussions on the impacts and potential solutions to displacement-related problems.

The survey team examining village relocation comprised international staff from the Center for International Forestry Research (CIFOR), local project staff from the National Agriculture and Forestry Research Institute (NAFRI) and Northern Agriculture and Forestry Research Center (NAFReC), including a Hmong translator and a representative from the District Agriculture and Forestry Office. The constant presence of a district representative and the short duration of field missions influenced the responses that we received from the villagers. This also proved to be the case in a workshop in which government officials outnumbered village representatives. However, despite these constraints, villagers expressed their dissatisfaction with the relocation process and proposed alternatives.

In the Tanzanian case, we draw on data collected by two of the contributors as a part of their PhD studies between 2003 and 2009.[2] The most important sources of data were individual and group interviews (semi-structured and structured), conducted with local people in two villages adjacent to the Derema Corridor. We also participated in two meetings of farmers affected by displacement and interviewed other actors involved in the process, including representatives of the local government, the forestry department and other organizations. In addition, to a lesser degree we draw from two household surveys, conducted in 2005 (48 respondents) and 2008 (139 respondents). They differed in coverage but were designed to identify patterns of resource use and access, and experiences of different social groups resulting from the establishment of the Derema Corridor. Moreover, previous research reports from the area, legal and policy documents, and grey literature were used as source material and as a means of cross-checking some of the information.

Displacement in Laos

Our three research villages in Laos are all close to the Nam Et-Phou Louey National Protected Area, which is high in biodiversity and valued for its population of tigers (ICEM, 2003). The district is also one of the 47 poorest districts in Laos and thus a priority area for development. The dominant land use, outside the protected area, is swidden agriculture. Consequently, the landscape mainly comprises a mosaic of swiddens and fallows with more permanent agricultural land uses, such as paddy rice fields, teak plantations and cash crops, as well as protected and conservation forest areas.

Livelihood strategies and social characteristics

Ethnically, Viengkham District is predominately Khmu (64 per cent), with the two minority groups being Hmong and Lao Loum. The village of Phadeng, the most remote site of the Laos component of the Landscape Mosaics project (Figure 4.1), is located at an elevation of 1100m and exclusively Hmong. In 2008, Phadeng consisted of 40 households. Village lands directly bordered on the Nam Et-Phou Louey National Protected Area. Consequently, park authorities and the Wildlife Conservation Society had engaged the village in conservation activities, such as border patrols and cardamom cultivation in the buffer zone. The villagers in Phadeng relied more heavily on livestock in their subsistence strategy than did the people of the two more accessible villages we studied. Prior to 2003, the villagers had cultivated opium for both local consumption and trade (Fitriana, 2008). Villagers were forced to increase their reliance on shifting cultivation, livestock and wild species after the implementation of the national opium eradication strategy.

The villagers of Phadeng originally came from Dien Bien Phu in Vietnam. Since the 1920s, Phadeng's people have periodically relocated their village, because of war or in search of better agricultural land, and have often dispersed in separate villages or hamlets. The most recent settlement of Phadeng was only fully consolidated in 2003, a time at which it had a population of 50 families. Between 2004 and 2008, at least ten families left the village (Fitriana, 2008). The out-migrations, in part, triggered the process that would lead to the displacement of the village.

In terms of food security and access to wild species, the villagers of Phadeng were relatively well off – income is a key issue in evaluating the effects of resettlement (High, 2008; Petit, 2008). But the village's inaccessibility meant that the inhabitants received little in terms of basic human services or extension support. This was evident in the lack of tap water, medical supplies and a working school, although a small primary school was managed by a local teacher. The distance to the main road also made access to education and health services in the district capital difficult. Some parents were able to provide for secondary education for their children in Viengkham.

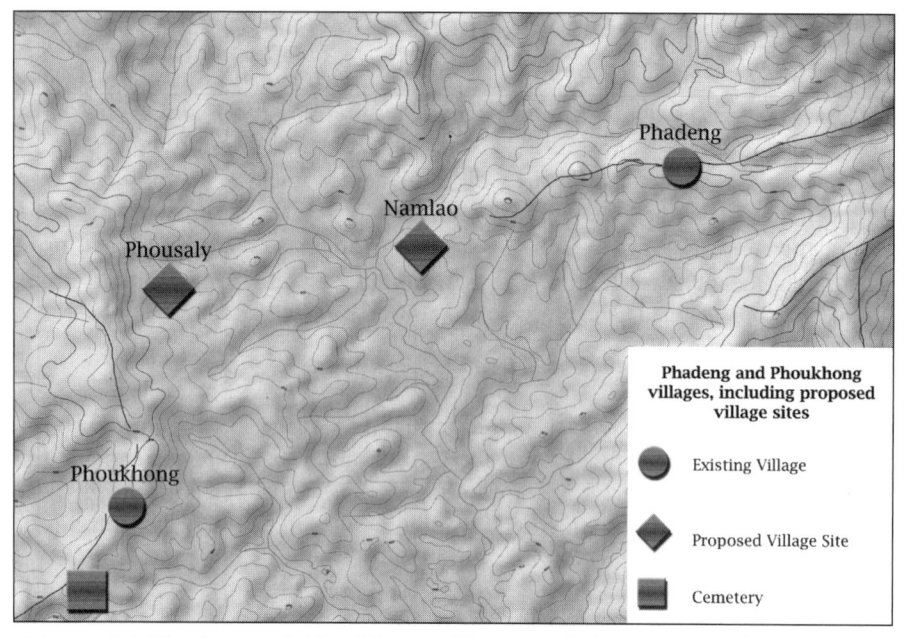

Figure 4.1 *Phadeng and Phoukhong villages, including proposed village sites*

Policy context and governance

Since the 1975 revolution, the Lao government has relocated remote populations nearer to infrastructure. But it was only during the late 1990s that government policies began to encourage the resettlement of remote villages under the auspices of the Rural Development Programme (Lestrelin, 2009; Evrard and Goudineau, 2004). Under these terms, resettlement has five justifications: reduction of shifting cultivation, eradication of opium, security (better control of remote populations), access to services (such as education and health care) and cultural integration of ethnic minorities (see Baird and Shoemaker, 2005; Fox et al, 2009; Messerli et al, 2009; Mertz et al, 2009). This important national policy is articulated around two concepts: the focal site strategy, in which relocated villages are provided with services; and village consolidation, which aims at merging small villages (with populations of less than 500 in lowland areas and 200 in upland areas) into bigger settlements (Baird and Shoemaker, 2005). In these relocation villages, Land Use Planning and Land Allocation programmes have been applied to stop deforestation, intensify agricultural production and improve land taxation (Evrard, 2004, in Lestrelin and Giordano, 2007).

Despite the development goals of these policies, the overall consequences have been negative (Fujita and Phengsopha, 2008; Ducourtieux et al, 2005). In the most extreme cases, there are high levels of mortality among the displaced (Baird and Shoemaker, 2005). In the majority of cases, conflicts arise between the resettled people and the original inhabitants, often over access to land and ethnic differences (see Petit, 2008; Ducourtieux et al, 2005). Another common effect is land degradation as the human population exceeds the carrying capacity of village land. By studying village relocation in Phonxay District, Luang Prabang Province, Jones et al (2004) highlight several flaws in the implementation of resettlement policies. A central conflict relates to the policy of reducing the number of villages. Because of the district's insufficient staff and budget to implement the process in a way that meets the development objectives, international NGOs are often asked to support the process (Baird and Shoemaker, 2005). Jones et al (2004) demonstrate that mediated solutions, recognizing livelihood needs of villagers and meeting national policy objectives, are possible. Such solutions, although less desirable than avoiding relocation, require greater participation of villagers and flexibility from district authorities in implementation.

Displacement process

In early 2007, villagers from Phadeng were informed that a planned road would link a nearby village, Phoukhong, to Namlao, and they were asked to move to Phoukhong. Namlao is located near a river, at the border between Phoukhong and Phadeng, and villagers grow upland rice and raise cattle. Because no one had moved by the end of 2008, district officials informed the villagers of Phadeng that their village had been merged with Phoukhong and they had to relocate their houses there. No real consultation took place with villagers. In 2009, Phadeng's official recognition was not renewed, and the village stamp was cancelled. At the time of our visits, in early 2009, a district officer joined our team of scientists and discussed with villagers (representatives of the main unions) the necessity of moving.

Phoukhong is a predominately Hmong village, accessible by a semi-permanent road, 30 minutes' drive from the main road and three hours' walk from Phadeng, its closest neighbour. The justifications for relocation included more access to services and markets, greater distance from the protected area (thus reducing pressures from swidden agriculture and livestock grazing) and compliance with minimum village size standards, as required by the village consolidation policy. The selection of the village may have seemed appealing to district authorities because of the shared ethnicity and pre-existing ties (intermarriage) between the two. The issues of where the villagers would settle and farm, however, were neither addressed nor negotiated.

Despite the order to relocate, the villagers still had not done so by February 2009. But officially the village of Phadeng had ceased to exist and the new official entity was the village of Phoukhong, consolidating the lands of both villages.

Perceptions and scenarios: catalyzing discussion

The decision to merge Phadeng and Phoukhong was made by Viengkham district officials and was not subject to debate. The official justification was that the merger would improve development and conservation outcomes, but aside from closer proximity to the road, the means for achieving the goals were unclear to us. In addition, the degree of participation of villagers in the decision-making and their perception of the process was not known. To address this, in February 2009 we began to explore the perceptions of the villagers and district-level actors involved in the displacement process, as the villagers were being asked to relocate.

During two field visits, we discussed scenarios about the relocation with the interested stakeholders in the presence of Viengkham district staff. We met first with the villagers of Phadeng and Phoukhong separately, and then with the district officials. In Phadeng we asked about the history of the relocation process, the wishes of the different groups in the village (whether they were for or against relocation), the choice of location and their concerns about possible consequences (e.g. land-use planning, village boundaries, integration with the new village and access to services). In Phoukhong, we interviewed key villagers about their willingness to merge with Phadeng and the consequences for land use.

In those discussions, we noted that villagers in favour of the relocation expected to gain better access to services, markets and job opportunities (see High, 2008). Those against it (mainly the elders) said that none of the proposed sites were on mountaintops (as is traditional for the Hmong), and they feared loss of access to their lands. The site originally chosen by the district government was, according to all interviewed villagers, not suitable for the relocation: it had insufficient land for resettlement, and it was far from the river and near a cemetery. Phadeng villagers preferred a different site, Phousaly, which was situated in close proximity to Phoukhong, but more suitable to their needs. At the end of the second visit, we prepared a document for the district authorities, explaining the different options for relocation, with their advantages and weaknesses. It became clear that staying put was not an option for the villagers. We proposed a workshop where villagers and district personnel could discuss these options. After the presentation of the results of these field missions, a representative from the district government went to Phoukhong and Phadeng to discuss settlement locations. Through consultation with each village committee, the intermediate location of Phousaly, halfway between Phoukhong and Phadeng, was chosen as the site of resettlement. This site met the district's objectives of relocating the village closer to the road and offered villagers of Phadeng sufficient land and access to water and valley bottoms for agriculture. The villages would administratively become one, although located an hour's walk apart. The villagers of Phadeng were then given from June until December 2009 to relocate to the new site.

Re-evaluating the new site through participatory mapping and scenario exercises

Upon learning that an intermediate solution had been found, the LM team felt excluded from this process and tried to promote a more participatory approach, seeking to re-engage in the displacement discussion. To clarify both what had occurred and to understand what our role in the process should be, we organized another field trip to the two villages and the district in June 2009. In doing so, we sought to understand the concerns and official positions of the villages and the district, and whether the people's desires had been met through the selection of the new site. Furthermore, we aimed to get a clearer understanding at the household level of what each of the settlement options meant for the villagers of Phadeng.

In Phoukhong, we spoke with the village council, in particular the head of the village elders committee. We discussed the pros and cons of the four settlement scenarios that had been identified with Phadeng residents – no relocation, relocation to Phoukhong and the new settlement options of Phousaly and Namlao (located within the village territory of Phadeng). These focus group discussions were complemented by a participatory mapping exercise, in which we asked the villagers to map out their current land uses and consider how such uses would be affected by the different settlement options. Although each discussion highlighted positive and negative outcomes, it was generally accepted that the new official choice of Phousaly was the best and that it could not be changed. The problem with that site for the people of Phoukhong – as with so many resettlement areas – was the sharing of land: for that site to be considered acceptable, the lands of Phoukhong and Phadeng had to be jointly managed. The area where the new settlement was to be located was a swidden agricultural area for the village of Phoukhong. To be satisfied with the relocation, the villagers needed a new joint land-use plan.

In the case of Phadeng, the (former) village council explained their understanding of the situation to us. They had negotiated with the district representative and then Phoukhong to move to the Phousaly site, and they were to begin moving in June and finish by December. The site was seen as the best option because of its proximity to the old village, the potential accessibility by road and access to water and good land. Some of the elders of the village preferred to stay in Phadeng or alternatively, in Namlao. The villagers agreed that the lands should be equally shared between the two villages. If the village moved to the new site, however, the members would need assistance with developing a joint land-use plan, tools and supplies (including food for the workers who would build the road) and agricultural extension support.

Not all the villagers were content with the negotiated option for the relocation. Eight families had already left the village in anticipation of the relocation. Six families had moved to Vientiane province, one to Xayaboury province and two to villages in Luang Prabang province. Some moved because they had relatives in the area; others wanted to start lowland paddy farming and have greater market access.

We then interviewed members of ten Phadeng households about how they imagined each of the four settlement options would affect their lives (positively or negatively and why), using the five capital assets of the Sustainable Livelihoods Framework (Department for International Development, 1999).

Table 4.1 *Results from household perception surveys in Phadeng*

Total	Positive attributes of site	Negative attributes of site	Balance (positive minus negative)	Households preferring each site
Phoukhong	111	79	32	0
Phousaly	124	66	58	5
Namlao	113	77	36	3
Phadeng	86	104	–18	2

The villagers were then asked for their overall settlement preference and their reasons for this course of action, whether they had any alternative suggestions, where they would farm and whether they would share their lands with Phoukhong.[3] Table 4.1 above shows the survey results.

Although Phousaly was seen as having the most positive attributes, only half of those interviewed nominated it as their preferred choice of settlement; the majority of those who preferred it said the relocation would lead to increased access to services and markets. Reasons for choosing Namlao or remaining in Phadeng were to be closer to swidden and grazing lands, and to benefit from greater access to non-timber forest products (NTFPs).

After the field trip, we met with the Viengkham district governor to present our results and discuss the district's official position on the relocation process. The governor explained that the new settlement location was chosen after the head of the district court, also ethnically Hmong,[4] was sent to discuss the options with the villages. After agreeing on the site, the villagers of Phadeng requested the governor's support for tin roofs and rice during the move, which would affect their rice harvest. The governor restated all the official reasons for the move, adding only that some families were still planting opium. The governor agreed that the workshop, which the team had planned to determine local people's views, was needed but with a different kind of agenda. Instead, he set out to explain these issues:

- government policy and the reasons for relocating the village;
- improving infrastructure for livelihoods;
- changing from shifting cultivation to cash crops; and
- protecting forests for sustainable use.[5]

Participatory negotiations for displacement

The workshop proceeded in July 2009 with discussions about the implications of the relocation for livelihoods and development, land-use planning and natural resource management.

Five representatives from each village (three officials, both men and women, and two elders, one man and one woman) attended the workshop. From the district, representatives of the district agriculture and forestry office, land management authority and the village cluster attended. A representative from the protected area also attended. Our Lao partners, NAFRI and

NAFReC, acted as facilitators for the workshop, offering no direct input during the discussions.

The workshop began with presentations from the district, villages and protected area. These were followed by focus group discussions about the livelihood and development implications of relocation, as well as potential solutions to related issues. Depending on the issue discussed, groups were mixed or composed of only villagers or only officials. For one topic (the planned road linking Phoukhong to Phousaly), we proposed a group with only villagers from Phadeng and Phoukhong. For issues relating to land-use planning and natural resource management, three-dimensional topographic maps were used to aid the discussion of a land-use plan for the combined villages. The discussions included villagers' considerations, the desire of the protected area's representative to improve park protection and the wishes of the district officials for better land management. At the conclusion of the workshop, a negotiated land-use plan was represented on the three-dimensional map, and a plan for how to meet the livelihood and development needs of the villages was developed.

Although we circulated the agreements to district and village participants after the workshop, they could not implement the land-use plan alone. We anticipate that it will be implemented under the Upland Research and Capacity Development Programme in 2010. Shortly after the workshop, Phadeng and Phoukhong villagers completed a road, accessible by motorbike, from Phousaly to Phoukhong. By the end of 2009, ten households had relocated to Phoukhong, and others were in the process of relocating or moving to other locations.

By February 2010, Phadeng had been completely abandoned. The last family left in December 2009, having waited for the mother to give birth. The resettlement of the villagers began in August 2009 and finished in December. By our February 2010 visit, most of the houses had been destroyed, some because villagers reused their materials for the new settlement or for the Sanam Namlao (the hamlet near Namlao River), some because of roaming cattle, buffalo and goats. About 14 of the 40 houses were still standing, though some were already in bad condition. A few were still locked, their owners protecting goods remaining inside, intending to move these later, on their relocation to Phousaly. Villagers still go to Phadeng two or three times a week, to take care of the livestock.

There was no plan to make Phadeng into a *sanam* (a small hamlet of field huts near people's fields). Villagers explained that they had made swiddens too recently there for reuse, and they saw no immediate need for a *sanam*.

In the meantime, the Sanam Namlao had developed. Before resettlement, there were only three houses there. We saw a group of 11 houses near the Namlao River, and villagers said there were about ten more, spread along the river. The majority of people at Sanam Namlao already had a house in the new location, Phousaly. Only four households had yet to build one. Namlao is about to become the main *sanam* for Phadeng villagers because it is still located in their former territory, and so villagers do not need to ask permission from Phoukhong villagers to open new swiddens.

In Phousaly, by February 2010, 20 houses were already finished. One small school was standing, with a new teacher providing education to grade three. One villager from Phoukhong had settled in Phousaly because he was the owner of the fish pond located in the new settlement site. The people of Phoukhong do not recognize the name Phousaly; for them the area is just 'the fishpond'. The district supported villagers with basic tools (no explosives, which the villagers had anticipated), and the Food and Agriculture Organization of the United Nations (FAO) provided rice for workers. The villagers began the roadwork in July 2009 and finished a month later. One person per household was designated to participate, receiving between 9.5kg and 15kg of rice per person per day, depending on the difficulty of the task. Villagers said the district had promised but not yet delivered 20 roofing pieces per family. Water was no longer an issue, as it had been in Phadeng, and villagers were apparently happy with their new situation, along with access to the market (via the new road for motorbikes) and health care (three vaccination campaigns had been conducted since August 2009). Two houses had electricity from micro-hydropower. The main issue was the problem of unclear zoning and selecting sites for swiddens.

Phoukhong villagers were still using their traditional lands for swiddens at the time of our visit. However, beginning in 2011, they plan to plant in old Phadeng territory. One element of concern for Phadeng villagers living in Phousaly was that they still needed to obtain permission from Phoukhong to open a new field. They would like to agree on clearer zoning and be able to plant upland rice according to this zoning, without having to ask permission. For this reason, they were still using their old territory. The original inhabitants of Phoukhong, on the other hand, were more concerned about the village boundaries with neighbouring villages than joint land-use zoning with the merged village. Four households from Phadeng moved to Vientiane, and one to Houay Choy. The rest were still residing between Phousaly and Namlao, and every day people from Phousaly travel to the *sanam*.

Displacement in Derema Corridor[6]

The East Usambara Mountains are part of the Eastern Arc mountain range, ranked high in biodiversity conservation value (see Chapter 5). The landscape consists of a mosaic of forests, small-scale agriculture and commercial estates, tree plantations and villages. Table 4.2 presents the land uses in the early 1990s. Since then, agricultural areas are likely to have increased substantially, mainly at the expense of unreserved forests (Jaclyn Hall, unpublished data).

The Derema area is mainly upland rainforest (above 800m). In 2000, it was considered to be one of the few remaining large, unreserved tracts of forest, as noted in studies by the Forestry and Beekeeping Division and others (FBD et al, 2004; Newmark, 1993, 2002). Nevertheless, in the early 1990s, 86

Table 4.2 *Land uses in East Usambaras, 1993*

Main land-use class	Hectares	Percentage
Forest	42,121	50.4
Agriculture	35,909	43.0
Peasant farming	31,716	88.3
Tea	2363	6.6
Cocoa	1107	3.1
Sisal	535	1.5
Woodlands	4113	4.9
Grassland	345	0.4
Ponds and rivers	101	0.1
Barren land	393	0.5
Settlements	620	0.7
Total	83,601	100

Note: The survey does not cover the total landscape.
Source: Hyytiäinen (1995)

per cent of the planned Derema forest reserve[7] of 790ha was classified as 'cultivation under forest' (Johansson and Sandy, 1996, p31). In the farming system commonly used by local small-scale farmers, some indigenous trees have been retained in the *mashamba* ('farms' in Swahili), as shade trees for the crops and for other purposes. From a distance, then, the area looks like a lightly disturbed forest.

The dominant ethnic group in the East Usambara Mountains is the Shambaa. A number of other groups are also present as a result of immigration, and many villages are socially mixed. Swahili, Tanzania's official language, is spoken by virtually everyone and is most commonly used to communicate public matters in community meetings. The main source of living is small-scale farming, including cultivation of subsistence crops, such as maize and yams, and cash crops, especially spices and sugar cane. Historically, local people have depended on forests in many ways – as a source of fuelwood, building material, food, other forest products and agricultural land. Today, access to forests is increasingly limited because much of it has been either converted to agricultural land or placed within the boundaries of protected areas. Agroforests have largely replaced forests as the most important source of forest products (Rantala, unpublished data), although the people living adjacent to the forests still obtain fuelwood and some minor products from the forest.

Msasa IBC and Kwezitu were among the five villages that were directly affected by the establishment of the Derema Corridor. In total, the five villages had an estimated population of about 7900 people (URT, 2006), but not all of the villagers held land in Derema. Of the population of 2200 in Msasa IBC, about 26 per cent signed up for compensation for lost access to land in the corridor; in Kwezitu, less than 10 per cent of the population of 2311 made such a request. From the perspective of the villagers, the area offered an important source of income and food from subsistence and spice crops, as well as future agricultural land, fuelwood, timber and other forest products. In addition to farming, other sources of livelihood in the villages include small businesses and wage labour in the tea fields and factories. Some farmers have also benefited from the development interventions introduced by conservation projects in past years, such as butterfly farming and the commercialization of the seeds of an indigenous tree species, *Allanblackia stuhlmannii*.

To address the threat of forest conversion to smallholder agriculture, forest management in the East Usambaras today is conservation-oriented, especially regarding the remaining catchment forests in the higher altitudes. Much of the upland forest has been protected as government forest reserves or nature reserves. Since the end of the 1980s, the Tanzanian government has focused on the enforcement of forest regulations with the help of donor-funded conservation projects. At the same time, especially from the 1990s onwards, efforts have been made to involve the communities in conservation, for instance through integrated conservation and development projects and the establishment of village-level organizations involved in conservation (see Stocking and Perkin, 1992; Woodcock, 2002; Vihemäki, 2009).

Following pilot projects in the East Usambaras and elsewhere, the National Forest Policy of 1998, introduced by the Ministry of Natural Resources and Tourism (MNRT, 1998), marked a policy shift, scaling up participatory approaches to forest management. The Forest Act of 2002 reinforced this shift (see Chapter 5). Yet exclusionary approaches also remain on the forest conservation agenda. In the East Usambaras, forest control is guided by the Eastern Arc Mountain Forests Strategy, which seeks to expand protected areas to conserve biodiversity (MNRT, 2006).

Efforts to mitigate the ecological effects of forest fragmentation in the East Usambara Mountains include plans for conservation corridors between the strictly protected forest reserves to enhance landscape connectivity (Tye, 1993). The Derema Corridor was considered the most urgent for its role in connecting the Amani Nature Reserve to the south with forest blocks to the north (Figure 4.2). Conserving Derema was first suggested in the 1970s (Iversen, 1991), but the process of establishing the corridor did not start until the late 1990s, as part of a large donor-funded conservation intervention, the East Usambara Conservation Area Management Programme (EUCAMP). The corridor plan included only farms and unprotected forests, and excluded settlements.

Initial reactions to corridor plan

Before the start of EUCAMP's engagement, some villagers were reluctant to give up their agricultural land for forest conservation. The secretary of Msasa IBC stated in the early 1990s that villagers would meet further reservations with resistance (Mwalubandu et al, 1991). The situation had not changed much when negotiations over the Derema Corridor began in early 2000. The plan of EUCAMP, however, was to conduct this conservation initiative in a participatory way, in the sense that local people would be consulted and their livelihoods would not be endangered by the exercise (Vihemäki, 2009).

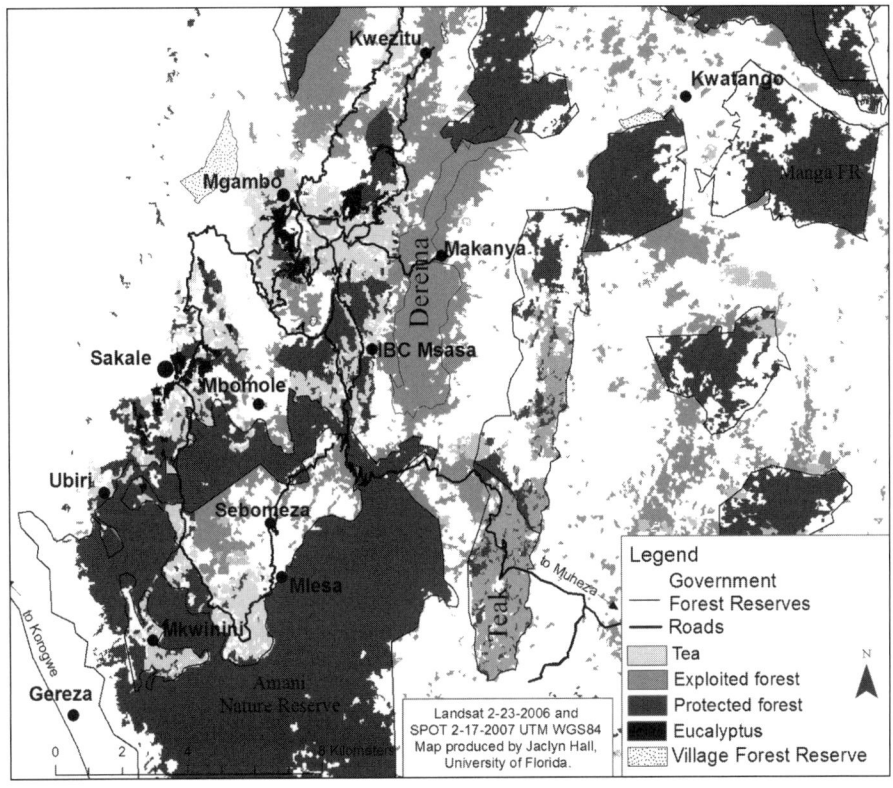

Figure 4.2 *Location of Derema Corridor and study villages in East Usambaras*
Source: Map produced by Jaclyn Hall (2010)

The status of land rights in the Derema Corridor was not clear, as is evident from documents and controversial accounts given by actors involved (Vihemäki, 2009). Local villagers considered themselves to be owners of some of the land that was to be included in the corridor. For instance, an elderly man living in Makanya argued that the forests had previously been owned by two tea companies, Karimjee and Bombay Burmah, but were later given to the villagers. After the start of national village-making policies in the 1970s, land was commonly allocated to the farmers by village authorities. The village authorities and village environmental committee members also suggested that some of the land previously owned by the Derema tea estate had been awarded to the citizens. Villagers' claims to land came to be considered customary rights in this process (URT, 2006).

Customary land rights on village land are protected as private land rights, even if not registered (see Chapter 5). Nevertheless, the president can transfer village land to the category of reserved land for public benefit (Village Land Act, 1999). In such a case, land legislation defines the procedures for informing, consulting with and obtaining the consent of local communities, and a right to compensation. Communities can even withhold consent for land transfers that cover areas of less than 250ha. Although the process of establishing the Derema Corridor was launched before the current legislation came into effect, EUCAMP was advised early on to adhere to the new land law.

Social impact assessment

At the start of the process in 2000, a social impact assessment (SIA) was commissioned by EUCAMP and conducted by researchers from the University of Dar es Salaam. Its objective was to inform planners about the potential social impacts of the establishment of the corridor, and propose measures to mitigate the negative effects and enhance the positive ones (Jambiya and Sosovele, 2000). It also served to inform local people about the plan for establishing the corridor and probably raised expectations and fears. Data was collected mainly through questionnaires administered to almost 300 farmers in the five directly affected villages. In addition, about 70 other interviews and group discussions among the farmers and village leaders were conducted. But at that time, the actual number and identity of people to be affected were not known.

The attitudes of the farmers towards the initiative were mostly negative at the beginning (Jambiya and Sosovele, 2000; authors' fieldwork). The secretary of the Village Environmental Committee of Msasa IBC explained that the farmers had been consulted about their willingness to conserve the area but were reluctant to give up their land because they depended on it for their livelihoods. Some farmers also expressed their relative powerlessness if the government decided to establish a forest reserve. A small proportion of farmers (13 per cent) were initially positive about the conservation of the forest and shifting to alternative areas (Jambiya and Sosovele, 2000). Despite such local reservations, the conservation value of the forest was said by the SIA to be widely recognized by the villagers.[8] When asked about the use of potential compensation money, most anticipated that if it was sufficient, they would spend it on land (reported in the United Republic of Tanzania, URT, 2006).

The reactions of the villagers were partly shaped by their previous experiences of conservation-related displacement. During the establishment of the Amani Nature Reserve, the farmers whose land was appropriated were compensated neither according to the initial plans nor to their expectations (Jambiya and Sosovele, 2001; Vihemäki, 2009). There were also gender differences in the initial positions of the villagers. According to Jambiya and Sosovele (2000), women were more concerned than men about the change in access to NTFPs, such as fuelwood, herbal medicines and wild vegetables. Some of the women also feared that men would dominate the decision-making process about the use of the compensation money.

After the SIA survey, a feedback workshop was organized, involving 24 representatives of the villages and two from the ward level (Jambiya and Sosovele, 2000).[9] The aim was to share the findings with the stakeholders and make recommendations on how best to proceed. From the documentation, it appears that the position of the villagers had shifted at least in the official arena of discussion: their representatives were now by and large positive about the conservation plan, according to the SIA report (Jambiya and Sosovele, 2000).

The general conclusion of the workshop was that Derema must be reserved. However, the level of consensus among people whose farms were to be appropriated appeared to be lower than the documents suggest. Only a limited number of villagers' representatives participated in the workshop, mainly village leaders. In 2008, the majority of the farmers we interviewed, whose land had been appropriated for the corridor, stated that they had not participated in the decision-making process behind the land allocation (Rantala and Vihemäki, forthcoming). The expressed lack of participation was probably partly due to their dissatisfaction with the compensation and consequences of the process, but it also appears that many agreed to give up land because of the perceived lack of alternatives.

Although the planners stressed the integration of local people in the planning and management of the corridor, there was not much space in the process for negotiating overall land-use options. For instance, the SIA focused on the option of conservation through protected area establishment and did not explore other alternatives (see Sjöholm et al, 2001). The option of protecting the area through the establishment of village forest reserves instead of a strictly protected government reserve was discussed in a later workshop, but the idea was reportedly rejected by the community representatives. The decision on how local communities would participate in the management of the corridor was left open until the process was completed (URT, 2006).

Local responses and strategies

After the SIA, several meetings and negotiations among the farmers, government representatives and project staff were held with regard to how to go about establishing the Derema Corridor. Those who had fields in the corridor agreed to give up their area, and were promised compensation for the value of their crops (see FBD et al, 2004). The compensation rate was not known at that point. The determination of crop values was a complex exercise, affected by the change in land law. In principle, the valuation approach took into account the value of the lost production from main crops and fruit trees for three years, until new plants could grow to maturity at new sites (URT, 2006).

The expectation of receiving monetary compensation was probably an important reason why farmers agreed to the conservation plan, but agreement was also influenced by a perceived lack of alternatives. In the post-SIA workshop, it was suggested that the villagers should be 'educated' about the importance of giving up their land by experts or village leaders (see Jambiya and Sosovele, 2000). According to a forest official involved in the process, some pressure was also put on the farmers to accept the plan. An elderly lady, a village council member, explained, 'We accepted since we knew that when the government decides to do something, no one can refuse, and this came from the government' (16/03/2005, Msasa IBC). Her comment illustrates a clear disparity in the negotiation power, as experienced by the affected people.

Opposition to the plan or its conditions had not disappeared completely. When the Derema Corridor was being demarcated and the crops growing in the boundary were being slashed, some farmers in Makanya resisted the process and wanted the boundary location to be changed. As a result, the police intervened. To settle the conflict, a meeting was arranged by EUCAMP, and the process continued (Pohjonen, 2002).

The project had anticipated applying the old land law to calculate the compensation and reserved funds accordingly. But the government instructed the project to use the new land law instead, causing confusion and delay (URT, 2006). Compensation for the lost crop production in the narrow boundary of the corridor was set at a high rate, raising expectations about similar compensation in the remaining area. The estimated yield of cardamom per plant used in the compensation calculations was based on estimates by the farmers, and was much higher than the normal yield (personal communication, T. Reyes). Only 172 farmers with crops along the boundary received compensation. Ironically, the boundary compensation created an incentive to plant new crops *inside* the corridor. Both villagers and outsiders started to plant crops in the uncompensated fields and previously uncultivated areas in the hope of being compensated (see Vihemäki, 2009). This strategic action by the farmers increased the overall costs of establishing the corridor.

Many of the farmers argued that they agreed to give up their lands upon promise of payment within six months of finalizing crop evaluation (authors' fieldwork, 2003, 2005). At the time of the crop valuation, people were told not to plant new crops or actively tend to the areas they had given up but they could still harvest crops there. The lack of maintenance reduced the yield and resulting income from their fields.

The compensation rates for the crops inside the boundaries were reassessed several times. The process of paying the compensations for these crops started only in mid-2005, three years after the closure of EUCAMP; the farmers had been waiting and following up on the compensation for several years. Over the course of time, the negative social and economic consequences of the intervention became evident.

In early 2005, the unpaid compensation for crops was a major concern for the farmers who had given up farms and not accessed alternative farmland. Many complained about loss of income. People made their living by cultivating the remaining fields or by other less 'traditional' means.[10] The affected people often used the Swahili expression *tumeathirika*, meaning 'we are negatively affected' or 'hurt'. In addition, they argued that they had no information on the amounts that they were to be compensated, suggesting inadequate communication. Contrasting discourses among the affected farmers about the future of the corridor were presented to the visiting researchers. Some requested to be paid compensation for crops, whereas others asked to be given their old fields back (Vihemäki, 2009). In addition, some farmers used the concept of rights to question the way they had been treated in the decision-making process, saying their rights had been bypassed.

The flow of information about compensation and the process as a whole after EUCAMP's pullout was neither appropriate nor adequate from the perspective of the affected farmers. After waiting for crop compensation and becoming increasingly uncertain about the outcome of the process, the farmers organized themselves to follow up. They sent several delegations to the government offices in Dar es Salaam, as well as those at district and regional levels, to meet with officials and policy-makers. This was largely a result of the farmers' disappointment with government authorities and the negative consequences for livelihoods. Yet their collective action was probably also spurred by the initial promises of a participatory and transparent approach (Jambiya and Sosovele, 2000). The outcomes of their efforts were often unsuccessful, as they were asked to return several times and wait, adding to their frustration. However, before the presidential elections in 2005, the issue of the unfinished compensation process was raised in parliament (URT, 2006), and the broader awareness created pressure to settle the matter.

The accounts of many of the Derema villagers also illustrate their experience of the loose commitment of the government to promoting social development, common also in other areas of Tanzania. Hydén (1994) argues that concentration of power during the post-independence period in Tanzania led to a general loss of social capital in the society, partly as a result of unfulfilled governmental promises and leaders' 'self-righteous' behaviour. In the Derema villages, the farmers' complaints about being 'cheated' by the government illustrate similar distrust towards the authorities, although the farmers also tried to engage with them by conducting follow-up. False information was sometimes given by officials or politicians, contributing to confusion and distrust among villagers, some of whom suggested that the government officials might be 'eating the money,' an expression used for corruption (Vihemäki, 2009). In addition, some feared that the forest department might introduce wildlife into the corridor after it was gazetted. The information gaps had created space for rumours, adding to the confusion.

In 2004, the Ministry of Natural Resources and Tourism (MNRT) approached a World Bank national conservation project and requested funds to complete the compensation process in Derema (URT, 2006). In relation to this initiative, a World Bank supervision mission for a nationwide forest conservation project visited the area, including representatives from the United Nations Development Programme (UNDP), World Wide Fund for Nature (WWF), Tanzanian government and donors. They 'heard loudly from Msasa IBC villagers how annoyed they were by the payment delays' (URT, 2006, p26). As a result of new organizations' involvement, a requested resettlement action plan in line with the World Bank's Operational Policy 4.12 on Involuntary Resettlement was submitted for World Bank approval in 2006. During the preparation of the action plan, government authorities were consulted and meetings were organized in each of the five villages, in which village leaders and some farmers participated (URT, 2006). The number of participants in the meetings was not documented.

In the meantime, towards the end of 2005, half of the compensation (the rate of which had been re-estimated and reduced after the initial payouts) was awarded to the farmers, partly in anticipation of the upcoming presidential elections (see URT, 2006). The farmers accepted the money without agreeing that the payments represented 50 per cent of the sums due (URT, 2006) – evidence of strategic action to keep the negotiation process open, in hope of securing better compensation. During the verification process that preceded the payments, some farmers in Msasa IBC expressed strong discontent. The action plan report (URT, 2006, p29) describes the situation at the time of the payments as follows: 'In Msasa IBC villagers forced the team to stop the verification exercise, and police had to "rescue" the team.' In early 2008, farmers in Msasa IBC also argued that at the time of the 2005 payments there had been disputes between the farmers and the authorities. In addition, some had not turned up to collect the money (URT, 2006; authors' fieldwork) because they rejected partial compensation, reflecting their opposition to the way in which the process was handled.

As a part of the resettlement action plan, a re-evaluation of the remaining payments was made. This reduced the estimated production per plant from 5kg of dry cardamom to 3kg (URT, 2006). The level of final compensation was to be lower than for the boundary crops, although interest payments for the delay were to be added to the sums.

The action plan and allocation of World Bank money resulted in the establishment of a facilitation and monitoring process, with WWF becoming the primary facilitator. A compensation and development office led by a coordinator was also established in the district capital. Alternative farming land at sisal estates in the lowlands for those wishing to relocate was also to be made available with the agreement of the district commissioner (World Bank, no date). However, the land had been previously allocated to farmers who had lost land to the Amani Nature Reserve and was not available. The alternative land was not yet available by the end of 2009, although the coordinator had made progress in securing new land for the affected farmers.

Access to final compensation payments was still a major concern for many in early 2008. The remaining compensation, with the money originating from the World Bank loan, was paid gradually, between February and May 2008. Before that, in early 2008, the Derema farmers' representatives were planning a demonstration at the district headquarters, to be paid the remaining money. One argued that the affected farmers had earlier accepted the lower rate of compensation crops, even though they considered the rate unfair. Despite the alleged consensus and the adjusted rates, some of the farmers still criticized the level of compensation and the process. Furthermore, the village chair of Msasa IBC stressed the poverty effects. The most anxious individuals even raised the option of suing the government.

Soon after the final payments, access to alternative land in the lowlands became an urgent issue for the affected people, although many complained that there was no money left for the improvement of new farms because of the delay in land allocation. They expressed resentment towards the authorities over the inadequate handling of the land issue. The farmers now also demanded cash compensation for the land itself, in addition to the crops.

The reactions and strategies of the Derema farmers suggest that efforts to avoid negative outcomes and involve affected people in the planning stage failed to make the process socially acceptable. In fact, the negotiations, meetings and expectations of compensation probably fuelled the farmers' collective action. Despite the broad dissatisfaction, not all affected farmers turned out to be 'losers' in the end. Some received profitable compensation and invested in new land or other assets (Rantala and Vihemäki, forthcoming). Yet, they were a clear minority among the affected people, and they also criticized the overall process and its effects.

Discussion

Policies that lead to displacement represent the most severe consequence of the great disparities in power between villagers and other actors involved in landscape management. The distance between decision-makers and those affected, as well as the lack of involvement of those affected by the decision, often means that a 'no-change' option is not possible. In such situations, how can we achieve the best or 'least bad' outcomes for people's livelihoods?

The two case studies described different pathways of communicating the livelihood concerns of the affected people to the actors implementing the displacement process, and other means used to mitigate harm (Table 4.3). The experiences suggest that the degree to which the process can render the outcomes socially acceptable depends on the context, such as conditions of land tenure, people's experiences from other displacements and variations in negotiating power. Other central factors shaping the social outcomes include the commitment of the organizations initiating the displacement to address and monitor the long-term social implications, willingness to negotiate conditions of displacement, flow and accuracy of information and level of representation of the different groups.

In the Lao case, we described interlinked pathways of communication – the first in presenting empirical research and scenarios to decision-makers, which then opened the pathway for the second phase, direct community participation through a workshop. Although we do not argue that the best solution was achieved, we think that through enhanced communication and ultimately some participation, a situation better than the initial option and past examples of displacement in Laos was attained. Several factors contributed to the success of the communication strategies in achieving a mediated outcome. The most significant factor was that policy objectives for Lao displacement were first for development, and second for conservation. Even though the process was imple-

Table 4.3 *Communication with affected people in two displacement processes*

Stage	Laos	Tanzania
Initial consultation	Study team conducted perception studies and participatory mapping in two communities to understand perceived impacts and alternatives. The team shared results with district authorities, but there was no official consultation process	Social impact assessment was conducted in 2000, including survey and interviews of villagers (370 people)
Information sharing prior to displacement	Village authorities were officially notified of governmental intent to relocate village	Social impact assessment created expectations and fears among farmers; findings were shared with 24 village representatives in workshop; farmers were educated about importance of conservation; promises were made about compensation for lost income
Negotiations on conditions of displacement	Study team facilitated workshop between district and village officials using participatory and 3D mapping to address livelihoods and land use	Meetings of local government and project with affected farmers or their representatives were held to reach agreement; village representatives participated in demarcation of corridor; meeting was held to solve conflict on boundary location
Consultations during process	Villagers rarely consulted after displacement began	Farmers were consulted about crop prices to assess compensation rates; communication was irregular and information flow limited between affected people and government, 2003–2005; conservation organizations and donors made brief missions to Derema villages, 2004; WWF coordinator facilitated process, 2006–2009; half- to one-day meetings were held with village leaders and villagers during preparation of resettlement plan, 2006
Land-use planning for new or protected area	Participatory land-use planning is scheduled for 2010	Workshops were organized in 2000–2001 to decide management model for corridor, but planning was incomplete; decision on management of area was postponed

mented without appropriate consultations, surveys and resources, the overarching objective provided a mechanism for altering the outcome. The second most important factor was the timing: the process was in the initial phase when the project team presented its research on the perceived impacts of displacement. The third factor was that the project team was given legitimacy as an independent and credible conduit of information between villagers and the district, and later became a facilitator of participatory discussions at a workshop on the displacement process. The fourth factor was the low population density of the area and the official state ownership of the land (even when customarily owned by local people). These conditions gave the district a greater degree of flexibility in reaching a mediated solution. Project intervention was not able to overcome the district's lack of resources to provide improved services to the displaced villagers, however.

In Tanzania, despite the efforts to avoid negative social outcomes and fairly compensate farmers for the lost land, there was ultimately a high level of dissatisfaction. The situation also appeared to be prone to conflicts. Consent for the plan to establish the Derema Corridor seemed partly imposed; the village leaders and representatives who reached agreement to protect the reserve did not represent the views of all groups. Some also accepted the plan, as in Laos, thinking that they had no alternatives. In addition, the delays in providing compensation contributed to livelihood losses and increased negative attitudes. Furthermore, some of the promises made by the representatives of the government, such as providing access to alternative farmland, had still not been met by the end of 2009.

Among the factors that complicated the process and interfered with efforts to improve social acceptability in Derema were the limited time frame of the project that launched the process, poor planning and initial misunderstanding of the context. The changes in land legislation and related difficulties in implementing the displacement also posed severe challenges. Because of the many agencies involved and unclear authority relations, the affected farmers often had difficulties getting information about the process, including which agencies to approach with their concerns and claims.

An interesting difference vis-à-vis the Lao case, however, is the development of collective action in Tanzania, indicating these farmers' greater perceived negotiating power. This could partly be an outcome of initial consultations and negotiations, as well as the expectations of compensation. In the Lao case, however, two factors hindered the development of such collective action. The first is the political context, and the second is the marginalized position of the Hmong in Laos (Petit, 2008; Drouot, 1999; Milloy and Payne, 1997; Ovesen, 2004).

Conclusions

In situations where displacement cannot be avoided, can we achieve a 'least bad' solution and minimize social harm? Enhancing flows of information and direct communication between those implementing the process and the displaced is likely to make displacement less harmful to the displaced, but it does not always guarantee the best outcome. The outcomes depend largely on the context, such as the actual space for negotiating the conditions of displacement, availability of alternative lands or resources, and negotiation power of the groups involved. Making displacement less adverse and more inclusive requires careful planning and understanding of the context, commitment from organizations involved, a significant amount of time and access to resources.

Our experience suggests that avoiding or mitigating negative social consequences is likely to be contingent on several factors:

- an intermediary whose legitimacy is accepted by different parties to catalyze discussions;
- flexibility, in terms of available land, financial and human resources, on the part of implementing agencies to create a mutually acceptable alternative;
- the stage of the process and receptivity of decision-makers to launch or reinitiate discussions on the displacement;
- trust between the parties involved, as well as accurate and continuous information shared between parties;
- sustained commitment by the implementing agency; and
- adequate information about the existing institutional and material conditions that shape people's lives, including rights to natural resources.

The intervention of external parties, such as researchers, conservationists or NGOs, can supplement a government's efforts to mitigate negative impacts of displacement and settle disputes. Yet, any long-term solution requires governmental reform at all levels. Particular areas where changes are needed include conventional good governance issues such as transparency and accountability. But perhaps more important are changes in attitudes and organizational cultures. Government bureaucracies are hierarchical; working with local communities – crucial in displacement – requires a different, more horizontal and risk-taking mindset, as well as greater respect for the rural peoples, their claims and rights, and their aspirations. Truly participatory processes cannot begin with explanations of government policies, the provision of false 'choices' that ignore the people's wishes, and 'facipulation' (facilitation and manipulation) that pushes people in predetermined directions. Finally, our research leads us to conclude that to improve social acceptability, officials need to gain skills and be provided with adequate resources in social analysis and learning, participatory planning, facilitation, conflict management and resolution, as well as inclusive negotiation.

Notes

1. Social scientific evidence shows that 'restricting access' to resources vital for livelihoods is similar to economic displacement (Cernea, 2005).
2. Rantala was connected institutionally to the Landscape Mosaics project but conducted her research in Derema independently. Vihemäki was at the time of conducting her research affiliated with the Institute for Development Studies of the University of Helsinki.
3. Despite our best efforts, we were able to interview only one female household representative. Many women were working in their fields at the time of interviews.
4. Petit (2008), using personal histories, documents the important role that Hmong political leaders can play for their fellow ethnic group members.

5. Cf the older but still relevant description of the warping of 'participation' in Laos (Arnst, 1997).
6. This section includes materials that will be more fully described and analysed in Rantala and Vihemäki (forthcoming).
7. The planned size and the name of the proposed conservation area have varied in different documents produced during the process. The final size of the Derema Corridor was 968ha (Burgess, N., personal communication, 08/09/2010).
8. Many local people associate forest cover with the availability of rain and water. This is also widely conceived as a justification for forest conservation (see Chapter 5; Vihemäki, 2009).
9. The workshop discussions are documented in the annex of the SIA, Summary of the Conclusions of Muheza Stakeholders' Workshop.
10. For instance, some young men from Derema villages had started to mine gold in 2004–2005. They argued it was due to the lack of alternatives after their families' farms were appropriated.

References

Agrawal, A. and Redford, K. (2009) 'Conservation and Displacement: An Overview', *Conservation and Society*, vol 7, no 1, pp1–10

Arnst, R. (1997) 'International Development Versus the Participation of Indigenous Peoples', in D. McCaskill and K. Kampe (eds) *Development or Domestication? Indigenous Peoples of Southeast Asia*, Silkworm Books, Bangkok, pp441–454

Baird, I. G. and Shoemaker, B. (2005) 'Aiding or Abetting? Internal Resettlement and International Aid Agencies in the Lao PDR', Probe International, Toronto

Baird, I. G. and Shoemaker, B. (2007) 'Unsettling Experiences: Internal Resettlement and International Aid Agencies in Laos', *Development and Change*, vol 38, no 5, pp865–888

Brockington, D. and Igoe, D. (2006) 'Eviction for Conservation: A Global Overview', *Conservation and Society*, vol 3, no 4, pp424–470

Cernea, M. (1993) 'Anthropological and Sociological Research for Policy Development on Population Displacement', in M. Cernea and S. Guggenheim (eds) *Anthropological Approaches to Resettlement: Policy, Practice, and Theory*, Westview Press, Boulder, Colorado

Cernea, M. (2005) '"Restriction of Access" is Displacement: A Broader Concept and Policy', *Forced Migration Review*, vol 23, May, pp48–49

Cernea, M. (2008) 'Reforming the Foundations of Involuntary Resettlement: Introduction', in M. Cernea and H. M. Mathur (eds) *Can Compensation Prevent Impoverishment? Reforming Resettlement through Investments*, Oxford University Press, New Delhi

Department for International Development (DFID) (1999) 'Sustainable Guidance Sheets: Framework', DFID, London

Drouot, G. (1999) 'Pouvoir et Minorités Ethniques au Laos: De la Reconnaissance Constitutionnelle à la Participation Effective à L'Exercice du Pouvoir [Power and Ethnic Minorities in Laos: From Constitutional Recognition to Effective Participation to the Exercise of Power]', *Mousson: Social Science Research on Southeast Asia*, vol 99, pp53–74

Ducourtieux, O., Laffort, J.-R. and Sacklokham, S. (2005) 'Land Policy and Farming Practices in Laos', *Development and Change*, vol 36, no 3, pp499–526

Evrard O. (2004) 'La mise en oeuvre de la reforme foncière au Laos: impacts sociaux et effets sur les conditions de vie en milieu rural' [Implementation of Tenure Reform in Laos: Social Impacts and Effects on Life Conditions in the Rural Milieu], LSP Documents de Travail no 8, FAO, Rome

Evrard, O. and Goudineau, Y. (2004) 'Planned Resettlement, Unexpected Migrations and Cultural Trauma in Laos', *Development and Change,* vol 35, no 5, pp937–962

FBD, WWF, TFCG and IUCN (2004) 'Securing the Derema Forest Corridor in the East Usambara Forests of Tanzania: The Most Important Biodiversity Corridor in the Eastern Arc and Coastal Forests Hotspot', Draft Paper, Dar Es Salaam, June

Fitriana, Y.R. (2008) 'Landscape and Farming System in Transition: Case Study in Viengkham District, Luang Prabang Province, Lao PDR', Agronomy and Agro-Food Program, Institut des Regions Chaudes-SupAgro, Montpellier, France

Fox, J., Fujita, Y., Ngidang, D., Peluso, N., Potter, L., Sakuntaladewi, N., Sturgeon, J. and Thomas, D. (2009) 'Policies, Political-Economy and Swidden in Southeast Asia', *Human Ecology,* vol 37, pp305–322

Freeman, L., Lewis, J., Borreill-Freeman, S., Wiedmer, C., Carter, J., Clot, N. and Tchoumba, B. (2008) 'Free, Prior and Informed Consent: Implications for Sustainable Forest Management in the Congo Basin', Paper Presented at Workshop on Forest Governance and Decentralization in Africa, 8–11 April 2008, Durban, South Africa

Fujita, Y. and Phengsopha, K. (2008) 'The Gap Between Policy and Practice in Lao PDR', in C. J. P. Colfer, G. R. Dahal and D. Capistrano (eds) *Lessons from Forest Decentralization: Money, Justice and the Quest for Good Governance in Asia-Pacific,* Earthscan/CIFOR, London, pp117–132

High, H. (2008) 'The Implications of Aspirations: Reconsidering Resettlement in Laos', *Critical Asian Studies,* vol 40, no 4, pp531–550

High, H., Baird, I. G., Barney, K., Vandergeest, P. and Shoemaker, B. (2009) 'Internal Resettlement in Laos', *Critical Asian Studies,* vol 41, no 4, pp605–620

Hydén, G. (1994) The Role of Social Capital in African Development', in H. Marcussen (ed) *Improved Natural Resource Management. The Role of the State Versus that of the Local Community,* Occasional Paper 12, Institute of Development Studies, Roskilde University

Hyytiäinen, K. (1995) 'Land-use Classification and Mapping for the East Usambara Mountains', Technical Paper 12, East Usambara Catchment Forest Project, FBD and Finnish Forest and Park Service, Dar es Salaam and Vantaa

ICEM (2003) *Regional Report on Protected Areas and Development: Review of Protected Areas and Development in the Lower Mekong River Region,* Indooroopilly, Queensland, Australia

Iversen, T. (1991) *The Usambara Mountains, NE Tanzania: History, Vegetation and Conservation,* Uppsala Universitet, Uppsala, Sweden

Jambiya, G. and Sosovele, H. (2000) 'Social Impact Assessment (SIA): Derema and Wildlife Corridors – EUCAMP' (unpublished consultancy report), Dar es Salaam

Jambiya, G. and Sosovele, H. (2001) *Conservation and Poverty: The Case of Amani Nature Reserve,* Research on Poverty Alleviation (REPOA), Dar es Salaam

Johansson, S. and Sandy, R. (1996) 'Protected Areas and Public Lands – Land Use in the East Usambara Mountains', Technical Paper 28, East Usambara Catchment Forest Project, FBD and Finnish Forest and Park Service, Dar es Salaam and Vantaa

Jones, P., Sysomvang, S. and Amphaychith, H. (2004) 'Village Land Use and

Livelihoods Issues Associated with Shifting Cultivation, Village Relocation and Village Merging Programmes in the Uplands of Phonxay District, Luangprabang Province', in B. Bouahom, A. Glendinning, S. Nilsson and M. Victor (eds) *Poverty Reduction and Shifting Cultivation Stabilization in the Uplands of Lao PDR: Technologies, Approaches and Methods for Improving Upland Livelihoods,* Proceedings of Workshop held in Luang Prabang, Lao PDR, January 27–30 2004, National Agriculture and Forestry Research Institute, Vientiane, Lao PDR

Lasgorceix, A. and Kothari, A. (2009) 'Displacement and Relocation of Protected Areas: A Synthesis and Analysis of Case Studies', *Economic and Political Weekly,* vol xliv, no 49, p37–47

Lestrelin, G. (2009) 'Land Degradation in the Lao PDR: Discourses and Policy', *Land Use Policy,* doi:10.1016/j.landusepol.2009.06.005

Lestrelin, G., and Giordano, M. (2007) 'Upland development policy, livelihood change and land degradation: interactions from a Laotian village', *Land Degradation and Development,* vol 18, pp55–76

Mertz, O., Leisz, S. J., Heinimann, A., Rerkasem, K., Thiha, Dressler, W., Pham, V. C., Vu, K. C., Schmidt-Vogt, D., Colfer, C. J. P. and Epprecht, M. (2009) 'Who Counts? Demography of Swidden Cultivators in Southeast Asia', *Human Ecology,* vol 37, no 3, pp281–290

Messerli, P., Heinimann, A. and Epprecht, M. (2009) 'Finding Homogeneity in Heterogeneity: A New Approach to Quantifying Landscape Mosaics Developed for the Lao PDR', *Human Ecology,* vol 37, no 3, pp291–304

Milloy, M. J. and Payne, M. (1997) 'My Way and the Highway: Ethnic People and Development in the Lao PDR', in D. McCaskill and K. Kampe (eds) *Development or Domestication? Indigenous Peoples of Southeast Asia,* Silkworm Books, Bangkok, pp398–440

MNRT (Ministry of Natural Resources and Tourism) (1998) National Forest Policy, United Republic of Tanzania, MNRT, Dar es Salaam

MNRT (2006) Eastern Arc Mountains Forest Strategy, FBD, December 1 2006 version

Mwalubandu, C., Ngaiza, A., Kazungu, E., Ramadhani, H. and Temba, P. (1991) 'Whose Trees? Learning from the Past: The East Usambara Catchment Forest Project, Tanzania', in N. Atampugre (ed) *Whose Trees?,* Panos Institute, London, pp58–99

Newmark, W. (1993) 'The Role and Design of Wildlife Corridors with Examples from Tanzania', *Ambio,* vol 22, no 8, pp500–504

Newmark, W. (2002) 'Conserving Biodiversity in East African Forests: A Study of the Eastern Arc Mountains', *Ecological Studies,* vol 155

Ovesen, J. (2004) 'All Lao? Minorities in the Lao People's Democratic Republic', in C. R. Duncan (ed) *Civilizing the Margins: Southeast Asian Government Policies for the Development of Minorities,* Cornell University Press, Ithaca, New York, pp214–240

Petit, P. (2008) 'Rethinking Internal Migrations in Lao PDR: The Resettlement Process Under Micro-analysis', *Anthropological Forum,* vol 18 no 2, pp117–138

Pohjonen, V. (2002) 'Synopsis of Derema Ecological Corridor' (unpublished), accessed from Indofor/Savcor, Helsinki

Rantala, S. and Vihemäki, H. (forthcoming) 'Forest Conservation and Human Displacement: Lessons from the Establishment of the Derema Corridor, Tanzania', in I. Mustalahti (ed) *Features of Forestry Aid: Analyses of Finnish Forestry Assistance,* Department of Political and Economic Studies, University of Helsinki, Finland

Schmidt-Soltau, K. and Brockington, D. (2007) 'Protected Areas and Resettlement: What Scope for Voluntary Relocation?' *World Development*, vol 35, no 12, pp2182–2202

Sjöholm, H., Malimbwi, R. E., Turunen, J. and Wily, L.A. (2001) 'Mid-term Review of the East Usambara Conservation Area Management Programme (EUCAMP), Phase III', Widgari Consultants, FBD and Finnish Forest and Park Service

Stocking, M. and Perkin, S. (1992) 'Conservation-with-development: An Application of the Concept in the Usambara Mountains, Tanzania', Transactions of Institute of British Geographers, vol 17, pp337–349

Tye, A. (1993) 'Establishment of Forest Corridors and Other Protected Forest Areas in East Usambara', Conclusions of Workshop held in Tanga, 15 July, and Action Plan, EUCDP, (unpublished)

URT (United Republic of Tanzania) (2006) 'Derema Forest Corridor: East Usambara Mountains, Resettlement Action Plan for Farm Plots Displaced for Biodiversity Conservation in the Derema Forest Corridor', Report for World Bank, Ministry of Natural Resources and Tourism, Dar es Salaam, www.easternarc.or.tz, accessed 3 May 2010

Vihemäki, H. (2009) 'Participation or Further Exclusion? Contestations over Forest Conservation and Control in the East Usambara Mountains, Tanzania', PhD thesis in Development Studies, Interkont Books 18, Institute of Development Studies, University of Helsinki, Finland

Woodcock, K. (2002) *Changing Roles in Natural Forest Management: Stakeholders' Roles in the Eastern Arc Mountains, Tanzania*, Ashgate, Burlington

World Bank, 'Derema Corridor: the East Usambara Mountains' (unpublished), Dar Es Salaam

5

Changing Landscapes, Transforming Institutions: Local Management of Natural Resources in the East Usambara Mountains, Tanzania

Salla Rantala and Emmanuel Lyimo

Sustainable governance of landscapes requires the integration of differing interests of multiple stakeholders at various levels of policy formulation and implementation (see Chapters 1 and 11). Decentralization of natural resource governance is often seen as a useful tool in achieving this. In many countries, decentralization has given rural communities the chance to make decisions about the management of natural resources in their domain and to formalize customary management rules and practices. Some suggest that customary management systems that have evolved over long periods in response to location-specific conditions make communities better resource managers than centralized agencies (Deininger, 2003), whereas others stress the flexibility and responsiveness of local management systems (Hartanto, 2009; Marfo et al, 2010), which can be indispensable characteristics in rapidly changing tropical landscapes. It has also been suggested that democratic decentralization, through greater participation, helps to better match public decisions to local needs and aspirations, and to increase equity in the use of public resources (cf Ribot, 2003). In other words, decentralization is heavily laden with expectations of improved effectiveness and equity in local resource management, in turn enhancing the governance of whole regions and landscapes.

Those assumptions can be highlighted through an examination of the processes whereby individuals and groups establish or re-establish rights to important productive resources in their domain, often following decades of centralized management. Rights authorize their holder to use, manage and benefit from resources, but such rights exist only in cases where there is a social mechanism that allocates duties and binds individuals to them (Bromley, 1991). Statutory and/or customary rules serve as such mechanisms. Customary institutions refer to self-organized rule systems that are rooted in shared social experiences and histories of communities (Benjamin, 2008). In Africa, statutory and customary legal constellations guiding resource management have coexisted since the colonial administration and in many places continue to do so.

Through decentralization, state natural resource management systems may create new community-based institutions that are designed by external agents and implemented at local level (Benjamin, 2008). The relationship between customary and new community-based institutions may range from ignorance to subordination to effective accommodation (Marfo et al, 2010). Integration or 'meshing' (Marfo et al, 2010) of customary law with state and other institutions brings the situation closer to nested institutions, where each set of rules and objectives of management is linked to the rules and objectives of the next level in multilevel governance systems. This is said to enhance the legitimacy of local rules across levels of governance, making them less frequently challenged in courts and administrative and legislative settings (Ostrom, 1999). A situation where customary and modern institutions exist independently or are mutually ignored is explained by legal pluralism, which recognizes the coexistence of various normative orders at different levels, with their own bases of validity and legitimacy; that is, the validity of local rules is independent of state recognition (Griffiths, 2002).

In legal pluralist settings, actors can manipulate the discrepancies between the statutory and the customary to influence the direction of political and economic change, one strategy being the selective use or invention of custom (Wiber, 1990; Yngstrom, 2002). As such, legal pluralism raises important questions of power: how the 'customary' is embedded in local networks of power and information, but also influenced from the outside by the same factors (Griffiths, 2002). At the same time, the nested institutions approach appears to make assumptions about the integrity, coherence and uniformity of state law, which may be similarly treated with scepticism (cf Griffiths, 2002). In both cases, the performance of decentralized governance in terms of effectiveness and equity ultimately depends on the actors to whom rule-making powers have been devolved and their accountability to their constituents (Ribot, 2003).

In general, customs in precolonial African societies may have been loosely defined and flexible (Rangers, 1983; cited in Dore, 2001). This is especially true for modern-day Tanzania, where customary does not necessarily equal 'traditional'. Statutory regimes began subordinating customary natural resource management systems during the colonial era. After independence, the 'Operation vijiji', or village-making policies of the 1970s, involved shifting millions of people into new village communities around the country (Shao, 1986). Since then, local rules have been shaped by community-based decision-making and administration not necessarily according to traditions but often in response to directives from central or district government. Customary law in many Tanzanian villages today may thus be described as 'prevailing norms' (Alden Wily, 2003, p11). To a certain extent, they are acknowledged and supported by the current national legislation following a decade of decentralized policies in land, forestry and wildlife.

The East Usambara Mountains in northeastern Tanzania are an ecologically and culturally diverse landscape. The area is one of the world's biodiversity hotspots and home to several endemic species of plants and animals (Rodgers and Homewood, 1982; Burgess et al, 2007). The total population is estimated to be more than 130,000 people (Tanzania Forest Conservation Group, 2008) across 61 villages (www.easternarc.or.tz/eusam). The village populations today consist of people originating from numerous ethnic groups, including the dominant Shambaa, who are believed to be the original inhabitants, as well as other ethnic groups that moved to the area to work as labourers in tea plantations or pursue other economic opportunities, and later obtained farmland in the surrounding villages. The current mosaic of forest and agricultural land uses has been shaped over time by diverse local actors with varying economic, political, social and cultural interests, under the influence of colonial and post-independence government policies, market fluctuations of agricultural and forest commodities, and global policy shifts between development and conservation. It seems plausible to assume that local governance of natural resources today reflects many of those influences, as much as the customary management practices of the Shambaa.

This chapter studies the development from the customary to modern community-based institutions in the East Usambara Mountains and its implications for assessing the performance of decentralized natural resource governance. In describing the current practices of land, forest and tree rights administration within the study communities, the focus rests on factors that affect the outcomes of decentralization in terms of elasticity, effectiveness and equity. These factors include the actors holding rule-making and implementation powers, the direction of their accountability and the influence of a broader structural and relational context and continuum.

We focus on the legal constellations pertaining to land and forest, which are central in the governance of the East Usambaran landscape, where forest conservation objectives and growing local needs for agricultural land have been contested for a long time. Land and forest are also often referred to simultaneously in both modern state law and customary law, making it practical to treat the two side by side in the analysis.

First, the current decentralized governance of natural resources in terms of the relationship between statutory and customary local management systems in Tanzania is placed in the context of legal documents, scholarly analyses and observations. The study proceeds with a historical overview of customary management of natural resources in the East Usambaras, drawing on secondary literature and, to a lesser extent, empirical fieldwork. In the next section, empirical results concerning current local practices of land, forest and tree rights administration in two villages are provided. Finally, the results are synthesized to highlight the key issues.

Fieldwork for this study was carried out within the governance domain of the Landscape Mosaics project (see Chapter 1) between April 2008 and September 2009. The research first focused on the formal structures of landscape governance versus local practices in the East Usambaras generally, as well as those in the three villages selected as representative local landscapes. The methods included qualitative interviews with individuals and groups, a household survey and participant observation. For the current study, the previous data was complemented mainly through group discussions with elders, village leaders and village land and forest committees, to answer specific questions regarding the relationship between customary and modern institutions. At this stage, the fieldwork focused on the villages of Misalai in the East Usambaran uplands and Kwatango in the lowlands.[1] In these so-called natural villages, people were already living and managing natural resources before the 1970s, when many new villages were created. In terms of forest rights, we also draw on fieldwork conducted in other villages of the East Usambaras.

Decentralization in Tanzania

The framework of central control over natural resources was established in Tanzania during colonial rule and reinforced in the first decades of independence. The laws and policies of the state placed the control over resources such as forests, land and wildlife in the hands of the government and ruptured many of the rights of local people to the resources. The notion of the sovereignty of state law became dominant, subordinating local customary laws.

Decentralization was introduced in the 1990s to address many of the observed shortcomings of centralized natural resource control. The driving force of the movement towards more decentralized and devolved forms of governance in the past two decades has been the quest for increased efficiency and equity, including more direct benefits from natural resources for the rural population and ultimately poverty reduction (Kallonga et al, 2003).

Table 5.1 lists the major national laws and policies that have set the stage for scaling up decentralization of land and forest management nationwide.

An aspect of the Local Government Reform Programme is political decentralization, the devolution of powers to locally elected councils and committees at district, ward and village levels. Local governments are autonomous of the central government but subordinate to the parliament (URT, 1998a).

Table 5.1 *Laws and policy processes in decentralization of land, forest and tree governance in mainland Tanzania*

Sector	Law or policy	Year of enactment
Local Governance	Local Government Authority (District and Urban) Act	1982
	Local Government Reform Programme	2000
Land	Land Act	1999
	Village Land Act	1999
	National Land Policy	1997
Forest	Forest Act	2002
	National Forest Policy	1998[2]

The village is the lowest recognized unit of government where natural resources are managed on village land, meaning the legal category of land inside surveyed village boundaries. Primary decision-making bodies are the village assembly, a periodic meeting that is open to all villagers above 18 years of age, and the village council, whose 25 members are the chairpersons of sub-villages (*vitongoji*) and representatives elected by the village assembly. The village council is the manager of land, forest and tree resources on village land, and in general accountable to the village assembly. Chairpersons of the councils and village executive officers sit in the ward development council. The central decision-making body at the district level is the district council, with one elected member from each ward in the district, three members appointed by the minister for local government, members elected by the district council among party organizations and village chairpersons, and any member of parliament representing constituencies in the district (Local Government [District Authorities] Act, 1982).

In principle, the relationships between levels are to be administrative, technical and/or consultative and advisory in nature (URT, 1998a). Nevertheless, the decentralized land and forest laws maintain top-down elements through which central government may continue to exercise power over village natural resource management. For example, many of the land and forest management documents drawn up at village level require approval and signature by the district council or central government agencies. On the other hand, mechanisms of downward accountability of the village council are limited to the theoretical approval or disapproval by the village assembly of decisions made in some issues, but not all (Alden Wily, 2003; Sundet, 2005). Elected members of the village council must be registered members of political parties, although not of the ruling party, Chama Cha Mapinduzi (CCM, Party of the Revolution) as previously (Alden Wily, 2003). In many areas, including the East Usambaras, support for CCM remains strong and most village council members belong to its ranks, forming a direct political link from the central government to the village level.

The Land Act (1999) and Village Land Act (1999) set out the current statutory provisions for land tenure in Tanzania. As a colonial legacy, all land (soil) is vested in the president of the republic on behalf of the people, whereas citizens may hold certain rights to land. The principal way for the Tanzanian rural population to access land is through customary rights. In principle, the Village Land Act recognizes customary rights as existing and secure rights, even when not registered, but it provides an opportunity to enhance the security through registration of customary rights. Village councils can apply for a certificate of village land issued by the land commissioner once the village boundaries are determined and agreed upon with neighbouring entities. This gives them the authority to grant certificates of customary occupancy to individuals and groups. Customary rights to land are treated as private land rights, which can be transferred, bought and sold, leased and inherited (Alden Wily, 2003).

Whereas the certificate of village land is not needed for the village councils to manage land and other natural resources according to the state and customary laws, the formal establishment of village boundaries and the certificate are likely to decrease the space for different interpretations on where general land ends and village land starts. General land refers to a residual category that is not reserved or village land, but the 1999 Land Act includes a caveat pointed out by Alden Wily (2003): 'general land' is defined to also include unused or unoccupied village land. Thus the Land Act definition could be used as a pretext to exclude villagers from considerable areas of common property. Furthermore, in informal discussions it has become apparent that forestry officials continue to interpret unregistered village land as general land, which has implications for the exclusion of forest resources from villagers' domain. This view has been adopted in, for example, the National Framework for Reduced Emissions from Deforestation and Forest Degradation (REDD) (URT, 2009).

Despite the provisions that recognize customary land tenure, 'customary' remains undefined in the law. Reference is made to the Interpretation of Laws and General Clauses Act (1972), where customary law is defined as a prevailingly accepted set of rules by the community whereby rights and duties are acquired or imposed; thus, customary law can also include modern rules. Nevertheless, Section 20(2) of the Village Land Act (1999) implies that customary laws require historical precedence, but only so long as the practices 'do not deny women, children or persons with disabilities lawful access to ownership, occupation and use of lands' (Alden Wily, 2003, p12). Section 20(4) simply states that the applicable customary law shall be 'in the case of a village not established as a result of Operation vijiji the customary law which has hitherto been applicable in that village'. Perhaps more importantly, how one demonstrates pre-existing customary rights to land is not clearly established in the laws. A 'peaceful, open and uninterrupted' occupancy for a minimum of 12 years justifies an application for a customary right of occupancy. However, whether the right is granted is also subject to the village council's assessment of the applicant's skills to develop the land. Village land that is free of use or undeveloped may be considered communal land or even general land, which is outside the council's domain and managed by the land commissioner (Village Land Act 1999, Section 57(1) and 23(1); Sundet, 2005).

Villages may set up independent dispute management systems. An adjudication committee can be elected by the village assembly to determine boundaries and interests in land, and a village land council may be appointed by the village council to settle land disputes. The land council mediates disputes in accordance with 'any customary functions of mediation' or 'natural justice in so far as any customary principles of mediation do provide for them', meaning a requirement for a fair procedure, or any training they may have received (Village Land Act 1999, Section 61(4)).

The National Forest Policy and forest law build on the legal framework of communal land tenure administered by the village councils and provide an opportunity for community-based forest management (URT, 1998b). Two main schemes to devolve forest rights to the local level through participatory forest management were introduced into national law in the Forest Act of 2002, of which Community Based Forest Management (CBFM) concerns forest management on village land. Although the law lists several types of forest on village land,[3] it is within the declared village land forest reserves where villagers hold most extensive rights to forest resources. The CBFM policy focuses on the demarcation and setting aside of these reserves, which are usually managed by a village forest committee or environment committee (URT, 2007). Villagers, through the council as the trustee, have the right to make and enforce rules about the harvesting and management of the reserve, exclude others, monitor resource use and sanction violators. They may harvest timber and forest products, collect fines and collect and retain forest royalties (Blomley, 2006). To take official legal effect, a village forest reserve requires a management plan and related bylaws approved by the village assembly and the district council.

Although the national CBFM guidelines recognize that forests may have been sustainably managed by customary practices in many places, the designation of communal forests as village forest reserves is justified as necessary to secure their 'protection by the national law' and to counter threats of forest conversion to other land uses (URT, 2007, p3). The Forest Act (2002) makes only a few references to the role of customary management of forests, and this is in relation to the reserves. The role of customary law in defining the rights and responsibilities of villagers in respect to a village forest reserve is equated to those defined in the village forest management plan, bylaws and other rules and agreements (Section 40). The law specifically states the procedures concerning the investigation of existing rights and consultation, and compensation in the case of the transfer of land under customary ownership to a central government forest reserve. The law even requires a review of existing rights in case a declared village forest reserve is to be upgraded to a gazetted village forest reserve. But it is silent on how customary rights to forest and land should be taken into account in the initial stages of the establishment of a village forest reserve on village land.

The National Forest Policy of 1998 identified a gap in the institutional framework for management of private forests (URT, 1998b). One of the stated objectives of the Forest Act (2002) is to encourage participatory 'planning, management, use and conservation of forest resources through the development of individual and community rights, whether derived from customary law or under this Act' (Section 3(b)). The objective notwithstanding, the law does not mention customary or local management of forest and trees in privately held or unreserved communal village areas. This gap, coupled with the ambiguous interpretations of village and general lands, is reflected in the management of the so-called reserved trees – the valuable timber or endangered tree species that have been protected by state law since pre-independence (Woodcock, 2002). Forest officials continue to interpret the forest law in such a way that the reserved trees are owned and managed by the government regardless of the tenure of the land they stand on, as was the case according to the old law. The only exemption is when the reserved trees grow within an established village forest reserve, in which case the rights to these trees are re-allocated from the government to the village council (Blomley, 2006; URT, 2007). In fact, an alternative interpretation of the Forest Act leads to the conclusion that these trees are always managed by the village council when on village land, whether within a reserve or not – if village land can be unambiguously defined.

Thus, the Tanzanian CBFM policy largely follows the reserve-centric model of central government forest management applied since colonial times, with vaguely subordinating customary management to serve this approach, but without a clearly assigned role. The next part discusses how this approach came to be dominant.

Customary Management of Natural Resources in the East Usambara Mountains

The Usambara Mountains probably experienced three periods of settlement: 100–400 CE, 900–1100 and from 1600–1700 to the present day (Schmidt, 1989). The largest ethnic group in the area has been the Shambaa, who share traditions regarding the arrival of, and consented to being governed by, the ruling class of Kilindi around 1740, resulting in the development of an extensive kingdom encompassing the West and East Usambara Mountains in the 19th century (Feierman, 1974; Iversen, 1991).

It is perhaps impossible to describe a pristine past where the Shambaa managed natural resources free of any external influences, given that the earliest written accounts of local natural resource management were by German explorers pursuing the same rich resources (cf Iversen, 1991). Similarly, historical accounts based on oral traditions should be treated with some caution, since oral traditions have been accused of being selective, ethnocentric, elite-centric and hardly free of later influences (Koponen, 1988). Nevertheless, in the following we borrow mainly from the detailed works of Feierman (1974) and Woodcock (2002) to identify the main characteristics of land, forest and tree rights administration of the precolonial Shambaa. By precolonial, we refer to the time of the Shambaa kingdom (1740–1890s), on which most research has focused; earlier traditions are likely to have been different (Feierman, 1974).

In the precolonial era, land was classified into uncultivated, forested wilderness areas, further divided into sub-categories of different types of forest and bushland, as well as various types of cultivated land cover. Wilderness areas held regenerative and healing powers associated with water or rain. Several forest areas, typically on hilltops or ridges, were associated with rituals for making rain or 'healing the land', such as the Mlinga peak that looms over Kwatango village. The Kilindi king was believed to hold the power to bring rain and was vested most rights over land in the political sense in exchange for use of his benevolent power (Feierman, 1974; Woodcock, 2002). He distributed parts of the kingdom to his descendants to rule (Winans, 1962; Feierman, 1974). The Kilindi leaders also held the right to make rules about the ritual forests. Other Shambaa leaders had access and rights for ritual purposes, and the responsibility to enforce rules and sanction violators. To community members, ritual forests offered a place that hosted private initiation ceremonies and, to an extent, enabled access to forest products (Woodcock, 2002).

Other forest areas not associated with rituals were managed by various lineages[4] (patrilineal descent groups, Winans, 1962; Feierman, 1974). These forests were also seen as a regenerator of life, providing water, land and forest products. Leaders held the rights to make and enforce rules over the forest, to distribute land to community members for cultivation and to permit or inhibit felling of trees. Community members had the rights to access and use forest products and services. Felling of trees or clearing of forest for agriculture required sacrifices to calm the spirits of ancestors who resided in the trees (Woodcock, 2002).

Rights to use and manage the land could be bought and sold privately, although not the land itself, which was communally owned. Woodcock (2002) states that rights to land were held indefinitely and were not subject to the use of the land. However, Dobson (1940, p14), who studied the land tenure of the Shambaa in the West Usambaras during British colonial rule, observed that 'there is no obligation for a man to cultivate his land', but that he was more likely to lose it if he did not, as in some cases private land rights were terminable by the chiefs.

Lineages had informal councils who handled affairs within the lineage. The eldest male of each commoner lineage was the representative to the council of elders of the local Kilindi chief, *zumbe*, who in practice held little power over affairs within lineages (Dobson, 1940; Winans, 1962, 1964). Winans (1964) mentions that politically ambitious commoners could manoeuver for bureaucratic positions in local governance and influence it. Several lineages resided in villages side by side (Feierman, 1974). The British administration later grouped these lineages under a single native authority headman of a village (Winans, 1964). The council of elders was recognized by the administration and received a salary from the national treasury, although its main source of income was from presents received for settlement of disputes (Dobson, 1940). Village headmen were under the *zumbe* and his elders, and wielded little power other than collecting taxes, but were later delegated tasks such as soil and tree conservation (Dobson, 1940).

Descent among the Shambaa was traced through patrilineal kinship and men inherited the land of their fathers (Winans, 1962; Feierman, 1974). According to Woodcock (2002), in precolonial times Shambaa women had access and use rights to the land of their lineage, which they had to give up upon marrying into another lineage. If the woman later divorced or became a widow, she could return to her paternal home and regain her access and use rights to her father's land. Dobson (1940) observed that Shambaa women could inherit their fathers' land but in much smaller portions than their brothers. Ideally, a father would distribute his land before his death, but the allocations could also be changed posthumously (Winans, 1964). Women could not inherit their husbands' land, which would go to the children or the husbands' brothers if there were no children or the children were too small to take care of the land. In polygamous families, children inherited the farms their mother worked (Dobson, 1940; Feierman, 1974).

Tree tenure in the precolonial era was tied to land rights; individuals had rights to use and dispose of both planted and conserved trees on their parcels of land. Trees also increased the price paid for land. On communal land, the one who planted trees had the right to use them; wild trees were communally owned. Community members had access to trees on individual fallow land for firewood collection. Women's rights to trees correlated with their land rights, revocable upon marriage or divorce (Woodcock, 2002). In Kwatango village, tree planting correlated with the security of land tenure: if a person was allowed to plant a coconut tree on a parcel, he would hold that piece of land indefinitely; if not, the land would only be considered borrowed (group discussion, Kwatango village land council, September 2009).

Local elders recall how beliefs and taboos also acted to protect the communities' forests, since breaking them were seen to lead to disasters such as drought, famine, disease or death. For example, only dry wood was used for firewood. Large trees were left to remain in the fields and forest to provide shade, fodder and wild foods. Certain tree species were protected, regardless of tenure, because they housed both malevolent and benevolent spirits (Woodcock 2002; authors' fieldwork). Specific tree species mentioned by village elders as having been protected by local taboos include *mbuyu* (baobab) and *kanita*; these were big trees underneath which the elders used to worship. If somebody cut down a *kanita* or *mbuyu*, the person was expected to become blind, disabled, or even to die. 'Hisa's uncle did not obey this rule. He cut down *kanita* and became blind for his entire life' (group discussion with elders, Misalai village, June 2009).

Feierman's (1974) study highlights a dualism in the relationship of the Shambaa to land and forest. Men inherited not only their fathers' property but also their fathers' skills and knowledge related to agriculture, soils, rainfall, medicinal plants and traditions regarding forests. This was necessary for survival, but it was also believed that without knowledge of the rites and spells, men would die (Feierman, 1974). Although the close relationship with wilderness as a provider of life was central to the Shambaa existence, there were also more material concerns: each man was supposed to marry as many wives as possible and provide for each of them and their children, as well as dedicate himself to acquiring material wealth while the wives took care of daily subsistence farming. Each wife was given her own garden to farm. As the population grew, more and more land was needed to supplement fathers' inheritances that would have otherwise been divided into smaller and smaller parcels. Dobson (1940) observes that Shambaa farms were never divided into uneconomically small pieces. Feierman (1974) notes that in precolonial times, young men could acquire additional land by clearing forest fairly easily. The boundaries of communities appear to have been permeable, and newcomers were welcome to join the villages (Feierman, 1974). Based on archaeological findings and species composition in landscape patterns, Hamilton (1989) suspects that human beings have influenced virtually all forest areas in the East Usambara Mountains to some extent. In the more densely populated West Usambaras, where the capital of the Shambaa kingdom was located, large tracts of forest on steep slopes had been converted to farmland by the arrival of the Germans (Iversen 1991).

The colonial era in the East Usambaras ruptured the traditional land and forest tenure system (Hamilton and Mwasha, 1989a; Woodcock, 2002). All land was handed over to the Germans 'for all time' in a treaty agreed to by a local Shambaa chief in 1885. Land was then divided into forest reserves, private estates and public land (Hamilton and Mwasha, 1989a). By 1912, there were eight forest reserves in the East Usambaras, and most of today's 20-odd reserves were established pre-independence; few have been added since (Iversen, 1991). The British continued to adhere to this system after World War I. The British administration was especially concerned with the rigorous enforcement of top-down, exclusionary conservation of forest in reserves for water, climate regulation and soil erosion control (Hamilton and Mwasha, 1989b). Yet, forestry business and export interests are likely to have motivated management in practice (Woodcock 2002). On public land, local people continued to hold access and use rights to land and forest, with the exception of the state-owned reserved tree species – but without management responsibilities (Woodcock, 2002).

The post-independence government sought to reverse the alienation of the local population from land in estates and reserves. An early 1960s act declared all underdeveloped land state property, and according to Hamilton and Mwasha (1989c) led to the expansion of public land and agricultural encroachment, especially by newcomers, into the reserves of the East Usambaras. Immigration from other parts of Tanzania had already increased because of financial opportunities in the tea and sisal estates. In the 1930s, Shambaa leaders were struggling to retain the land under their administration exclusively for the Shambaa. Furthermore, they tried to discourage the sale of private land rights due to the perceived misuse of land by newcomers from other ethnic groups, who bought land with no intention to settle, made what money they could, then sold it (Dobson, 1940). By the 1940s, traditional leaders had lost much of their authority under British rule, and people no longer sought their consent for felling trees (Woodcock, 2002). Many workers in commercial logging and pit-sawing operations came from other regions with no cultural or spiritual relationship to the local forests. Woodcock (2002) concludes that much of the customary management of natural resources in the East Usambaras eroded following the disruption of customary land tenure regimes and the simultaneous weakening of enforcement of statutory regimes. Forests in reserves were in theory closed to the people, but in practice often converted to de facto open-access regimes (Woodcock, 2002). In discussions, villagers commonly refer to the period before the establishment of village councils in the 1970s, or even before the decentralization initiatives of the 1990s, as the time when 'there were no rules', for instance in terms of the conversion of forest to farmland or the felling of trees.

Research on natural resources in the East Usambaras has been carried out since German times (Iversen, 1991). In the early 1980s, awareness of the unique biodiversity of the mountain forests (Rodgers and Homewood, 1982) finally ended the commercial logging operations. The management of the landscape began to be geared towards conservation. Conservation projects such as the East Usambara Catchment Forest Project (EUCFP), funded by the Government of Finland, and the East Usambara Conservation and Agricultural Development Project, of the International Union for Conservation of Nature (IUCN), dominated the management of the landscape from the late 1980s to the early 2000s. Whereas further reservation of forests and biological research formed the core of these projects, community participation started to gain importance during the 1990s. Environmental education or 'sensitization' activities and community forestry were promoted (EUCFP, 1995; Tye, 1995; East Usambara Conservation Area Management Programme or EUCAMP, 2002). The shift towards increased community participation was probably driven by the perceived shortcomings of centralized management, with the aim of transferring forest management responsibilities to local communities, but rights reallocation was more cautious (cf Ellman, 1996). The relationship of the project and the villagers was still very much top-down (cf Frontier Tanzania, 2002), although local customary management was given limited attention in some early management plans for village forest reserves (Veltheim and Kijazi, 2002).

Since the mid-2000s, NGOs such as the Tanzania Forest Conservation Group (TFCG) and World Wide Fund for Nature (WWF) have embraced the possibilities provided by decentralized land and forest policies to promote community-owned management of natural resources across the landscape. The central government remains preoccupied with the forest reserves, and the local government (district) has inadequate resources to implement its role in landscape management.

Current Practices in the East Usambaran Villages

Context: The study villages

Today, more than half the population in the villages of Misalai and Kwatango belongs to the Shambaa ethnic group, consisting of descendants of settlers from both West and East Usambara Mountains. Two-thirds of them were born in the villages, and the rest are from other villages in these mountains. The other half of the population is ethnically very mixed, the Bondei and the Zigua being the second-largest ethnic groups. Despite expectations of higher ethnic diversity in the upland villages, where most immigration has been directed, the demographics are very similar in these two study villages.[5]

Kwatango, in the lowlands, is a remote village; the road to the village is in poor condition for most of the year, severely limiting people's access to public services and economic integration with regional urban areas. The population density is low, at about 0.2 persons/ha (National Census, 2002). According to a village land-use planning exercise carried out in 2008, only 25 per cent of the 3650ha village area is cultivated actively (WWF, 2009a), and communal land is available for allocation to villagers and even private outside investors. Virtually all inhabitants earn their living from farming, the most common farm size being four acres. The most important cash and subsistence crops are maize, bananas and groundnuts. A number of orange and teak trees have been planted but are not yet producing.

Misalai is a densely populated upland village, with 3.4 persons/ha (National Census, 2002), squeezed between tea estates, forest reserves and other populous villages. Population densities appear to correlate with rainfall and are higher in the uplands of the East Usambaras than in the lowlands, as in other parts of Tanzania (cf Shao, 1986). Eighty-seven per cent of the village area of 635ha is under cultivation (WWF, 2009b); reportedly, all available farmland has been distributed among the inhabitants. The most common farm size is two acres. Two-thirds of the surveyed population report farming as their most important livelihood activity; wage labour in the tea fields or factory is the most important livelihood for the rest. Yet nearly half of the population is involved in the tea industry, at least in the form of seasonal labour. The most important cash crops are cardamom and sugarcane; yams, beans, maize, cassava and bananas are cultivated for purposes of subsistence and cash income.

Land rights

Land management in the study villages, as well as in other East Usambaran villages, is centred on the village council, which has replaced the role of the traditional authorities. Misalai and Kwatango villages have not yet obtained their certificates of village land, although the boundaries have been surveyed. In Kwatango, a complicated boundary dispute with a neighbouring village stands in the way of drawing the village borders and formalizing villagers' land rights.

Both villages have started to apply elements of the Village Land Act (1999) through the election of 'land committees' for adjudication. Kwatango also has had a land council for the settlement of disputes since 2004. A dispute typically involves farm borders, especially where land has been fallow for a long time. In such a case, the land council seeks advice from village elders who know to which family the land has traditionally belonged. The formal institution that is the land council thus draws on the knowledge of the elders, whose traditional role was as a 'council of elders' to be consulted in various communal matters – but a formal 'council of elders' no longer exists, nor do the elders have an official position in the village governance structure.

Despite the village council's formal control over all village land, households manage their farms privately; villagers most commonly state that they 'own' their farmland. Virtually no one (96 per cent of the household survey respondents in 2008) has a land title or a certificate for a customary right of occupancy. Land is bought and sold according to customary ownership. In most cases, the land has been obtained through inheritance. Purchase is the second most common way of obtaining land in Misalai, where the only way for a newcomer to acquire land is to find a willing seller of a private parcel; anyone who approached the village council for public land would do so in vain. The council needs to be consulted on any private sale of land in both villages, indicating an attempt to exercise control over private land management.

Both villages are currently enforcing a bylaw that allows the village council to reallocate private fallow land to those in need of land. Kwatango leaders date this bylaw to the 1980s, whereas in the more densely populated West Usambaras the practice seems to have been in place in the 1930s (cf Dobson, 1940). Growing needs for farmland may have contributed to the later adoption of the rule in the East Usambaran villages, yet the current competing demands for land not only include those of farmers. In Misalai, this rule was applied in the transfer of some private farmland for the establishment of a village land forest reserve. In Kwatango, where availability of land is not yet an issue, the bylaw has been passed by the council, because of damage to crops caused by pest animals residing in fallow areas. The owner is given up to six months to clear his farm before the land is reallocated. In Misalai, no notice period was given to the farmers who lost part of their land to the village forest reserve. Also in other villages, the establishment of village forest reserves has involved disputes concerning borders, with the village council or forest committee challenging the legitimacy of certain farms 'encroaching' into the forest (eg group discussion, Kwezitu village forest committee, September 2009). In addition, the village forest committee of Misalai told farmers that unless they used areas adjacent to the reserve, more land would be taken for conservation, creating pressure to clear uncultivated private land and a disincentive to retain private forest or fallow areas.

A shift in local approaches to land management in Kwatango is said to have started around the time the new Land Act and Village Land Act were passed (1999), but the need for it arose earlier, when newcomers started to move into the village:

Now there is a mixture of different people because they come to ask for land, but in the past people lived like relatives, they received farms from each other. ... Therefore, if a conflict arose between relatives, it was difficult to report it ... because it was an action of taking your relative before the government; you could be seen like you have crossed a big line. So that is why things were resolved at home. (Group discussion, Kwatango village land council, September 2009)

Under colonial rule, Kwatango remained largely inaccessible although officially within the sphere of influence of a local *zumbe*, or subchief. *Zumbe* represented another layer in the local dispute management system; he was not usually involved but rather used to frighten people, as a last resort when a person disobeyed the decision of the elders. For example, if a person refused to pay the required chicken for the elders' mediation of a border dispute, he would be sent to *zumbe*. 'When you are sent to *zumbe*, you are not going to pay a chicken anymore, there is a risk of being sent to jail.' Land disputes today are also generally solved privately or through land council mediation with the help of elders 'because [people] have seen that many times when conflicts are sent to court, it becomes worse with the judge's involvement. By ourselves, with the help of elders and law books, it is easy to solve them.' Traditional rules and government rules coexist in a hierarchical manner. 'Now, when old rules fail, we start applying the government rules' (group discussion, Kwatango village land council, September 2009).

Women's land rights appear to be an area where modern statutory institutions have not significantly replaced customary practices. Farmland accessed by households is mostly said to be owned by men, although on average 25 per cent of the female survey respondents reported owning land themselves as opposed to accessing land that is owned by the spouse, jointly owned or owned by another family member. In Misalai village the proportion was nearly 40 per cent. A woman was more likely to own land personally if she was the head of a single-parent household. It was explained that Shambaa women sometimes receive an allocation of land within the family from their husbands, so that they take care of their own parcels. Nevertheless, when it comes to the sale of the harvest or the land itself, a woman's ownership may be questioned, and she is often sidelined in the decision-making or income from the sale.

Although traditionally a divorced or widowed woman could leave her husband's village and return home (cf Woodcock, 2002), in practice this is complicated. A levirate marriage tradition continues, especially among those Shambaa who are recent immigrants from the West Usambara. That is, a trustworthy brother, selected by the family, takes over a deceased man's land and his wife and small children, thus reinforcing the lineage continuity (cf Winans, 1962). Often a widowed mother acts as a caretaker of the land on behalf of her small children. If a widow wishes to move, she can only do so after her children are old enough to take care of the farm themselves, unless she is willing to leave the children behind. If the children follow their mother to her paternal home, land conflicts with the maternal uncles may ensue. In a recent case, a young widow decided to leave her late husband's village following a dispute with the in-laws. This Shambaa woman returned to Misalai, her father's village, and began cultivating her maternal grandfather's land, which she requested from her mother, who had inherited the land from her father. Her children, however, could not join her and were sent to school in a third location. Another widow, the second wife of a polygamous husband, refused to marry a brother-in-law, whereas the first wife's children inherited the farms she had worked hard on. Subsequently, all support was cut off from the husband's family. For the children's sake, she did not want to leave the village. With the help of her elder children, she purchased a new farm, which now supports the family. In a third case, a local Shambaa woman was farming the land she had been allocated by her father with her husband, who was an immigrant worker in the tea estates and did not possess land. After the father's death, the woman's brothers chased her away from the land. Because the family is too poor to buy new land, she now farms along the edges of the tea fields (personal communication, B. Powell, November 2009).

For women of all ethnic groups, the most common way of obtaining personally owned land is through inheritance. The young widow whose mother and aunts had been distributed land by their father explained: 'The Shambaa do not consider gender, all children can inherit land. They think, what if my daughter is divorced and needs land?' (interview, Misalai, September 2009). This perception is not shared by everyone; in many cases, female children are not considered when a father's inheritance is distributed, they receive only a small parcel, or male relatives use their traditional right to reallocate the father's inheritance posthumously. In most cases, women still access farming land through their husbands.

Forest rights

Land rights are essential for farming-based livelihoods and thus continually claimed and contested by local actors; in contrast, forest rights are potentially an area where local rights to the natural resources can now be re-established or reinvented after several decades of centralized forest management.

Many villages in the East Usambara Mountains are establishing village forest reserves to protect remaining patches of forest on village land. The promotion of village forest reserves as a means of safeguarding forest connectivity is one of the major activities of the East Usambara Forest Landscape Restoration Project, implemented by TFCG in partnership with WWF since 2004. With the project's facilitation, nine villages across the landscape have established reserves since 2006 (personal communication, E. Mtui, April 2009).

In interviews and informal discussions, most villagers readily list good reasons for the establishment of forest reserves and forest conservation in

Table 5.2 *Forest uses according to village bylaws in Kwatango and Misalai*

Activity	Kwatango	Misalai
Collecting firewood	Allowed	Allowed to collect dry firewood for own use
Collecting vegetables and mushrooms	Allowed	Allowed to collect for own use; for sale requires permit and payment
Collecting traditional medicine	Allowed to collect for own use; for sale requires permit	Allowed to collect for own use; for sale requires permit and payment
Charcoal-making	Prohibited	Prohibited
Crop cultivation	Prohibited	Prohibited
Cutting trees for timber or building poles	Prohibited	Prohibited
Grazing	Prohibited	Prohibited
Hunting	Prohibited	Prohibited
Beekeeping	Requires permit	Requires permit and payment; natural honey collection prohibited
Research	Requires permit	Requires permit and payment
Rituals, sacrifices	No mention	Allowed
Tourism	Requires permit	Requires permit and payment

Source: Kwatango Village Forest Management Plan, August 2005, and Misalai Draft Village Forest Management Plan, March 2008

general, almost invariably related to forest environmental services rather than direct use benefits. Most commonly, a healthy forest is associated with a favourable climate, secure rainfall and provision of water. The main purpose of forests as a land use is conservation, villagers say, and the management plans are mostly geared towards conservation, regeneration and replanting. Direct utilization of the village forest is limited to the collection of wild foods and other non-timber forest products (NTFPs) (Table 5.2).

The process of village forest reserve establishment and bylaw formulation has followed a similar course in all the villages. TFCG first approached the village council, which then introduced the proposal to the village assembly, where a planning committee to prepare the forest management plan and bylaws was elected. Many members of initial planning committees later became members of village forest committees in charge of the reserve. When asked about the origins of the bylaws, village leaders and forest committee members stress the role of 'villagers themselves' in their formulation, although 'education given by TFCG' is invariably mentioned as an influence. Further probing reveals that sometimes the forest bylaws were copied from existing village bylaws or national laws, with the two often confused:

> ...these bylaws we have taken from the village laws. Village government had bylaws. Even me, when I was growing up, I was told that there is a government tree and also there is a tree that I can harvest without the government bothering me. So we were already told by wazee [elders] that there are government trees. After coming back to the village office [...] the village chairman

and his village executive officer opened the msahafu[6] *and said, use these bylaws because they have already been approved and they are here. And those bylaws are what we use to squeeze them [punish offenders].* (Secretary of Misalai forest committee, September 2009)

In practice, the village leaders and forest committee have usually prepared the bylaws and presented them to the village assembly for approval with few comments or contributions by other villagers.

Once approved by the assembly, the management plan is revised by the district lawyer, in accordance with the land laws, before submission to the district council. The bylaws are then given a 'legal jargon' format annexed to the management plan. It is striking that even if villages' forest management plans differ, at least in terms of details, the bylaw sections of most plans (all of them dated 2005) are exactly the same – and in most cases bear a slight mismatch with the rest of the plan. Which bylaws are then actually applied in village forest management is thus not clear from the management plan.

According to the elders of Misalai and Kwatango, there are similarities between old local rules and current village forest bylaws, such as the prohibitions on tree cutting, grazing in the forest and starting a bush fire. In Kwatango, the traditional role of wild vegetables in the local diet was a determining factor for allowing their collection in the village forest reserve, and customary rules in general have been perhaps emphasized more here than in other villages (group discussion, Kwatango land council, September 2009).

Despite the similarities in the actual rules, the elders see a difference between the traditional and current systems of managing forests: in the olden days, the rules were obeyed better because of the taboos, the breaking of which had irrevocable consequences, such as death. In the views of the elders, fines are less effective. Few current village leaders and forest committee members disagree, even though some say that since people today are more educated, they also obey rules better. A change in the way in which offences are dealt with is deemed inevitable because many of the beliefs of the elders have eroded, as the following example illustrates:

…they [elders] were saying there was a tree which was forbidden to cut, but now we do cut it. It was called mgambo. *Once you cut it, it would bleed. And when you rest for a while in order to catch breath, it has recovered, you see?*[7] *It was really that way. But now I can take an axe and cut* mgambo *and not have any effect, and the tree falls down.* (Secretary of Misalai village forest committee, September 2009)

The erosion of old rules is partly explained by the in-migration of newcomers, who do not share the same beliefs and refuse to obey the local traditional rules, thereby corrupting also the locals' obedience of those rules:

> *Here the* mvule *[Milicia excelsa] was 300 years old, but the Hehe*[8] *were the ones who came to destroy* mvule, *when they came is when they started to cut them down for timber. Also the first people to get to Mlinga [a sacred place for the Shambaa] were the Hehe, ask anybody, they are the ones who went there to process timber. Now, they have influenced Shambaa people, also the Shambaa have misbehaved due to the interaction, but in the past a Shambaa did not know how to prepare timber with a saw. So* mvule *had no problem.* (Kwatango village chairperson, September 2009)

Consequently, there is a need to enforce rules that everybody can share – and the government rules can function as such:[9]

> *Those old rules are used by those who know each other. For those who are different from you, it is not easy to follow your rules. For example, if I ask you to pay a chicken for uprooting, then you tell me that 'I don't know such a rule, I won't provide a chicken'. It is then you switch to those rules which we all obey, the government ones.* (Member of Kwatango land council, September 2009)

Some think that the general population growth has forced the shift from locally applied rules, harvesting quotas and reconciliatory measures, which were effective as long as resources were abundant, to government rules and fines. Since the local environmental carrying capacity has been exceeded, they say, the only way to rescue the remaining forests is to apply strict rules and heavy-handed punishments for offenders. Some forest committee members also think that physical punishments like those in the time of *zumbe* would not suffice anymore, but a large fine – or even jail – works better than a literal stick:

> *… right now if I say I will give you 20 strikes for cutting that* mvule, *you will agree and go to cut it. But this law that if I cut one* mvule, *I pay 500,000; if I don't have it I will end up in jail. So people are afraid. They decide to keep quiet [obey rules]. That is the existing system.* (Secretary of Misalai forest committee, September 2009)

In practice, though, punishments are applied flexibly and according to circumstances. For example, in Kwezitu village a fine of TSH 50,000 (US$38) for cutting trees can be negotiated down to TSH 10,000 (US$7), still a hefty sum for a local subsistence farmer, depending on how the offender behaves before the forest committee. Apologies and mistakes, perhaps made by children, are looked upon favourably with first-timers, but strict punishments apply to those who are seen as purposefully breaking forest rules.

Rights to trees on farms

Village land in the East Usambaran uplands consists of a mosaic of different land uses, many of which have trees. An important land use is agroforestry, in which the primary forest is gradually cleared and, especially in the early stages, resembles a disturbed rainforest. But whenever a tree-dominated area is privately owned, even if it is left to regenerate as forest, it is still called *shamba* (farm), not *msitu* (forest). In the villages studied it is common that *msitu* is only used to refer to a reserved area, such as a village forest reserve. This may reflect the recent history of alienation of local people from forests within the central government reserves: describing an area as 'forest' might imply forfeiting one's rights to the natural resources and land in that area.

Firewood, wild vegetables, medicines and building poles are commonly harvested in *shambas*. There are no regulations concerning forest product harvesting outside the reserves, with the exception of logging. It is common for women to collect firewood or even building poles not only on their own land but also on fallows or farms owned by relatives and neighbours, according to the Shambaa tradition. Immigrants from other areas have had to conform to this practice. Few people attempt to restrict others from collecting firewood on their land, and such attempts are frowned upon. In Misalai, a story is told with great amusement of a woman who complained to her parents-in-law that she did not like their collecting firewood on her farm. To let her know what they thought about such 'bad words', the in-laws appeared at the woman's house with all the firewood they had collected and dumped it at her feet.

The village council regulates timber harvesting on public as well as private land. Felling of mature indigenous trees is not allowed by the village bylaws, even on private land. It is possible to obtain a permit from the council to cut a tree on one's farm for personal use, but harvesting timber for sale requires a permit, which is reportedly hard to get, from the district forest officer, subject to village council approval. When explaining why they do not allow tree regrowth on their land, farmers cite their lack of knowledge of application procedures and difficulty in obtaining permits to utilize the trees once mature. Villagers report that as a consequence of such strict regulations, many people get rid of the trees on their farms 'illegally' or without following the official procedures.

Woodcock (2002) has pointed out that the weak land rights of women in the precolonial era probably discouraged them from planting or caring for trees on their farms. Similarly, we observed that those people (often women) who access land by borrowing it think that they do not have the right to make a decision about planting trees.

Conclusions

In the villages studied, modern community-based institutions have largely replaced customary institutions, which are deemed too eroded to ensure sustainability of local natural resource management in the face of high population densities, demographically heterogeneous communities and scarce natural resources. Yet the current institutions should also be understood within the historical and socio-economic context of the landscape where they operate (see Hartanto, 2009) and in the light of a continuum where local management practices are shaped by the changing local structures of power and authority.

Some dimensions of local customary management have persisted, and a closer look reveals nuances of legal pluralist practices. Certain rules and beliefs continue to exist with a narrower base of legitimacy and validity among a subset of community members. Many traditional rules and practices have, however, been merged and confused over time in community-based decision-making that conforms to statutory laws and responds to the requirements of central government and other external agents, such as projects and conservation NGOs. For instance, although the rationale of village forest management has a new basis in national policy and the education provided by NGOs, conservation ideas may have special appeal to East Usambaran villagers because of old beliefs, such as the association of forests and rain.

The gaps between the statutory and customary – especially when the latter are easily challenged, as in the question of land rights – usually work to the advantage of the socially and politically more powerful actors in the villages, such as the village council or forest committee. This new class of community decision-makers, who have replaced the Kilindi chiefs and councils of elders, hold three basic types of powers in the decentralized system: legislative (creating rules), executive (making, implementing and enforcing decisions) and judicial (adjudicating disputes) (Ribot, 2003). The decision-making bodies have been democratically elected, but their subsequent accountability to the constituents is limited to the formal approval of decisions by the village assembly. Rule-making processes have in practice been finalized before being presented to the assembly. Any effect of broader participation is further diluted through, for example, the homogenization of village forest reserve bylaws across villages at the district approval stage. Responding to external political and ideological influences and requirements rather than to the needs of the majority may sometimes help 'nest' local management with higher-level priorities and increase its upward legitimacy, but it can also decrease local legitimacy.

What may have been inherited from the customary system is the use of fear in the exercise of power. The Kilindi leaders legitimized their power by claiming to control the rain, in the absence of which crops would fail and people would suffer, as well as through fear of physical punishments for those who broke the rules. Similarly, current leaders use the threat of monetary sanctions to deter unauthorized resource use, although the elder generation does not agree on the effectiveness of this strategy.

The distrust of community leaders towards their fellow villagers in questions of communal natural resource management may originate in the historical relationships between the communities and external agents, where it is still evident. A good example is the management of reserved trees, which forestry officials are reluctant to entrust to communities and continue to control through skewed interpretation of the laws. Insufficient knowledge of modern statutory laws, coupled with the influence of decades of conservation education, leaves villagers without means to contest officials' interpretation. The outcome is a strong disincentive for retaining trees on farms, where they are seen to compete with crops over which the farmer has more control for sale. This may eventually undermine the overall sustainability of landscape management by compromising environmental services that trees on farms can provide, such as biodiversity conservation, soil erosion control and water services.

At the same time, the fuzziness of local management may provide an opportunity for socially disadvantaged actors to challenge customary practices. Women's land rights remain considerably weaker than men's, but a closer examination reveals variations among families and more dynamic power relations that may affect land access. It is difficult to determine to what extent historical variation has been diluted in accounts of customary practice based on oral tradition, often biased to male informants. It is also possible that as improvements in women's status globally trickle down to rural areas, and Tanzanian policies since the socialist era acknowledge women's role in production in addition to reproduction, women may now find more ways to claim and defend their rights – and to go around the social norms if their rights are taken away. Those means might be limited by land titling, which has proved in many cases devastating for women's land rights (Yngstrom, 2002). Experience from a recent large conservation intervention in the East Usambaras that aimed at registering small-scale landholders, though using inadequate methods, points to the same result (Rantala and Vihemäki, forthcoming).

Some flexibility has been retained in the enforcement of village forest reserve bylaws. The application of sanctions is graduated, which may help to maintain effectiveness where individual costs (such as restricted access to forest resources) matter more to some actors than the communal benefits of forest environmental services. Increased visibility of sanctions to those who purposefully defy common rules may similarly enhance effectiveness of common resource management (Ostrom, 2005).

In sum, the case of the East Usambara Mountains supports the notion that communities' resource management institutions are, rather than adapted to static local conditions as a result of a perfecting evolution over long periods of time, adaptable and responsive to changing social and environmental conditions, and to the political and economic requirements that community leaders are inclined or forced to follow. The outcomes of decentralization are thus dependent on the structural and relational context and on the continuum within the landscape where the policies operate. They do not automatically produce equitable and effective local natural resource management institutions. Yet it may be rightly asked whether more just or legitimate governance systems would exist in the absence of decentralization. Centralized management clearly led to unsustainable and inequitable practices, and the customary management systems, which also embodied a variety of undemocratic structures and dynamics (cf Ribot, 2003; Benjamin, 2008) had already broken down. The current decentralized framework entails many good ingredients to support the development of more effective and equitable community management institutions, such as key land and forest rights that have been devolved to the village level with real powers to make and execute management decisions. Empowering a wider range of community actors to exercise those rights, as well as the available accountability mechanisms, might help improve local management outcomes, with a balance between conservation and development interests. Scaled up, these in turn might lead to more sustainable governance of entire landscapes.

Notes

1. Editor's note: This site reverses our usual link between remoteness and higher altitudes. In Tanzania, the lowland village was the more remote.
2. Currently being reviewed, with a new 'zero draft' National Forest Policy dated November 2008.
3. Village land forest reserves; community forest reserves, created out of village forests; non-reserved forests; and private forests of individuals and groups, based on customary or granted land rights.
4. Woodcock (2002) refers to forests managed by 'clans', but both Winans (1962) and Feierman (1974) stress the importance of lineages in the years since kingship replaced clanship as the basis of political organization, and discount the role of clans in the past 200 years. 'There is no ... clan ownership of property' (Winans, 1962, p35). The Swahili word *ukoo* is often translated as meaning both clan and lineage.
5. In the household survey carried out in September 2008 as part of the Landscape Mosaics project, 127 respondents from 82 randomly sampled households in Kwatango and Misalai villages (41 households in each) were interviewed. For each household (except single-parent households), both spouses were interviewed separately. Of the respondents, 69 (54 per cent) were women and 58 (46 per cent) were men. The median age for female respondents was 38, and for males, 41.
6. Literally Koran, used here to refer to a book. Later, a discussion with the village chairperson clarified that 'the book' was a national law book.
7. This relates to the beliefs held of the *mgambo* tree, said to be impossible to cut down because it would always recover, even if one spent the whole day trying to fell it.
8. An ethnic group from Iringa region in central Tanzania.
9. See also Meinzen-Dick and Pradhan, 2002.

References

Alden Wily, L. (2003) 'Community-based Land Tenure Management. Questions and Answers about Tanzania's New Village Land Act, 1999', IIED, Issue Paper no 120
Benjamin, C. E. (2008) 'Legal Pluralism and Decentralization: Natural Resource Management in Mali', *World Development*, vol 36, no 11, pp2255–2276
Blomley, T. (2006) 'Mainstreaming Participatory Forestry Within the Local Government Reform Process in Tanzania', IIED Gatekeeper Series 128

Bromley, D. W. (1991) *Environment and Economy. Property Rights and Public Policy*, Basil Blackwell, Oxford

Burgess, N., Butynski, T., Cordeiro, N., Doggart, N., Fjeldså, J., Howell, K., Kilahama, F., Loader, S., Lovett, J., Mbilinyi, B., Menegon, M., Moyer, D., Nashanda, E., Perkin, A., Stanley, W. and Stuart, S. (2007) 'The Biological Importance of the Eastern Arc Mountains of Tanzania and Kenya', *Biological Conservation*, vol 134, no 2, pp209–231

Deininger, K. (2003) 'Land Policies for Growth and Poverty Reduction', World Bank Policy Research Report, World Bank, Washington, DC

Dobson, E. B. (1940) 'Land Tenure of the Wasambaa', *Tanganyika Notes and Records*, vol 10, pp1–27

Dore, D. (2001) 'Transforming Traditional Institutions for Sustainable Natural Resource Management: History, Narratives and Evidence from Zimbabwe's Communal Areas', *African Studies Quarterly*, vol 5, no 3, www.africa.ufl.edu/asq/v5/v5i3a1.htm, accessed 16 February 2010

Ellman, A. (1996) 'Handing Over the Stick? Report of a Village Management and Farm Forestry Consultancy', East Usambara Catchment Forest Project, Technical Report 18

EUCAMP (2002) 'East Usambara Conservation Area Management Programme Completion Report of the Phase III (1999–2002)', Administrative Report 40

EUCFP (1995) 'East Usambara Catchment Forestry Project. Project Document Phase II: 1995–98, Vol I: Main Document', 2 January 1995

Feierman, S. (1974) *The Shambaa Kingdom. A History*, University of Wisconsin Press, Madison

Frontier Tanzania (2002) 'Environmental Education Programme. Environmental Activities within Local Communities, East Usambara Mountains', East Usambara Conservation Area Management Programme, Frontier Tanzania Environmental Research Report 92, Society for Environmental Exploration, UK, University of Dar es Salaam, Forestry and Beekeeping Division, Tanzania

Griffiths, A. (2002) 'Legal Pluralism', in R. Banakar and M. Travers (eds) *An Introduction to Law and Social Theory*, Hart Publishing, Oregon

Hamilton, A. C. (1989) 'History of Resource Utilization and Management. The Pre-Colonial period', in A. C. Hamilton and R. Bensted-Smith (eds) *Forest Conservation in the East Usambara Mountains*, IUCN Tropical Forest Programme, World Conservation Union, Gland and Forest Division, Ministry of Lands, Natural Resources and Tourism, Dar es Salaam

Hamilton, A. C. and Mwasha, I. V. (1989a) 'History of Resource Utilization and Management. Under German Rule', in A. C. Hamilton and R. Bensted-Smith (eds) *Forest Conservation in the East Usambara Mountains*, IUCN Tropical Forest Programme, World Conservation Union, Gland and Forest Division, Ministry of Lands, Natural Resources and Tourism, Dar es Salaam

Hamilton, A. C. and Mwasha, I. V. (1989b) 'History of Resource Utilization and Management. Under the British', in A. C. Hamilton and R. Bensted-Smith (eds) *Forest Conservation in the East Usambara Mountains*, IUCN Tropical Forest Programme, World Conservation Union, Gland and Forest Division, Ministry of Lands, Natural Resources and Tourism, Dar es Salaam

Hamilton, A. C. and Mwasha, I. V. (1989c) 'History of Resource Utilization and Management. After Independence', in A. C. Hamilton and R. Bensted-Smith (eds) *Forest Conservation in the East Usambara Mountains*, IUCN Tropical Forest

Programme, World Conservation Union, Gland and Forest Division, Ministry of Lands, Natural Resources and Tourism, Dar es Salaam

Hartanto, H. (2009) 'Adaptability of Customary Forest Institutions in Kerinci, Central Sumatra, Indonesia', PhD thesis, Monash University, Australia

Iversen, S. T. (1991) 'The Usambara Mountains, NE Tanzania: History, Vegetation and Conservation', Uppsala Universitet, Uppsala, Sweden

Kallonga, E., Ndoinyo, Y., Nelson, F. and Rodgers, A. (2003) 'Linking Natural Resource Management and Poverty Reduction', Tanzania Natural Resource Forum, Occasional Paper 1

Koponen, J. (1988) *People and Production in Late Precolonial Tanzania. History and Structures*, Monographs of the Finnish Society for Development Studies No 2, Transactions of the Finnish Anthropological Society No 23, Studia Historica 28, Gummerus, Jyväskylä, Finland

Marfo, E., Colfer, C. J. P., Kante, B. and Elias, S. (2010) 'From Discourse to Policy: The Practical Interface of Statutory and Customary Land and Forest Rights', in A. Larson, D. Barry, G. R. Dahal, and C. J. P. Colfer (eds) *Forests for People: Community Rights and Forest Tenure Reform*, Earthscan/CIFOR, London, pp69–89

Meinzen-Dick, R. S. and Pradhan, R (2002) 'Legal Pluralism and Dynamic Property Rights', CAPRi Working Paper 22, Washington, DC

National Census (2002). www.tanzania.go.tz/censusf.html, accessed 31 May 2010

Ostrom, E. (1999) 'Self-governance and Forest Resources', Occasional Paper 20, CIFOR, Bogor, Indonesia

Ostrom, E. (2005) *Understanding Institutional Diversity*, Princeton University Press, Princeton

Rantala, S. and Vihemäki, H. 'Forest Conservation and Human Displacement: Lessons from the Establishment of the Derema Corridor, Tanzania' in I. Mustalahti (ed) *Features of Forestry Aid: Analyses of Finnish Forestry Assistance* (forthcoming)

Ribot, J. C. (2003) 'African Decentralization. Local Actors, Powers and Accountability', United Nations Research Institute for Social Development Programme on Democracy, Governance and Human Rights, Paper No 8

Rodgers, W. A. and Homewood, K. M. (1982) 'Species Richness and Endemism in the Usambara Mountain Forests, Tanzania', *Biological Journal of the Linnean Society*, vol 18, pp197–242

Schmidt, P. R. (1989) 'Early Exploitation and Settlement in the Usambara Mountains', in A. C. Hamilton and R. Bensted-Smith (eds) *Forest Conservation in the East Usambara Mountains*, IUCN Tropical Forest Programme, World Conservation Union, Gland and Forest Division, Ministry of Lands, Natural Resources and Tourism, Dar es Salaam

Shao, J. (1986) 'The Villagization Program and the Disruption of the Ecological Balance in Tanzania', *Canadian Journal of African Studies / Revue Canadienne des Études Africaines,* vol 20, no 2, pp219–239

Sundet, G. (2005) 'The 1999 Land Act and Village Land Act. A Technical Analysis of the Practical Implications of the Acts', Working Draft, 28 February 2005, www.oxfam.org.uk/resources/learning/landrights/downloads/1999_land_act_and_village_land_act.rtf, accessed 16 February 2010

Tanzania Forest Conservation Group (TFCG) (2008) 'Training and Collection of Baseline Data for Nine Village Forest Reserves in the East Usambara Forest Landscape Restoration Project', East Usambara Forest Landscape Restoration

Project, www.tfcg.org/downloads/TFCG-East-Usambara-baseline-monitoring-report.pdf, accessed 16 February 2010

Tye, A. (1995) 'East Usambara Conservation and Development Project. Final Report', (unpublished report), IUCN, Dar es Salaam

URT (1998a) 'Policy Paper on Local Government Reform', Local Government Reform Programme, Ministry of Regional Administration and Local Government, Dar es Salaam, October 1998

URT (1998b) 'National Forest Policy', Ministry of Natural Resources and Tourism, Dar es Salaam, March 1998

URT (2007) 'Community-based Forest Management Guidelines for the Establishment of Village Land Forest Reserves and Community Forest Reserves', Ministry of Natural Resources and Tourism, Dar es Salaam, October 2007

URT (2009), 'National Framework for Reduced Emissions from Deforestation and Forest Degradation (REDD)', August 2009, www.reddtz.org/images/pdf/redd%20framework%2009_new.pdf, accessed 16 February 2010

Veltheim, T. and Kijazi, M. (2002) 'Participatory Forest Management in the East Usambaras', East Usambara Conservation Area Management Programme, Technical Paper 61

WWF (2009a) 'Mpango wa Matumizi Bora wa Ardhi, Kijiji cha Kwatango, Wilaya ya Muheza', unpublished report, WWF Tanzania Program Office, Dar es Salaam

WWF (2009b) 'Mpango wa Matumizi Bora wa Ardhi, Kijiji cha Misalai, Wilaya ya Muheza', unpublished report, WWF Tanzania Program Office, Dar es Salaam

Wiber, M. G. (1990) 'Who Profits from Custom? Jural Constraints on Land Accumulation and Social Stratification in Benguet Province, Northern Philippines', *Journal of Southeast Asian Studies,* vol 21, no 2, pp329–339

Winans, E. V. (1962) *Shambala. The Constitution of a Traditional State*, University of California Press, Berkeley and Los Angeles

Winans, E. V. (1964) 'The Shambala Family', in R. F. Gray and P. H. Gulliver (eds) *The Family Estate in Africa: Studies in the Role of Property in Family Structure and Lineage Continuity,* Routledge, London and New York

Woodcock, K. (2002) *Changing Roles in Natural Forest Management. Stakeholders' Roles in the Eastern Arc Mountains, Tanzania*, Ashgate Studies in Environmental Policy and Practice, Ashgate Publishing, Aldershot, UK

Yngstrom, I. (2002) 'Women, Wives and Land Rights in Africa: Situating Gender Beyond the Household in the Debate Over Land Policy and Changing Tenure Systems', *Oxford Development Studies,* vol 30, no 1, pp21–40

6

Traditional Use of Forest Fragments in Manompana, Madagascar

Zora Lea Urech, Mihajamanana Rabenilalana, Jean-Pierre Sorg and Hans Rudolf Felber

Biologically, Madagascar is one of the world's most important biodiversity hotspots (Myers et al, 2000). But forest resources in Madagascar are disappearing at an alarming rate, and pressure on them remains high (Harper et al, 2008). Missing data makes it difficult to calculate the country's exact deforestation rate, but the area of evergreen rainforest alone is decreasing every year by an estimated 102,000ha (Dufils, 2003) and in our specific research area by 2.5 per cent (Rakotomavo, 2009). The reasons for the high rate of deforestation are complex and sometimes specific to certain localities (Jarosz, 1993). Other authors have differentiated between direct causes, such as agriculture, timber or fuelwood, and indirect reasons, such as migration, government policies and property rights (Casse et al, 2004). It is, however, indisputable that shifting agriculture is a major reason for the deforestation on the eastern slopes (Pfund, 2000; Razafy and Andrianantenaina, 1999), whereas timber and fuelwood play only minor roles. Madagascar's rural population depends heavily on mountain rice, known as *tavy*, which is often cultivated in forest clearings (Messerli, 2000; Pfund, 2000). Because of the practice of shifting cultivation, large and closed forest areas are transformed into a patchwork of smaller fragments that make up a mosaic landscape with alternating cultivated agricultural and fallow parcels.

Despite many efforts from international organizations and the Malagasy government, deforestation has not been halted (McConnell and Sweeney, 2005). We could say, somewhat abstractly, that there are two movements in Madagascar concerning the preservation of forest resources. One movement, 'conservationist', mainly directed from outside the country, would like to create protected areas to prevent the local populations from using the natural resources, whereas the other movement seeks to preserve the forest by means of participatory and sustainable management practices (Sorg, 2006).

Experience has shown that the concept of protection is not enough by itself to preserve natural resources in the long run and over large areas (Bertrand, 1999). In addition, protection and the creation of national parks and other protected zones are controversial because they have a negative effect on people's livelihoods (Adams et al, 2004; Keller, 2008; Sunderlin, 2003). Moreover, the direct and centralized management of the forest by the state's forest service has not been able to guarantee sustainable management and address the needs of the population at the same time.

In 1996, therefore, a new decentralization law was introduced in Madagascar that handed over the management of natural resources to the communities (*Transfer de Gestion*). This new law, 96-025 (*La Gestion Locale Sécurisée* or GELOSE), mandated communities to negotiate with the Forest Administration services and with a voluntary association of community residents known as *Communauté de Base* or COBA (Tsitohae and Montagne, 2009; Raik, 2007). This tripartite contract allowed the communities and COBA to use the natural resources (Montagne and Bertrand, 2006), and was the first instance since colonial times in which institutional regulations took local populations into account (Bertrand, 1999). Yet the implementation of the contracts for the sustainable use of forest resources has remained difficult. The state's main interest is that COBA protect the forests without necessarily drawing any real economic advantages from them (Hockley and Andriamarovololona, 2007; Razafy, 2004).

Seeking a better system, in 2006 the United States Agency for International Development (USAID), in collaboration with the Forest Administration, developed a new concept, KoloAla, which means 'maintain the forest'. KoloAla is based on the same principle of management transfer, and attempts to combine the preservation of the forests and the enhancement of livelihoods with a decentralized and sustainable use of forest resources by the communities. The principle of the tripartite negotiated contract remains and the forest is managed by COBA, but whereas the previous principle was motivated internally, the KoloAla approach originated more in the international community. A project team works with the local community in its planning and implementation over a long period. At the time of writing, a total of 390,000ha of forest has been designated into KoloAla zones, including the following three (USAID, 2009):

- a production zone for timber harvesting, where the timber or the user rights can be sold;

- a zone where the population has the rights to use the forest for subsistence; and
- a zone for strict nature conservation without user rights for the local communities.

Earlier experience with the transfer of management clearly shows how important it is to integrate the social context and customary rights of the local population into new management forms (Randrianasolo, 2000). Such processes of devolution are often in danger of ignoring or overlooking existing and functioning regulations or important local actors (Barry and Meinzen-Dick, 2008). There is already a danger that the concept will be too strongly influenced by international ideas or fleshed out by them. The local project, implementing KoloAla in the Manompana community, with close and lasting cooperation from the local population, has allowed for a certain degree of flexibility, so that social rules and needs might flow into the future management of forest resources.

Research Objectives and Methods

Unbroken forest massifs on the east coast have become increasingly fragmented (Consiglio et al, 2006; Harper et al, 2008). Until now, however, the research, development and conservation activities in Madagascar have focused on larger, mostly cohesive forest areas, which have more biodiversity than fragments and are therefore of higher national and international interest (Malanson and Armstrong, 1996). Little is known about the role and influence of forest fragments in local livelihood systems, although preliminary evidence suggested that they be taken into account in the design of governance arrangements for the development and sustainable management of forested landscapes.

The aim of our research project has been to reach a comprehensive understanding of the importance of these forest fragments for the livelihoods of the local population. To do this, we carried out ecological, botanical and socioeconomic analyses. In this chapter, we look at the part of the research that explains the issue of governance. Governance consists of complex mechanisms, processes and institutions through which citizens articulate and exercise their legal rights. It takes place at all levels – local, national, regional and global (Schmidt and Stadtmüller, 2009). With the results of our research and the knowledge gained about local processes, customary rights and the exercise thereof, in terms of the forest and its resources, we make recommendations about how they can be integrated into national mechanisms and institutions. If such local resource rights are not taken into consideration, forest management reforms could be sowing the seeds for future conflicts (Barry and Meinzen-Dick, 2008).

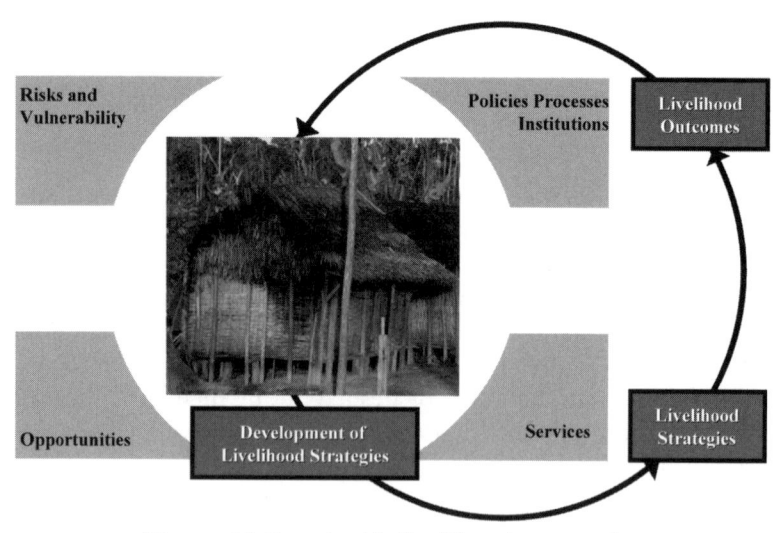

Figure 6.1 *Sustainable livelihood approach*

Interview methods

As a basis and general framework for our socio-economic analyses, we employ the sustainable livelihood approach (Baumgartner and Högger, 2004; Figure 6.1). This approach allows us to recognize the complexity of local people's livelihood systems and strategies at individual, family and community levels in relation to forest fragments. Livelihoods are obviously influenced by their context, including existing risks, opportunities, services and policies, and institutions that in our case affect forest resources as well. Above all, in this chapter we investigate the relationship between the context of policies, processes and institutions, and the local livelihood systems.

Diverse methods were employed. We began with open-ended discussions with several households in each of our villages to get an overview and to answer villagers' questions about our work. We then conducted the first semi-structured household interviews. In these interviews we gathered more specific information about the importance, perceptions, products collected from and uses of forest fragments (110 interviews in total). To deepen the information on importance and perceptions, we conducted scoring exercises with focus groups, which we carried out with people representing different wealth levels and separated by gender. To complete the information about the most important products, their use and management rules, we also accompanied villagers for transect walks into the forest. There, in particular, we identified plant species and were able to discuss resource management and governance in a more informal manner. To complete results, we held discussions with people who had specific knowledge or a special role in the social context. These included older persons, loggers, village chiefs and other traditional authorities. In all, during the two field periods of the project, we spent 11 months in the four villages.

Study area

The study area is located in a forest corridor in the district of Soanierana-Ivongo on the east coast of Madagascar. This forest corridor of about 38,000 ha, classified as evergreen humid forest at low altitude (0–800m) (Koechlin et al, 1997), is part of a larger landscape mosaic dominated by agricultural activities. It connects two protected areas: the special reserve of Ambahatovaky in the southwest and Mananara Nord Biosphere Reserve in the north. Within the entire area covered by these corridors, we differentiate between large forested areas, the 'forest massif', and the belt of fragments of forest that surround the larger expanses. As with many other forest areas in Madagascar, the borders of the cohesive forest are increasingly infringed on by the local population, who are mainly Betsimisaraka, an ethnic group (Cole, 2001) of primarily smallholder farmers. Their main agricultural system is swidden, which involves cutting forest to plant mountain rice (rain-fed rice), also known as *tavy* and, in the dialect spoken in our research area, as *jinja*. After one year of cultivating rice, farmers leave the land fallow or plant different crops, such as manioc and sweet potatoes, and sometimes an agroforestry system including a combination of mainly clove trees, coffee and annual crops. This results in a dynamic agricultural mosaic where changes occur at a number of temporal scales (Bennett et al, 2006). In these mosaics, forest fragments are integrated mostly as a reserve for future agricultural land (Tsitohae and Montagne, 2009). The swidden system is employed in forested landscapes, consistent with low population densities (Pfund, 2000; Erdmann, 2003). Because of Madagascar's population growth, however, fallow periods are becoming shorter, which leads to a loss of soil fertility (Styger et al, 1999). Whereas in 1970, the mean fallow period was between eight and 15 years, in our research area it has now been reduced to between five and eight years, still higher than in other regions in Madagascar.

As a consequence of population growth and decreasing soil fertility, the population enlarges the agricultural area each year at the expense of the forested area. What remains is a small and continuously shrinking forest corridor of a massif surrounded by a growing belt of fragmented forest.

For our research, we worked in four villages situated around the remaining forest corridor (Figure 6.2). The territories of the four villages have different forest cover and differ in their distance to the forest massif as well (Table 6.1). The villages Ambofampana and Maromitety are close to the massif in a remote area, whereas the villages Bevalaina and Antsahabe are farther from the massif in a territory of lower forest cover and are less remote. The differences in forest cover and distance to the massif allowed us to analyse the influence of forest loss on the human-forest interface.

The four villages not only differ with regard to their access to forest resources but also concerning their access to markets. The proximity of roads and markets led to degradation of forest resources. We therefore tried, as far as possible, to distinguish between changes to the system of livelihoods that were

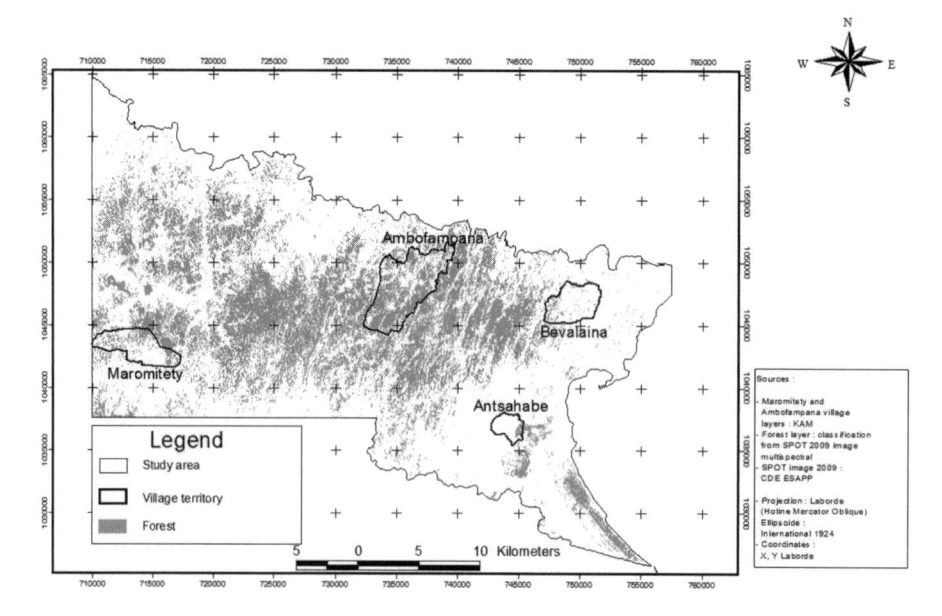

Figure 6.2 *Research area, Madagascar*

a result of the distance to the forest and those that were a result of the distance to the markets. Above all, in this chapter, we look at the influence exerted by the distance to the forest and its resources.

Forest fragmentation: definition and process

Forest fragmentation can be described as a dynamic process in which the habitat is progressively reduced into smaller patches that become more isolated and are increasingly affected by edge effects (Forman and Codron, 1986; Reed et al, 1996; Franklin, 2001; McGarigal and Cushman, 2002). Habitat fragmentation denotes a particular spatial process of land conversion. In the strict sense of the word, fragmentation refers to breaking a whole into smaller pieces while controlling for changes in the amount of habitat (Collinge, 2009).

Table 6.1 *Characteristics of four field research villages, Madagascar*

Characteristic	Ambofampana	Maromitety	Bevalaina	Antsahabe
Distance to forest massif (walking hours)	0.25	0.5	2	3
Forest cover (percentage)	86	75	43	21
Households	27	26	110	65
Population density (persons/km^2)	5	10.5	44	73
Ethnic groups	Betsimisaraka	Betsimisaraka	Betsimisaraka Sakalava	Betsimisaraka
Distance to nearest market (walking hours)	8	10	2.5	1.5

Deforestation in the tropics often creates matrices of human-managed areas and secondary vegetation with fragments of primary forest (Benitez-Maldivo and Martinez-Ramos, 2003). This process of habitat fragmentation is believed to create problems for the existence of the organisms living there in two ways: (1) the limited dispersal due to the higher degree of isolation reduces the probability that fragmented habitats can be colonized (Hanski, 1997), and (2) the reduced population size due to the smaller habitats increases the probability of extinction (Williamson, 1981; Diamond, 1984; Schoener and Spiller, 1995; Lande, 1998).

For this research, fragmentation is understood as a dynamic process whose effects depend on the species, type of landscape, spatial scale and geographical area. Deforestation results in habitat loss and isolation of remaining forest fragments.

Generally, the fragmentation starts with the creation of gaps in the habitat or continuous forest cover. The number and extent of gaps increase until the gaps either dominate the environment or modify the habitat in such a way that, in a changed, largely open landscape, only a few fragments of forest remain that represent the original vegetation cover. According to Jaeger (2000), the fragmentation process includes both a reduction of the habitat area and an increase of the proportion of the edge-influenced habitat.

The habitat fragmentation process encompasses three components (Andren, 1994):

- pure loss of habitat, leading to species extinction;
- reduction of patch size; and
- increasing spatial isolation of the remnant habitats.

Customary Resource Rights in Forest Massif and Forest Fragments

Here, we discuss the customary rights regarding three main roles of the forest: (1) land reserves for future agricultural activities; (2) a source of non-timber forest products (NTFPs); and (3) a source of timber. In a subsequent section, we focus on gender aspects, using the specific example of *Pandanus* spp. In Madagascar, the state is usually the legal owner of primary forest (Razafy, 2004). However, within the customary rights of the Betsimisaraka, forest ownership refers to people's right to use the land for agriculture in the future and not for the natural resources that it contains. In the following section, we consistently use the traditional understanding and customary interpretation of property rights and ownership rather than the legal one.

Traditional property rights

Following traditional rights, the inhabitants of a given area – independent of their village of residence, ethnicity or family affiliation – are generally allowed to collect timber and any NTFPs in all forests. Existing rules on certain species in our region are almost always kinship specific and not based on general regulations, but on more specific taboos. In this context the term kinship includes ancestors and their now-living descendants, as well as their future descendants (Keller, 2008). Contrary to such taboos for specific forest species, customary rights to convert forests into agricultural land apply in the whole study area. The traditional rule that gives the one who clears a forest the traditional landownership for the cleared area is well known by all farmers, regionally respected and applied. The existence of these traditional rules in the study area can be explained by the high importance of forest as a land reserve for agriculture, in contrast to the significantly lower importance of timber and NTFPs. In other regions of Madagascar we also find evidence that traditional landownership after deforestation falls to the person who clears a piece of forest, who is then the recognized owner (Keller, 2008; Oxby, 1985; Lindenmann, 2008). In this section we want to point out the differences between the forest massif and fragments in the Manompana corridor in terms of rights and regulations.

Forest massif

In the forest massif, a farmer who clears a part of the forest will subsequently be considered the rightful owner of the converted land. In other words, ownership is interpreted as the right to cultivate the land. According to our interviews with village authorities and farmers, any person, irrespective of village origin or kinship, is allowed to clear forest in principle. Following traditional customs, this household becomes the owner of the cleared area. The system enables a family to control a large area by clearing and appropriating the land, and thus securing their future rights to cultivate it (Messerli, 2002). This is an important aspect of the wish for descendants to whom one can give the land, and who can in turn offer it to their descendants; it is a feature that is deeply anchored in rural Malagasy culture (Keller, 2008).

Even though villagers or people from other regions do not have to ask permission to clear in the massif, they typically inform the *Mpisikidy*. The *Mpisikidy* is one of the village elders who is perceived to have the ability to contact spirits and ancestors, and to interpret their messages to the villagers. The role of the *Mpisikidy* is not to decide whether clearance is allowed. Instead, he contacts the forest spirits and the ancestors in a specific ceremony. The aim of the ceremony is to announce to the spirits that the family will be living in the forest and to ask the ancestors about future taboos for this particular household. The *Mpisikidy* is the person with the ability to contact the spirits; and he will inform the family of any traditional taboos, called *fady*, with respect to the future agricultural land. The family need only consult the *Mpisikidy* the first time they burn in the massif, not if the area to be cleared is a fragment surrounded by secondary vegetation or agricultural land.

Once the family in question has cleared the area of forest, the household has the right to fell other trees in the nearby forest until other families manifest

> ## Box 6.1 The origin of an *ala fady*
>
> A man from the village Ambofampana once searched for forest products in a forest near the village to build his house. When he did not come back in the evening, his family went looking for him and found him dead in the forest fragment. As they did not know why he died, the family asked the *Mpisikidy* to contact the ancestors and spirits and ask for the reason of his death. The *Mpisikidy* told the family that the forest spirits had been disturbed and were angry because he searched for products in this forest. It seemed that the spirits of this forest did not like it. Therefore from this day on it was forbidden for all descendants of the dead person to search for products or to burn the trees in this particular forest.

their interest. When this happens, the concerned families will discuss and agree on the limits of their respective ownership. Usually the limits follow topographic features such as rivers, small valleys or crests. To avoid potential conflict with others who may want to practise swidden agriculture in the surrounding forest, deforestation occurs as far as possible in circles around a remaining piece of forest, in order to separate it from the massif.

The only cases where clearing parts of the massif may be disallowed are *ala fady*, which are protected by a taboo (Keller, 2009). For example, it might happen that a forest is protected from clearance by a *fady* because of a particular experience a person had in this forest (see Box 6.1). But this *fady* concerns neither the whole village nor the whole region. Mostly *ala fady* are kinship specific and not known or respected by other family groups. But not all *fady* are family specific; some apply to entire village communities (Jones et al, 2008, Jarosz, 1994). For instance, in one of our villages, in the forest where the dead repose before being transported to their ancestral village, it is taboo to fell the trees. Nevertheless, family specific *fadys* are more frequent where the forest is concerned. In consequence, almost all once-*ala fady* forests have been burned by other families, which did not have to respect the *fady*. In other regions along the east coast it has been reported that parts of the forest cover were protected by forest specific traditions (*fomba*) and kept *fady* until new arrivals, migrants from other families, arrived and changed the usage (Tsitohae and Montagne, 2009). We also observed this in our study region.

Forest fragments

As soon as the forest massif is reduced to fragments and surrounded by agricultural land, the right to clear the remaining forest is limited to the owners of the adjacent agricultural land. If there is more than one farmer with agricultural land surrounding the fragment, then all are owners of the fragment. However, in these fragments any person can collect or exploit all forest products (NTFPs and timber) without asking the forest owners' permission. The traditional interpretation of ownership concerns only the land, not the resources growing naturally on it. The fragments represent the land reserve for agricultural purposes. However, as the family still sees itself as the rightful owner, even though they possess only the right to use the land, we speak here of so-called private forests, in the traditional sense. In the legal sense, however, the forest is still owned by the state. These private forest fragments will be distributed evenly among the children, regardless of sex or age, upon inheritance. Before the official distribution by the parents, children cannot become landowners of the forest. The deforestation of a forest fragment by a child does not create the right of future landownership. This is an important difference from the forest massif, where the deforested land always implies ownership, because the massif is not owned by a person.

We see that the main difference between the massif and the deforested fragments is that access to the pertinent soil reserve in forest massifs is free, whereas access to fragments counts as owned land. The traditional property rights regime of open access to forest and soil reserves in the massif makes it impossible for a village to control deforestation there. Villagers are not able to defend their forest as long as the forest is not fragmented, because it is not seen as either village land or private property. The needs of the population can be met by existing resources and the forest is sufficient to gain land for cultivation in future. Only if forests are fragmented is it possible to defend them against further deforestation following the customary right of private forest, but still only at an individual or a family level, not at a village level. The population perceives that private property of forest fragments is not associated with maintaining the forest, but with a land reserve for future descendants of the kinship group (Keller, 2008). With respect to forest fragments the decision of what will happen with them in the future remains with the family. On the village level it can normally not be decided whether a fragment should be protected or not. Therefore the concepts of families and kinship groups should always be at the centre when it comes to identifying the different actors involved in forest management (Tsitohae and Montagne, 2009).

Fady and *sandrana* for non-timber forest products

Over the entire research period we looked for local rules pertaining to the harvest of NTFPs. Regionally, the use of NTFPs is based on open access for anyone, regardless of kinship or geographical origin. All farmers can collect products in all forests, even in foreign village territories. The traditional system of private forests is not applied to forest resources. There do not seem to be any general rules, at either regional or village levels. However, the system of taboos and prohibitions may form an important traditional base for governance practices, and this was a central point in our research. In this section we describe the existing taboos (*fady*) and prohibitions (*sandrana*) concerning forest products and the potential to use these for conservation purposes. Certainly, anthropologists often rightfully complain that the Betsimisarakas' 'culture' is reduced to taboos (Keller, 2009); nevertheless, in relation to the collection of NTFPs in our study area, taboos seemed to be crucial elements in a broader cultural explanation.

Fady can be classified as taboos for certain days, taboos for specific species and taboos for applications of methods – for example, the taboo to 'fish with a net in a particular river'. The system of *fady* has also been described as an important protection mechanism for endemic species and habitats in Madagascar (Jones et al, 2008), as a prohibition to use particular species. Nevertheless, in our research area the system of *fady* seems to be very complex and not a system that could guarantee the protection of species or forests.

Taboos can be said to differ between individuals, families, ethnic groups and villages (Jarosz, 1994). The mostly kinship-specific *fady* concerning forests and forest species in our study region can originate from a particular experience or dream of a family member involving a particular species or part of the forest. The experience or dream has to be interpreted by the *Mpisikidy*, who contacts the ancestors and forest spirits. He will set the *fady*, which is then usually respected by all the descendants of the person who had the experience or dream. Consequently, rules and practices almost always differ between kinship groups and even families because they are strongly linked to their ancestors. The temporal limitation to cut forests in Ambofampana is one example of a *fady* in a village. For one kinship group, it is forbidden to clear forest on Tuesdays and Thursdays, whereas for another, it is forbidden only on Tuesdays. An important factor in terms of governance is that a *fady* can also be changed with the permission of the *Mpisikidy*, if required by circumstances (Box 6.2). A family can also offer a zebu (cow) or a chicken to the ancestors to change a particular *fady*.

Contrary to the *fady*, a *sandrana* is a somewhat stricter taboo, like a prohibition that can never be changed; it remains valid throughout life and for the whole family or even kinship group. A *sandrana* is a general prohibition to hunt, eat or collect a certain species, often a forest species. It has no temporal limitation, but as with the *fady* mentioned before, the *sandrana* also differs from family to family and is not a cultural belief that generally protects species. Whereas, for example, the *Indri indri* (lemur) is a *sandrana* for one kinship group, another from the same village can hunt it. We have found only two

Box 6.2 Example of change in a *fady*

In the past, when forest still covered a larger area of Bevalaina and Antsahabe, baskets and mats were woven using leaves of the plants of the Pandanaceae family, the *Pandanus* species. In those days it was *fady* to bring products made of *penja* (*Lepironia mucronata*) to the rice fields. *Penja*, a reed, grows in swamps and is only used for certain products. But with increasing population growth, farmers had to enlarge the agricultural area and the forest borders were pushed back. *Pandanus* became rare and swamped areas increased. As a consequence, women began to seek alternatives for *Pandanus* and produced their baskets and mats using *penja*. It therefore became difficult to collect the rice crop, as farmers could not bring their baskets, made of *penja*, to the rice fields. For this reason the *Mpisikidy* decided to change the *fady* and from then on, it was no longer taboo to bring *penja* products to the rice fields. In Ambofampana, where, unlike in Bevalaina and Antsahabe, abundant forests and *Pandanus* still exist, the *fady* for *penja* products remains.

> ## Box 6.3 How the Drongo bird became a *Sandrana*
>
> The Drongo bird, also called by the Betsimisaraka 'the king of the birds', is one of the rare species prohibited from being eaten by the whole population. Taboos are mostly based on experiences of certain persons, and so it is for the Drongo. People tell the story that the bird once saved a woman and her child. According to the story, rebels arrived in the village intent on killing all inhabitants. One woman managed to escape from the village and tried to save herself and her child in a cave. When the rebels approached the cave, however, the child began to cry, almost revealing his and his mother's whereabouts. Then, suddenly, a Drongo, with his voice of a crying child, began to sing and led the rebels in the wrong direction, so they could not find the woman with her child; both were thus saved. The woman told all her descendants not to eat the Drongo anymore because it saved her and her baby's life. Since then this bird has been a *sandrana* and a protected species for the entire population. In this sense, it is a binding *sandrana* for all Betsimisaraka living on the east coast.

species that are *sandrana* for all persons of the ethnic group Betsimisaraka (Box 6.3) in our research area.

Comparing the villages near and far from the forest massif, we can observe differences with regard to their *fady*. With increasing forest fragmentation and therefore less forest cover, obviously forest resources become rarer and have to be replaced by other species, whose collection may have been *fady* in the past. By consequence, villages with low forest cover have had to adapt certain *fady*, whereas these *fady* still exist in villages near the forest (see Box 6.2). We saw no evidence, however, of villages that are far from the forest having more *fadys* relating to forest resources, as these resources have become rare. On the contrary, if the diversity of forest resources decreases, the *fady* are likely to be repealed.

Nevertheless, *fady* and *sandrana* can provide protection for a few particular species or specific forest areas, if they are valid for the whole ethnic group or the entire village. A *sandrana* can also signify at least a certain reduction of over-exploitation. But the role of *fady* and *sandrana* should not be overestimated, since there is no guarantee that they will continue to be upheld. As we have seen in Box 6.2, they can be changed, or they may simply not be respected by immigrants who do not have the same ancestors (Tsitohae and Montagne, 2009). Additionally, if we look at the diversity and particularity of *fady* among kinship groups, they appear complicated. It seems almost impossible to respect all different and family-specific rules concerning products and forest areas. Even though we found that the *fady* and *sandrana* are very complex and not a guarantee for species or forest protection, we share the opinion of other researchers that *sandrana* and *fady* need to be better respected as cultural aspects, and therefore included in future resource management plans.

Use of timber products

Timber products are primarily used for house construction and making tools. In the two villages with high forest cover, no commercial wood exploitation is possible because of their remoteness. Therefore farmers use timber products only for subsistence purposes. In the two villages with low forest cover and a high degree of fragmentation, on the other hand, timber commercialization is very common, albeit on a small scale, because of the closer proximity to markets. Selling timber products can be an important source of income, especially in times of crisis or when harvests fail.

Nevertheless, in none of the four villages could we identify general rules or restrictions concerning timber harvesting. For NTFPs as well, the rule of general open access applies for timber harvesting. This means that even forest owners in a traditional sense are unable to prevent others from harvesting timber.

Involvement of women

Female and male members of a Betsimisaraka household both generally take important decisions, even though a traditional separation of gender roles exists. Women have the same rights as men, concerning inheritance and property to forest fragments and agricultural land. In this section, we focus on the necessary involvement of women in future forest management.

In our study site we recognized that in forest management activities organized by NGOs or by the state forest service, mostly men are involved. In general, men know the names of plants better than women and are usually in charge of the harvest of timber and NTFPs. Nevertheless, our results show that *Pandanus* spp., collected by women, are among the most important forest products (Figure 6.3). This identification of importance is based on local perceptions, the regular income that households earn by selling baskets and mats woven of *Pandanus*, and the high number of households using it (80 per cent). These products are of considerable importance, especially to poorer households, even though the mean income earned from *Pandanus* products per year and selling household is equivalent to only 5.3 Euros (13,425 Malagasy Ariary). Women explain its importance through its continuous availability during all seasons and in times of crisis or rice shortages. A household looking for additional income can produce and sell *Pandanus* products throughout the year.

We also observed changes in women's behaviour due to increasing forest fragmentation and decreasing diversity. In villages near the massif women collect plants alone in forests next to their village. Women are afraid to stray too far from their village, usually staying within a radius of one hour's walking distance. They are afraid of getting lost or meeting wild animals. There is also no need to travel farther because they still find plants of good quality near the village. However, women we talked to in villages far from the massif have had to change their harvesting practices because good plants have become rare near the village; now they usually go out in groups or with their husbands. Even so, the farther villages are from the forest massif, the less *Pandanus* they collect, because the plants are no longer available or too difficult to find. Nonetheless,

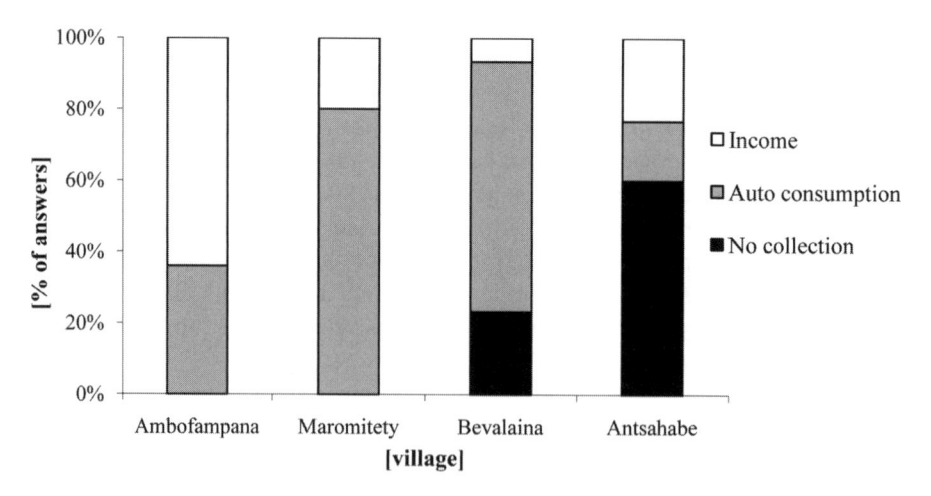

Figure 6.3 *The use of Pandanaceae*

women still depend on products made from *Pandanus*. Because of its growing rarity, this species has had to be replaced by other plant species, such as *penja* (*Lepironia mucronata*), which grow outside forests. Yet even where a substitute plant can be found for *Pandanus*, it remains the preferred plant because it is better suited for certain uses. For instance, rice dries more quickly on *Pandanus* mats than on alternative types of mats.

Awareness of Forest Depletion

Because the four studied villages differed in their distance and access to the forest massif, we could observe how villagers changed their perceptions towards forests and their resources along a gradient. With increasing distance away from the forest massif, fragments became a heightened source of contention among the farmers in neighbouring villages. In the following section we discuss the change in perception and why this has led to increased conflict. Achieving a heightened understanding of these perceptions and the origin of conflicts related to forest fragments should help to minimize conflicts in future resource governance.

The perception of forests is a broad topic covering a wide range of aspects. We looked particularly at people's awareness of the finite nature of forests as an exhaustible natural resource. For this analysis we asked households about their opinions regarding the formal legal prohibition of *tavy* in natural forests. All villagers were aware that *tavy* decreases the forest area. But their perception of the consequence of *tavy* on forest resources in the longer term was not evident at all. The results (Figure 6.4) highlight the difference between villages near the massif (Ambofampana and Maromitety) and the villages far away from it (Bevalaina and Antsahabe). In Ambofampana, for example, forest

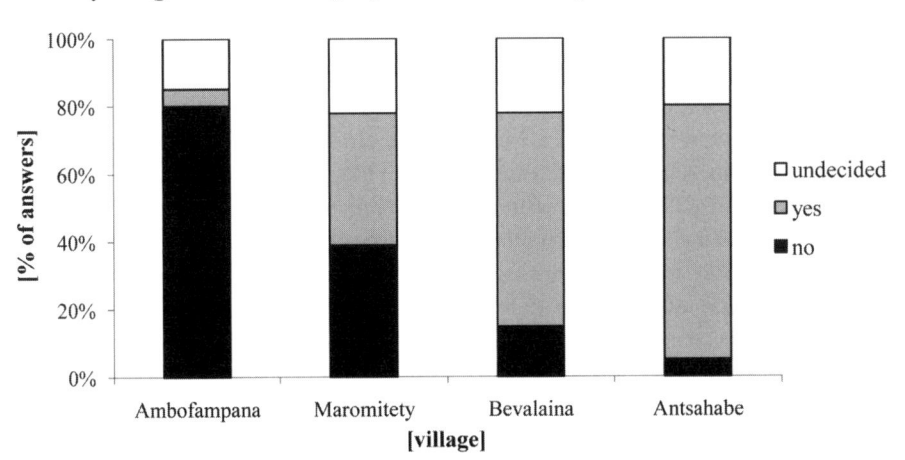

Do you agree with the legal prohibition of tavy on natural forest?

Figure 6.4 *Agreement with legal prohibition of* tavy *on natural forest*

resources were abundant, and people's perception was that the forest would never disappear. Therefore *tavy* was not seen as a serious problem. In contrast, in Antsahabe, where only a few fragments remain, the disappearance of forests had been observed over a longer period and the consequences were well known, so the farmers raised concerns. Figure 6.4 shows that as forest resources decrease, when only fragments remain, the system of *tavy* is increasingly perceived as a problem and a majority of farmers want it stopped.

Figure 6.4 also shows that in the villages nearest the massif (Ambofampana, Maromitety) and the villages farthest away (Bevalaina, Antsahabe), the answers were more homogeneous. But in the two villages in between, opinions were more divided. The differences were primarily between poorer households, which did not have enough land to satisfy their needs and rejected a prohibition on *tavy*, and richer households, which were less dependent on the forest as land reserves and wanted such a prohibition (p=0.022: Fisher's exact test).

In all four villages some people were undecided; these households were aware of the negative consequence *tavy* has on forest resources, but they also mentioned the lack of an alternative.

We conclude that people perceive forests more as an exhaustible resource when forests are already fragmented. As long as the massif is still close and mainly intact, it is not seen as a resource that should be used in a sustainable way. But awareness about the exhaustibility of forests in a village can lead to conflicts with other farmers, who continue with *tavy*.

Adaptation of Customary Rights to the National Legal System

The traditional system of forest use, including the practice of *tavy*, once seemed to be sustainable under conditions of low population density and high forest cover (Styger et al, 2007). In Madagascar this is no longer the case. The limited, kinship-related rules concerning NTFPs and forests, and the general open access for all forest products, also seem to have worked in the past, without generating conflict. In our research area this traditional system of open access still works in remote areas, where the forest massif remains extensive and villagers do not need to defend their forest against people from other villages. But as the massif becomes more fragmented and resources decrease, the traditional system no longer works and conflicts ensue. The local population in the deforested areas can no longer satisfy their needs from the natural resources of their own territory, and they begin to seek products in the territories of nearby villages. Following the traditional rule of open access, people are allowed to use the resources in other village territories. Villagers in areas with fragmented forests, however, are beginning to realize that the traditional rules no longer function; they are trying to find other ways to use their resources. The consequence is a mixture of customary rights with new rules, influenced by provisions of the national legal system.

In all four villages households were asked about problems they had concerning the current forest use in their village territory. Figure 6.5 shows the results. Some problems seem to be specific to a particular village. For example, in Maromitety, the fear that KoloAla would create a conservation site was noted several times (Figure 6.5). Maromitety lies near the special reserve Ambatovaky, which is already a conservation zone, outside our research area, where villagers have only limited access to forest resources (Randrianasolo,

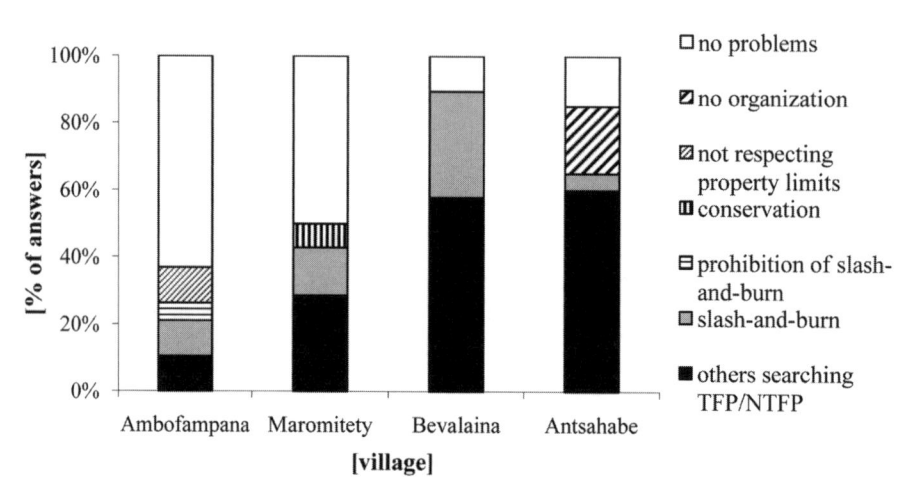

Figure 6.5 *Problems with current forest management*

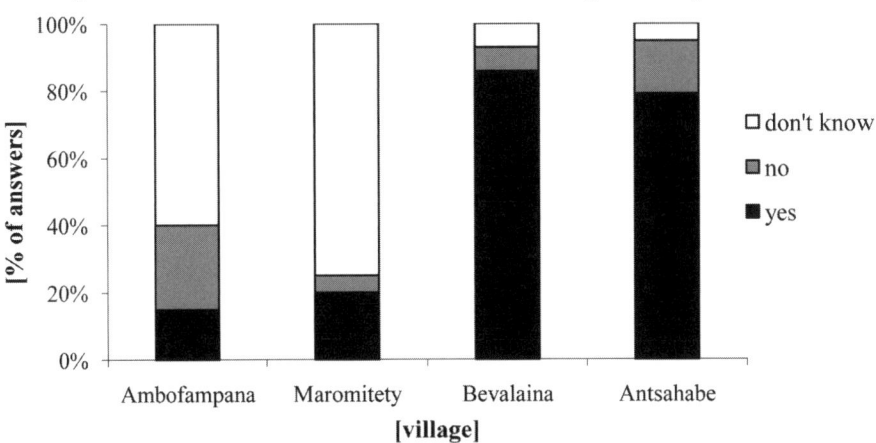

Do you have timber conflicts with other villages/foreigners?

Figure 6.6 *Timber conflicts*

2000). This conservation area does not affect Maromitety, but farmers have heard from other villages that their forest has been protected. Farmers in Maromitety seem to be confusing a protected area with the status of a KoloAla forest and are now afraid that they will be confronted with the same situation.

As illustrated in Figure 6.5, in all four villages, some informants said they had no problems with the current forest use. This was mostly the case in the two villages near the extensive forest massif. A majority of informants from these two villages did not consider it a problem if people from other villages used forest products in their forest, since resources were sufficiently abundant. On the other hand, a majority of informants in the two villages farthest from the forest massif mentioned forest use by people from outside their community as one of the most serious problems in their forest fragments.

Forest fragments thus become a source of conflict for villages as soon as their resources become scarce. Since people living in villages near the massif neither fear forest loss for the near future nor see it as a potential problem; they perceive no need at this time to change the rules concerning forest use. In Bevalaina and Antsahabe, on the other hand, where conflicts for timber products are common and increasing (Figure 6.6), interest in creating new rules within the national legal system and using formal conflict resolution mechanisms is high.

In Bevalaina, people are trying to change the traditional property rights regime of unlimited open access to a system with limited access rights. They have begun to introduce new rules, influenced by the national legal system, which are contrary to some customary rights. The owners of one forest fragment decided to establish a forest association to defend their fragment against any exploitation and collection of timber and NTFPs by outsiders. Anyone wishing to collect products from this particular fragment must now ask the association for permission. The customary right of open access, however, still applies to all other forest fragments. The association is not supported by the state for management transfer and has no backing from KoloAla or any other national management transfer programme; rather, it was founded following a local initiative with consensus among the farmers who are accepted as traditional owners of the forest fragment.

Naturally, owners of other fragments now want to defend their forest, too. When a farmer tried to prevent women from collecting *Pandanus* in his fragment, the traditional authorities told him to keep in mind the custom of open access to all forests not included in the forest association. Such situations have led to mistrust of the association by the village kin groups, and the rules governing the use of forest resources have become unclear.

Despite this mistrust and lack of clarity, we think this example of an association should be taken into account in the discussions about future systems for forest management. In many other projects of forest transfer to local communities, the potential for conflict arises from the incoherence between traditional ownership rights and new administrative limits. But in the case of Bevalaina the traditional owners of the concerned fragment are well recognized, and all are members of the association. Although other families seem to be jealous, the limits and rules of the association are well respected and the area has been conserved for six years.

Conclusion

Our research in Manompana was carried out in light of the new concept of KoloAla. Using KoloAla Manompana as a concrete example, we explain and illustrate how existing local systems can be incorporated into a decentralized forest management reform.

Generally speaking, existing local institutions, such as the forest association in Bevalaina, and systems of traditional rights concerning natural resources, should be incorporated into new forms of forest management as far as possible. An important finding is the difference between forest massifs and forest fragments, and its impact on customary rights. This needs to be integrated into future forest management. The traditional interpretation of forest ownership allows unlimited access to forest products but applies only to forest fragments.

This understanding of ownership is key to assigning areas to one of the three zones in the KoloAla system – wood production, subsistence use and conservation – in ways that minimize conflict. It would be advisable, for example, to include the wood production and conservation zones in plans for forest massifs, where the forest is not allocated to a family, thus avoiding some conflicts. Including in these zones any forest fragments that belong to one or more families, according to customary right, could lead to conflict within the village community. The profits from wood use would not directly benefit the forest owners but rather the entire community, which would mean also other kinship groups. The challenge would then be to persuade the current forest fragment owners to cede the entire management of their fragments to COBA.

Forest fragments should not be included in any conservation areas. Since allocation to the conservation zone absolutely forbids use, such designations would be accepted only with great difficulty by the current traditional owners. As local regulations respect the principle of free access to all forests, fragments are best suited to the category of subsistence use, so that the local population, as well as the traditional forest owner and other villagers, can still benefit from forest products.

A further reason why forest fragments should not be integrated into conservation or into wood production zones is that they play a central role in the livelihoods of the villages farther from the forest. Most families obtain vital products from these fragments, access to which is guaranteed by tradition.

Forest fragments are always in the possession of a family and important to the whole kinship group because they are land reserves for future living space and agricultural activity. This is why kinship groups with greater areas of forest are also those with the greatest influence and control over village affairs, and why the context of the different kinship groups in a village must be taken into account. For this reason, too, family members of both owners and non-owners of the forest must be represented in a COBA. Any system requiring that people discontinue the cultivation of forest fragments that they have traditionally claimed must be negotiated with the relevant families. This means that a reward system must be considered that can induce them to renounce their claim.

All kinship groups must be represented in a COBA because each has its own *fady*, *sandrana* and *ala fady*. Some *ala fady*, which forbid the taking of wood, are restricted to single families, and ignoring them can lead to conflict within a village community. By taking the situation of traditional forest owners into account, we want to move beyond the notion of supporting a private forest. Instead we want to show how community-based forest management, like that promoted by KoloAla, can best incorporate current forms of management.

Existing local *fady* and *sandrana* should be included in any future efforts because they are important local regulatory systems that go back to the peoples' ancestors. Although they should be taken into account, they rarely guarantee conservation activities, as they can be adapted to suit almost any situation and rarely apply to all kinship groups.

The KoloAla concept provides leeway to address NTFPs in management plans. We strongly recommend that this be done for the most important products. We have illustrated this point with Pandanaceae. *Pandanus*, for example, creates an important motivation to conserve forest fragments, particularly in regions near the forest massif where abundant forest resources are available and *Pandanus* of good quality can be found. The target group to motivate is primarily women: because *Pandanus* leaves of good quality cannot be found outside forests, women have an interest in conserving the remaining fragments. They also are aware of the decline in *Pandanus*, although few link the reduction to deforestation. It is therefore of the utmost importance to involve women as representatives in COBAs and to make the most of their knowledge and motivation for protecting *Pandanus* and other resources in future forest management activities.

Existing institutions, such as the traditionally founded forest association in Bevalaina, should be included in KoloAla and, as far as possible, left in their current form if their authority is to be recognized by the village community. External interventions that undermine existing institutions can quickly damage the KoloAla, as has been observed in earlier instances of management transfer.

We have seen that perceptions and opinions change with increasing distance to the forest massif and the concomitant decrease in forest resources: villages near the forest, and which therefore have forest in their territory, have little interest in conserving the forest. Meanwhile, villages that are farther away from the forest massif are greatly interested in its conservation. These different levels of interest can be a source of conflict, and this is why inhabitants from all villages need to be represented in the COBAs.

To sum up, local regulations and institutions governing the use of forest fragments exist, but certain customary rights to the forest are not compatible with national concepts. Some accommodation must be made, including a suitable reward system. As far as possible, reforms must take account of these customary rights and integrate priorities for NTFPs and specific *fadys* into the new management systems.

Acknowledgements

We would like to thank the organizations that have helped to fund our research. The research project has been supported by the Research Fellow Partnership Program (RFPP) and the Commission for Research Partnerships with Developing Countries (KFPE), both financed by the Swiss Agency for Development and Cooperation (SDC).

References

Adams, W., Aveling, R., Brockington, D., Dickson, B., Elliot, J., Hutton, J., Roe, D., Vira, B. and Wolmer, W. (2004) 'Biodiversity Conservation and the Eradication of Poverty', *Science,* vol 306, pp1146–1149

Andren, H. (1994) 'Effects of Habitat Fragmentation on Birds and Mammals in Landscapes with Different Proportions of Suitable Habitat. A Review', *Oikos,* vol 71, pp355–366

Barry, D. and Meinzen-Dick, R. (2008) 'The Invisible Map: Community Tenure Rights', Governing Shared Resources: Connecting Local Experience to Global Challenges: 12th Biennial Conference of the International Association for the Study of Commons, Cheltenham, England

Baumgartner, R. and Högger, R. (2004) *Search of Sustainable Livelihood Systems,* Sage Publications, New Delhi

Benitez-Maldivo, J. and Martinez-Ramos, M. (2003) Impact of Forest Fragmentation on Understory Plant Species Richness in Amazonia. *Conservation Biology*, vol 17, pp389–400

Bennett, A. F., Radford, J. Q. and Haslem, A. (2006) 'Properties of Land Mosaics: Implications for Nature Conservation in Agricultural Environments', *Biological Conservation*, vol 133, pp250–264

Bertrand, A. (1999) 'La Gestion Contractuelle, Pluraliste et Subsidiaire des Ressources Renouvelable' [Contractual, Pluralist and Subsidiary Management of Renewable Resources], *African Studies Quarterly*, www.africa.ufl.edu/asq/v3/v3i2a6.htm

Casse, T., Milhoj, A., Ranaivoson, S. and Romuald Randriamanarivo, J. (2004) 'Causes of Deforestation in Southwestern Madagascar: What do We Know?', *Forest Policy and Economics*, vol 6, pp33–48

Cole, J. (2001) *Forget Colonialism? Sacrifice and the Art of Memory in Madagascar*, University of California Press, Berkeley

Collinge, S. K. (2009) *Ecology of Fragmented Landscapes*, Johns Hopkins University Press, Baltimore

Consiglio, T., Schatz, G. E., McPherson, G., Lowry, P. P., Rabenantoandro, J., Rogers, Z. S., Rabevohitra, R. and Rabehevitra, D. (2006) 'Deforestation and Plant Diversity of Madagascar's Littoral Forests', *Conservation Biology*, vol 20, no 6, 1799–1803

Diamond, J. M. (1984) 'Normal Extinction of Isolated Populations', in N. Mh (ed) *Extinctions*, University of Chicago Press, Chicago

Dufils, J.-M. (2003) 'Remaining Forest Cover', in S. M. Goodmann and J. P. Benstead (eds) *The Natural History of Madagascar*, University of Chicago Press, Chicago and London

Erdmann, T. K. (2003) 'The Dilemma of Reducing Shifting Cultivation', in S. M. Goodmann and J. P. Benstead (eds) *The Natural History of Madagascar*, University of Chicago Press, Chicago and London

Forman, R. T. T. and Codron, M. (1986) *Landscape Ecology*, Wiley, New York

Franklin, S. E. (2001) *Remote Sensing for Sustainable Forest Management*, Lewis Publishers, Boca Raton, Florida

Hanski, I. (1997) 'Metapopulation Dynamics: From concepts and observations to predictive models', in Hanski, A. and Gilpin, M. E. (eds) *Metapopulation Biology*, Academic Press, San Diego, pp61–91

Harper, G. J., Steininger, M. K., Tucker, C. J., Juhn, D. and Hawkins, F. (2008) 'Fifty Years of Deforestation and Forest Fragmentation in Madagascar', *Environmental Conservation*, vol 34, pp325–333

Hockley, N. and Andriamarovololona, M. M. (2007) 'The Economics of Community Forest Management in Madagascar. Is there a Free Lunch?' An Analysis of *Transfert de Gestion*', USAID

Jaeger, J. A. G. (2000) 'Landscape Division, Splitting Index and Effective Mesh Size: New Measures of Landscape Fragmentation', *Landscape Ecology*, vol 15, pp115–130

Jarosz, L. (1993) 'Defining and Explaining Tropical Deforestation: Shifting Cultivation and Population Growth in Colonial Madagascar (1896–1940)', *Economic Geography*, vol 69, pp366–379

Jarosz, L. (1994) 'Taboo and the Time-work Experience in Madagascar', *Geographical Review*, vol 84, pp439–450

Jones, J. P. G., Andriamarovololona, M. M. and Hockley, N. (2008) 'The Importance of Taboos and Social Norms to Conservation in Madagascar', *Conservation Biology*, vol 22, pp976–986

Keller, E. (2008) 'The Banana Plant and the Moon: Conservation and the Malagasy Ethos of Life in Masoala, Madagascar', *American Ethnologist*, vol 35, pp650–664

Keller, E. (2009) 'The Danger of Misunderstanding "Culture"', *Madagascar Conservation and Development*, vol 4, pp82–85

Koechlin, J., Guillaumet, J.-L. and Morat, P. (1997) *Flore et Végétation de Madagascar* [Flora and Vegetation of Madagascar], Vaduz, V.A.R.G., Gantner Verlag

Lande, R. (1998) 'Anthropogenic, Ecological and Genetic Factors in Extinction and Conservation', *Population Ecology*, vol 40, pp259–269

Lindenmann, K. (2008) 'Representations of Forest. Social Anthropological Study of a Village on the Border of the Dry Deciduous Forest in Madagascar', Université de Neuchâtel, Neuchâtel

Malanson, G. P. and Armstrong, M. P. (1996) 'Dispersal Probability and Forest Diversity in a Fragmented Landscape', *Ecological Modelling*, vol 87, pp91–102

McConnell, W. J. and Sweeney, S. P. (2005) 'Challenges of Forest Governance in Madagascar', *The Geographical Journal*, vol 171, pp223–238

McGarigal, K. and Cushman, S. A. (2002) 'Comparative Evaluation of Experimental Approaches to the Study of Fragmentation Studies', *Ecological Applications*, vol 12, pp335–345

Messerli, P. (2000) 'Use of Sensitivity Analysis to Evaluate Key Factors for Improving Slash-and-burn Cultivation Systems on the Eastern Escarpment of Madagascar', *Mountain Research and Development*, vol 20, pp32–41

Messerli, P. (2002) 'Alternatives à la Culture sur Brûlis sur la Falaise Est de Madagascar. Stratégie en Vue d'une Gestion Plus Durable des Terres' [Alternatives to Slash and Burn Culture on the Eastern Escarpment of Madagascar. Strategy in View of More Sustainable Management of Lands], Centre pour le Développement et l'Environnement (CDE), Université de Berne, Berne

Montagne, P. and Bertrand, A. (2006) 'Histoire des Politiques Forestières au Niger, au Mali et à Madagascar' [Forest Political History of Niger, Mali and Madagascar], in A. Bertrand, P. Montagne and A. Karsenty (eds) *L'État et la Gestion Locale Durable des Forêts en Afrique Francophone et à Madagascar* [The State and Sustainable Local Forest Management in Francophone Africa and Madagascar], L'Harmattan, Paris

Myers, N., Mittermeier, R. A., Mittermeier, C. G., Fonseca, G. and Kent, J. (2000) 'Biodiversity Hotspots for Conservation Priorities', *Nature*, vol 403, pp853–858

Oxby, C. (1985) 'L'Agriculture en Forêt: Transformation de l'Utilisation des Terres et de la Société dans l'Est de Madagascar' [Agriculture in Forests: Transformation in the Use of Lands and of Society in the East of Madagascar], *Unasylva*, vol 148

Pfund, J.-L. (2000) *Culture sur Brulis et Gestion des Ressources Naturelles. Evolution et Perspectives de Trois Terroirs Ruraux du Versant est de Madagascar* [Slash and Burn Agriculture and the Management of Natural Resources. Evolution and Perspectives of Three Rural Territories on the Eastern Slope of Madagascar], ETH Zürich and EPFZ Lausanne, Zürich

Raik, D. (2007) 'Forest Management in Madagascar: A Historical Overview', *Madagascar Conservation and Development*, vol 2, pp5–9

Rakotomavo, A. (2009) 'Schéma d'Aménagement du Site KoloAla Manompana' [Management scheme for Manompanan KoloAla site], Manompana, AIM, Union Européenne

Randrianasolo, J. (2000) 'Rapport de Synthèse du Mandat: Capitalisation des Expériences en Gestion Contractualisée de Forêts à Madagascar' [Summary Report of the Mandate: Compilation of Experiences in Contract Management of Forests in Madagascar], in Intercooperation (ed), Intercooperation

Razafy, F. L. (2004) 'Les Intérêts des Différentes Acteurs dans la Gestion des Ressources Naturelles Forestières' [The Interests of Different Actors in the Management of Natural Forest Resources], *Schweizerische Zeitschrift für Forstwesen,* vol 155, pp89–96

Razafy, F. L. and Andrianantenaina, F. (1999) 'Gestion de Terroir et Tenur Foncière dans la Région de Beforona' [Management of Territories and Land Tenure in the Region of Beforona], Antananarivo, BEMA

Reed, R. A., Johnson-Barnard, J. and Baker, W. L. (1996) 'Contribution of Roads to Forest Fragmentation in the Rocky Mountains', *Conservation Biology,* vol 10, pp1098–1106

Schmidt, K. and Stadtmüller, T. (2009) 'Introduction. A Workshop Series on Forests, Landscape and Governance', in J. Carter, K. Schmidt, P. Robinson, T. Stadtmüller and A. Nizami (eds) *Forests, Landscapes and Governance; Multiple Actors, Multiple Roles.* Helvetas, Intercooperation, SDC

Schoener, T. W. and Spiller, D. A. (1995) 'Effects of Predators and Area on Invasion: An Experiment with Island Spiders', *Science,* vol 267, pp1811–1813

Sorg J.-P. (2006) 'Orientation Nouvelles de la Recherche dans la Zone des Forêts Denses Sèches à Madagascar' [A New Orientation to Research in the Dense Dry Forest Zone of Madagascar], in C. Schwitzer, S. Brandt, O. Ramilijaona, M. Rakotomalala Razanahoera, D. Achermand, T. Razakamananan and J. U. Ganzhorn (eds) *Proceeding of the German-Malagasy Research Cooperation in Life and Earth Sciences,* DAAD, DFG, Zoo Köln, Universität Hamburg

Styger, E., Rakotoarimanana, J., Rabevohitra, R. and Fernandes, E. (1999) 'Indigenous Fruit Trees of Madagascar: Potential Components of Agroforestry Systems to Improve Human Nutrition and Restore Biological Diversity', *Agroforestry Systems,* vol 46, pp289–310

Styger, E., Rakotondramasy, H. M., Pfeffer, M. J., Fernandes, E. C. M. and Bates, D. M. (2007) 'Influence of Slash-and-burn Farming Practices on Fallow Succession and Land Degradation in the Rainforest Region of Madagascar', *Agriculture, Ecosystems and Environment,* vol 119, pp257–269

Sunderlin, W. (2003) *Poverty Alleviation and Forest Conservation: A Proposed Conceptual Model,* CIFOR, Jakarta

Tsitohae, R. and Montagne, P. (2009) 'Le Foncier Forestier: Réserve de Terres Agricoles ou Véritable Espace à Gérer?' [Forest Tenure: An Agricultural Land Reserve or a True Space to Manage?], in F. Sandron (ed) *Population Rurale et Enjeux Fonciers à Madagascar* [Rural Population and Tenure Issues in Madagascar], Antananarivo, CITE et KARTHALA, Paris

USAID (2009) *For a Sound Management of Forest Resources. Outcomes of Five Years of Work,* USAID

Williamson, M. (1981) *Island Populations,* Oxford University Press, New York

7

The Role of Wild Species in the Governance of Tropical Forested Landscapes

Bronwen Powell, John Daniel Watts, Stella Asaha, Amandine Boucard, Laurène Feintrenie, Emmanuel Lyimo, Jacqueline L. Sunderland-Groves and Zora Lea Urech

Around the world, many landscapes surrounding protected areas are caught between two conflicting forces – the need to conserve local ecosystems and the biodiversity they contain, and the drive to develop. Often, communities living around forests and protected areas are economically very poor, by both regional and international standards (Fisher and Christopher, 2007). In seeking to understand the role of landscapes and various land-use types by studying the livelihoods of people living between protected areas and market centres, the Landscape Mosaics project examined an array of wild flora and fauna species as case studies highlighting the relationships between livelihoods and landscapes. These case studies, especially when combined, yield insights into past and evolving governance practices and illuminate complex interactions among forests, governance, culture, gender, ecosystem services and markets.

Villagers in the five project study sites (Cameroon, Tanzania, Madagascar, Indonesia and Laos) seek a better life, often through development that requires drastic changes in land use. By comparing different governance practices across sites, we can learn much to guide future efforts in achieving conservation and development, simultaneously. Close study of wild species can be used to analyse and communicate the complex social–ecological interactions within landscapes. Because they provide a common issue on which multiple actors can work together, individual species can play an important role in catalyzing the emergence of the necessary governance arrangements for development and sustainable management of entire landscapes. Although different species require unique management strategies, the identification of management issues is instructive for the development of viable and sustainable management and conservation solutions for the overall landscape. Focusing on individual, locally important species may be the key to unifying the interests of a wide range of stakeholders.

The five landscapes of the project are defined by administrative and customary boundaries, land use and land cover, including protected areas. These mosaics of land reveal many of the challenges of governing multifunctional landscapes (Tress and Tress, 2000). It is within these landscapes that local people seek their livelihoods, governed by local, national and global rules. Studying the use, management and sale of locally important species is a tangible way of exploring the relationships between people and their environment. The species presented below highlight how such relationships are shaped by multiple levels and forms of governance. In the case studies, we frame issues of governance within landscapes, with particular attention to place and scale.

Megafauna

Interactions between large wild animals and people living in and around forests have been widely studied. Many protected forest areas are intended to conserve megafauna that would otherwise be at risk from loss of habitat and hunting, whether for sale, consumption or pest control. Here we explore the varying roles of megafauna in the project sites in Cameroon, Tanzania and Indonesia.

Case 1: Cross River gorillas in Cameroon

The forested landscape inhabited by the remaining population of Cross River gorillas faces significant threat. Human population density in the region is among the highest in Africa, and to accommodate growing human demands, habitat is degraded through agriculture, burning for pastureland and timber extraction.

Resurrected as a distinct subspecies in 2000, Cross River gorillas (*Gorilla gorilla diehli*), called *meki* in the local Anyang language, are found in approximately 11 localities in Cameroon and Nigeria (Sarmiento and Oates, 2000; Oates et al, 2007). Genetic evidence suggests a recent significant decline in numbers, which is thought to have been caused by hunting, and current figures indicate a total remaining population of less than 300 individuals (Oates et al, 2007). A broad programme of conservation and research in Cross River gorilla habitat has been underway for more than a decade, and gorillas are central to government-level conservation planning in the area (Groves and Maisels, 1999; Oates et al, 2007).

Some cultural beliefs are associated with Cross River gorillas, and although gorillas have been traditionally hunted throughout their range, these beliefs have almost certainly contributed to their continued survival. For example, beliefs in parts of the Cameroonian range dictate that gorilla and chimpanzee meat not be sold and instead be divided among the hunters' families and the wider community. Since no financial benefit can thus accrue from ape hunting, hunters reportedly have often ignored opportunities to kill gorillas and conserved their ammunition for other species. Furthermore, communities surrounding the Kagwene Mountain, located in the eastern part of the population range, believe gorillas are human ancestors and hence must not be killed. Consequently, that locality is home to possibly one of the largest remaining groups (Sunderland-Groves et al, 2009). Undoubtedly, traditional beliefs such as these have been beneficial to the species, although they do not offer complete protection, and other beliefs may counter these. Many communities, for example, use gorilla body parts as traditional medicine, such as pressing gorilla arm or leg bones against a human baby's limbs to enhance strength. Skulls are kept in the house as trophies, and the hunter who kills a gorilla gains elevated social status. Typically, the weapon used is a locally made gun that is powerful enough to kill a gorilla but leaves the hunter defenceless if he misses the target. Many hunters report being attacked, and some bear serious scars from wounds inflicted during such incidents.

Hunting gorillas is internationally and nationally illegal, and initiatives to preserve the remaining Cross River gorillas and their habitat have been ongoing. In 2008, two new protected areas were recognized. The trade-off between conservation and local livelihoods could not be clearer: protected areas offer more secure environments for gorillas, but decrease local people's access to forest products. Restrictions on access to natural resources, such as bush mango, an important non-timber forest product (NTFP) (see Case 5), will ultimately harm livelihoods. Gorilla experts note that if gorillas are to survive in the region, additional areas will require official protected status. However, if protection cannot be balanced with attention to local livelihoods, long-term protection will be difficult to achieve.

Case 2: Bush Pigs in Tanzania

Dependent on subsistence agriculture, the Shambaa people of the Usambara Mountains have struggled with pest mammals for as long as oral history recounts. One of the most commonly cited pest species is the bush pig (*Potamochoerus porcus*), known as *nguruwe pori*. Although almost never seen, especially outside heavily forested areas, the bush pig is considered a menace, and few hesitate to express hope for extirpation of the species. Hunting bush pigs is an important way for men to gain social status but is pursued as an endeavour to exterminate the species. Despite reported falls in the pig population, locals still report significant crop losses. There is a strong perception that fields close to homes suffer less damage than those near forest reserves; some farmers even report abandoning fields close to reserved forests.

For more than 50 years, a group of men in Kwatango village has been cooperating to hunt bush pigs with dogs and spears; local people refer to the group as a cooperative. The men pay close attention to patterns of crop damage, have detailed knowledge of behaviour and ecology and are proud of their successes: 'We go to hunt in groups, we select people [to join us] who are brave. It is too dangerous an animal to go alone' (village man, Kwatango, Tanzania). Given that hunting the pig remains legal in Tanzania and simultaneously provides meat for consumption and rids the area of pests, the view that this is an appropriate solution is no surprise. Although trade in bush meat is not as developed as in West Africa, wild meat is a delicacy and may provide important nutrients often lacking in local diets. Aside from the Islamic prohibition on pork (about 50 per cent of the population is Muslim), only minor taboos exist against consumption of bush pig.

Feierman (1974) asserts that the bush pig is central to the Shambaa people's representation and understanding of their local environment. Many magical uses are reported, and bush pigs frequently appear in local stories and myths. Feierman (1974) recounts the story of the first king of the Shambaa people, the great hunter Mbegha, who was made king because of his powers as a bush pig hunter, both giving the people meat and killing the age-old agricultural threat. In some stories, people and ancestors take on the shape of bush pigs, although these myths are less prevalent than the transformation stories in other parts of Africa.

Although scientific information on actual populations is sparse, bush pigs are undoubtedly threatened. Whereas all hunting is strictly prohibited in neighbouring Kenya, the hunting of many species is only prohibited in reserves in Tanzania. No government policies currently protect this species, and traditional local practices offer only limited support for its conservation. The social and cultural importance of the bush pig hunt and the value of its meat may prove key to developing a sustainable relationship between the Shambaa people and a species that has raided their fields for generations.

Case 3: Wild pigs and tigers in Indonesia

In Indonesia agroforests act as buffer zones between villages and protected forest areas. Such zones establish an ecological transition between closed forest ecosystems and open fields, and prevent the fragmentation of forest cover. Although agroforests are important to local people for various reasons, biodiversity is not among them (Feintrenie and Levang, 2009). In fact, some of the preserved biodiversity, especially the large and sometimes dangerous wildlife, is far from welcome to farmers. Numerous pests plague rubber agroforests, including wild pigs, tigers, elephants and monkeys.

Wild pigs are important local pests; they are both greatly feared and highly destructive to agroforest productivity and thus affect local people's livelihoods. Of the original 500 trees per hectare of rubber seedlings usually planted in a newly established agroforest, more than 60 per cent are lost before the trees reach maturity, mostly to wild pigs, which are fond of young rubber trees. Pigs are a particular problem because they also feed on rice, cassava and other crops that are not top cash earners, and are thus less closely guarded. Local farmers cite the main drawback to the cultivation of agroforests as the fact that they attract mammals (Therville et al, 2010). Pigs are the main reason that farmers give for planting food crops close to home, rather than intercropping in rubber agroforests (Lehébel-Péron et al, 2010). Farmers, especially women, fear that pigs and also tigers will attack them while they are working.

Risk and damage caused by such pests are considered unpreventable. Local people feel unable to act to defend their farms from them. Under Islamic dietary laws, important to the 80 per cent of the local population who are Muslim, all the major mammal pest species are *haram*, or taboo to eat. Many Muslims even prefer to avoid obtaining financial benefit from or coming into contact with pigs. Hunting most pest species is illegal, but government laws are minimally enforced, and the primary deterrents to hunting are social and religious (Levang and Sitorus, 2006). Fences built around the young rubber tree plots are not sufficient to reduce the damage. While local conservation NGOs emphasize the potential of megafauna species for attracting tourism, local farmers remain helpless against the significant destruction these animals cause.

Wild animals as pests or pot fillers

The above case studies from Tanzania and Indonesia highlight the significant damage that megafauna can cause to local people's livelihoods. Understandably, pest species are the source of much conflict between local people and conservation efforts. In both Indonesia and Tanzania farmers report modifying their agricultural practices to limit losses to income and food security due to crop destruction.

Across the three cases, we see the different degrees to which megafauna are viewed as pests. In Cameroon, discussion of Cross River gorillas as agricultural pests is limited; people tend to focus more on the importance of the apes as a source of meat and trophies as a sign of hunters' bravery. In Tanzania, bush pigs are viewed as both pests and meat. In Indonesia, where religious restrictions prevent their consumption, wild pigs are viewed solely as pests.

The degree to which these animals are seen as pests seems to be inversely related to the level of hunting rights and practices. In Cameroon, hunting is a traditional way of life and the primary source of meat. In the Tanzanian site, where wild animal populations have been relatively low for the last century and traditional subsistence is based on agriculture, hunting, although legal, is a secondary activity. In Indonesia, hunting is rare because of the religious prohibitions against profiting from or having contact with *haram* species. Whether agricultural damage is in fact less in areas where hunting is allowed is impossible to determine without further research; perhaps people simply view such damage as less of a problem (or they discuss it less) because they value the pest species for its meat. Alternatively, damage levels could vary according to population densities of species. Pig populations seem relatively higher in Indonesia than in Tanzania. Better understanding of such dynamics would require an improved understanding of the pest population's size and its proximity to human habitation.

Regardless of their roles as pests or pot fillers, many of these animals are associated with very strong cultural beliefs and traditions. The Indonesia case is an exception, perhaps because the traditional cultural beliefs were diminished with the introduction of Islam. Pigs have a very high and positive cultural importance among many ethnic groups in other areas of Indonesia and Papua New Guinea (see Rappaport, 1968).

Many authors have suggested that cultural beliefs could be built upon to enhance local people's support for conservation (see Cocks, 2006), and the two case studies from Cameroon and Tanzania hold potential for the application of this idea. Although indigenous practices are not always positive, many people do not associate strongly with the abstract concept of biodiversity conservation, and given governments' limited ability to enforce legislation and policy, conservation based on cultural and religious values may be more sustainable (Cocks, 2006; Infield, 2001). Cultural values have been largely ignored in conservation practice (Infield, 2001), but a paper by Sheikh (2006) presents a conservation education project in western Karakorum, Pakistan, which was much enhanced through the involvement of local religious leaders, who helped to orientate the conservation messages within the local belief system and allowed sensitive issues to be addressed.

In the three case studies, it seems that most of the practices to improve the management of species of megafauna, such as banning hunting and setting aside protected areas, originated from national or even global actors. As has been reported in the past, in each of the cases presented here, communities have had a limited role in initiating or developing current management plans (where they exist) and thus lack a sense of ownership. Although less effective in addressing habitat destruction (one of the biggest threats to most megafauna), cultural and religious beliefs may in some cases provide a starting point for drawing local people into more inclusive and decentralized governance practices. Competition for habitat between expanding, poor, local communities and wildlife populations will undoubtedly intensify over the coming years, and both conservationists and local communities will have to be involved in the identification of sustainable solutions.

Commercialized Plant Species

Commercialization of non-timber forest products (NTFPs) is often touted as the best way to increase local people's monetary incentives for forest conservation and sustainable management of their environments. Here a look at NTFPs

Case 4: *Peuak meuak* in Laos

In Laos, 85 per cent of people live in rural areas and rely on agriculture and forest products for their livelihoods (MAF and STEA, 2003). In the project site, one of the most important NTFPs is the bark of *Boehmeria malabarica* (Urticaceae), a small bush that prefers moist, sunny locations. Collection of *peuak meuak*, as it is known locally, started around 2000, in response to demand from local traders doing business with Chinese merchants. Previously, local people did not pay attention to this species, and they have remained naive to foreign uses (incense sticks, mosquito repellents and glue); for them, it is simply a source of income (NAFRI, 2007). Over the past ten years, the demand and price for *peuak meuak* have remained stable. Today this NTFP is very important in communities with limited market access and virtually no other off-farm sources of income. However, uncontrolled harvesting has led to the rapid depletion of wild populations.

Peuak meuak is harvested throughout the year, but especially when people have lighter workloads. In household interviews, we found that women participated in collection more than men and that collection behaviour varies from one community to another. In some areas people collect individually and in other areas they collect in groups; amounts collected and frequencies of collection trips also vary greatly. Traders visit villages to purchase the bark and give local people information about price variations. A few locally based traders have started to appear; one, in Bouammi village, uses a motorbike to take large quantities to the less remote village of Muangmuay. There, other traders stockpile *peuak meuak* until the outside traders arrive. Customary rights grant open access to NTFPs that are essential to people's incomes, and since *Peuak meuak* has no local historical importance, it now falls in this category. When the product reaches the district capital of Viengkham, however, taxes are due to the Agriculture and Forestry Office (FAO, 2006).

Faced with high demand and falling wild harvests, outside traders have started dispensing advice on sustainable harvesting practices, such as not collecting the roots. Local people have discovered another solution; most have now turned to intensive management of *peuak meuak*, transplanting seedlings from the wild into private gardens. The extent of this trend is directly related to market access, associated with proximity to the road. Of families interviewed in the village closest to the road, 61 per cent of those selling the species (42 per cent of all households) reported having *peuak meuak* gardens. In a less accessible village, only one family of 45 has a *peuak meuak* garden, but others plan to start keeping gardens soon. In the most remote village, wild sources remain available and no domestication activities have been reported.

Harvesting practices are quite different when people collect in the wild, in comparison to their own fields. In the wild, to maximize the harvest the entire plant is extracted, whereas in personal gardens plants are harvested only after they have achieved optimal size (which takes one or two years), and the roots are left to regenerate. The provincial Agriculture and Forestry Office plans to start providing extension services for *peuak meuak*, although resources are not yet available (Nicholson et al, 2008). The transition from a free common resource to a managed private one could be a very interesting basis for engaging communities in the management of natural resources.

Case 5: Bush mango in Cameroon

Irvingia gabonensis and *Irvingia wombulu* (Irvingiaceae) are large trees commonly known as bush mango, or *ogbono*, in Cameroon and Nigeria (see Chapter 8). Although historically the seeds were not of high importance in the Takamanda area, they are now widely used as a condiment and soup thickener across West and Central Africa. Because of high market demand, about 75 to 80 per cent of the fruits collected by households in the Cameroon site villages are sold to generate income, and for some people the fruits provide their only income source annually. Harvests are highly variable, and an unproductive bush mango season often means that households tackle the year virtually without cash (Asaha et al, 2006; Tajoacha, 2008; Sunderland et al, 2002; Schmidt-Soltau, 2001). Bush mango collection and trade yield an average of 169,000 FCFA (about US$300) per household per season (Asaha et al, 2006).

About 85 per cent of the households within these communities gather bush mango directly from primary forests, at sites typically an hour's walk from the home. Less often, fruits are collected from scattered trees on secondary forest and farmland. The mature fruits of bush mango are picked from the forest floor and split open, and the cotyledons are removed and sun dried. In most cases this processing occurs in the forest. During harvesting, families may stay in forest 'bush houses' for more than a week. Women and children are particularly involved with collection; trade, however, is primarily men's responsibility.

Because production is highly variable, assessing the state of the resources and monitoring harvest intensity are difficult, but exploitation of bush mango seems, in theory, to be sustainable. Because of its local importance, there is considerable community-level control on its harvest. Importantly, outsiders are discouraged from harvesting; it is an activity reserved for people native to Takamanda. Encroachment of Nigerian communities into Takamanda for bush mango collection is a source of much conflict (Sunderland et al, 2002). Community members have open access to bush mango, in theory on a seasonal, first-come, first-served basis. In reality, however, families tend to harvest in the same area each year, close to their bush houses, and inferred resource 'ownership' is acknowledged (Sunderland et al, 2002).

All NTFPs located in farmland belong strictly to the farms' owners. Collecting here is regarded as stealing and is punishable by fines from the village council. By law, in Cameroon, all forested land and forest products belong to the central government, which nevertheless recognizes that communities living in and around forests are dependent on such products for their livelihoods. The law implies that access is allowed for consumption but not for sale. Although high market demand and local need for income have led to extensive reliance on bush mango, government regulations on NTFPs remain poorly understood by local people and are thus open to significant corruption, including illegal levying of 'taxes' by various government officials. This and the extremely limited market access are reported as the major constraints to increasing the income local people can obtain from bush mango and other NTFPs. In 2008, the Takamanda Forest Reserve was upgraded to a national park, and future management plans may include restrictions on NTFP collection (see Chapter 8). Because bush mango is a vital source of income, policy-makers must consider the effect of any restrictions on local livelihoods.

Case 6: Allanblackia in Tanzania

Villagers in the East Usambaras obtain income from the sale of a variety of forest products. The oil-yielding seeds of the tree *Allanblackia stuhlmannii* (Clusiaceae), locally known as *msambu*, are among the most important. Prior to World War I, *Allanblackia* had limited local use as timber and medicine. Villagers recall their first introduction to commercialization of the seeds, when an Indian man began purchasing the seeds on a small scale in the 1970s. In the 1980s, seeds were bought by a company and exported to Kenya (Ruffo, 1989). More recently, a partnership called the Novella Africa Initiative, involving the international company Unilever, International Union for Conservation of Nature (IUCN), World Agroforestry Center (ICRAF), Tanzania Forest Conservation Group (TFCG), Faida MaLi and various other NGOs, has led to the significant development of this product. When asked about traditional beliefs or stories surrounding this tree, people recount the story of the development of Faida MaLi and its trade.

Novella now buys *Allanblackia stuhlmannii* nuts for 250 TSH (about US $0.20) per kilogram. Both men and women collect seeds from fallen fruit, sometimes with the help of children. Most of the trees are found in tea company forests (from which anyone can harvest) or on individual farmland. Harvest is seasonal, from February to May.

The importance of *Allanblackia* as a source of income has led to the development of clear rules governing its use. Despite the absence of traditional management rules for *Allanblackia*, the strength of local government systems has allowed for new rules to be determined and disseminated efficiently. In Tanzania, every village government has a forest committee responsible for both setting new and enforcing old rules regarding the use and management of local trees. This has allowed previously unmanaged *Allanblackia* to become a success story for local community-based management. Marco Mgunga of Misalai village told us, 'There used to be many [*Allanblackia* trees], but now there are less after many were cut for timber by the Saw Mill Company in 1949'. The contrast between unrestricted historical harvesting and current

Figure 7.1 *Benjamin Njiku showing an immature fruit on a* msambu *tree on his farm in Shambangeda*

Source: Bronwen Powell

practice couldn't be greater. Musa Omar Kipingu of Shambangeda explained the situation under current rules, '*Msambu* is not allowed to be cut under any circumstances ... if you cut one young tree, you will be fined 5000 TSH [about US$4.50] and you have to plant ten trees. If you cut a mature tree, you will be sent to court.' Although these new rules are restrictive, most local people seem to support the current regulations. Magret Chilambo commented, 'Now [*msambu*] is a more valuable tree than in the past. It is a tree that should be taken care of'.

Case 7: Pandanaceae in Madagascar

Used to produce traditional crafts such as mats, baskets and hats, the leaves of *Pandanus* (Pandanaceae) are among the most important NTFPs in the Madagascar project site. All households report use of this species, and about 65 per cent sell *Pandanus* products to generate income. Only women collect, produce crafts from and trade the species. When collecting the leaves, women prefer to stay as close to their village as possible, no more than one hour's walk from their homes. Forests farther away from inhabited areas are referred to as 'the men's forests'. Women say they are afraid to 'get lost or meet wild animals' there.

Because only young, high-quality leaves are cut, allowing the rest of the plant to regenerate, harvesting practices should be sustainable and should not influence species distribution. However, the plants are not regenerating fast enough to meet demand, and high-quality leaves are reportedly becoming scarce. Venturing farther from home in search of leaves, women now harvest in groups or accompanied by men. More leaves are collected per trip, and harvesting is taking place even in forests outside village territories. We found that in villages with less forest cover (and thus presumably less forest access), more conflicts over NTFPs occur. Conflicts commonly occur when people from a village with limited forest area take advantage of customary laws, which grant open access to all NTFPs, and search for such products within the territories of villages with more tree cover: 'People from other villages are searching for products in our forests, and there is nothing we can do because our forests are open for all people' (villager, Manompana, Madagascar). Currently in Madagascar there are no governmental regulations on the quantity, season or location of harvesting *Pandanus* or any other NTFP. Neither are there historical or customary traditions that might help to ensure that this NTFP is used sustainably. The few taboos and traditional beliefs reported tend to vary significantly from family to family. Any person can collect in any forest without geographic, gender, ethnic or quantitative restrictions.

Because the quality of products from species other than *Pandanus* is inferior, the demand for *Pandanus* leaves from the forest remains high. The potential of *Pandanus* to catalyse the formation of novel NTFP regulations or forest governance practices may nevertheless be limited because when the local supply is exhausted, in addition to venturing farther to collect, women also switch to alternative plant species (*Lepironia mucronata*) available in marsh or swamp land. Although the current potential for *Pandanus* to increase women's involvement in resource governance has yet to be realized, this NTFP may provide an opportunity for the integration of women into future management planning as well as formal governance in general.

Changing importance of different land-use types

Of the several interesting themes that emerge from these four case studies in Laos, Cameroon, Tanzania and Madagascar, one is the lack of historical use of many commercialized species prior to the development of an outside market. For *peuak meuak* in Laos and *msambu* in Tanzania, local people are not even aware of the products' final destination or use. It is not surprising, then, to note that there are virtually no cultural or traditional beliefs attached to these species. The lack of customary management practices is understandable, given limited historical or cultural importance. Combined with often extremely rapid development of market demand, this has in most cases led to overharvesting and unsustainable exploitation (for example in Laos and Madagascar). The case of *msambu* in Tanzania is an exception; there, local governments have strong mechanisms for setting new regulations, which has allowed them to manage the species effectively. The case study from Tanzania also provides an excellent example of how cooperation among multilevel governance bodies has allowed for the successful decentralization or devolution of resource governance. The case studies from Madagascar and Cameroon highlight how lack of involvement and clarity from higher levels of government can impede the development of local governance practices. Markets and the state are two contextual factors that Agrawal (2003) argues were overlooked by earlier work on natural resource governance.

The importance of gender is apparent in these cases. In all four sites women are highly involved in harvesting, but in Laos, Cameroon and Tanzania it is the men who are responsible for trading and thus have control over the income generated. Because increased income in the hands of women improves family well-being more than it does in the hands of men (Kabeer, 2003), if commercialization of NTFPs is to play a positive role in local development (as defined by the Millennium Development Goals), women must have control of the income. Under the right circumstances, there is also potential for commercialized species to play a role not only in increasing women's roles in community-based resource management, but also in enhancing their participation in local governance in general. Perhaps future work on the governance of *Pandanus* in Madagascar will emerge as a success story.

Interestingly, in all four cases, local people have quickly adapted to decreased availability of wild populations. In Madagascar they switch from one species to another, and in Laos and Cameroon increasing rates of cultivation and intensive management are reported (research on domestication of *Allanblackia* is taking place, but has not yet been introduced in Tanzania). All of these practices act to remove these NTFPs from the forest, turning them into cash crops and breaking the link between income generated by the species and the forest ecosystem from which they originated. Other authors also note a common trend towards cultivation of NTFPs, when economically viable (Chamberlain et al, 2004; Belcher et al, 2005). Once these species are 'removed' from the forest, they no longer provide monetary incentives for forest conservation. Belcher and colleagues (2005) argue that because commercialization of NTFPs most often leads either to overharvesting or to cultivation, which often results in the clearance of forests and loss of biodiversity, it is increasingly hard to maintain hope that such commercialization is compatible with conservation goals. Our cases support this conclusion and suggest that NTFP commercialization in many cases lacks the ability to catalyze community-based management and ecosystem conservation.

Timber Species

Trees are the predominant feature in most tropical forested landscapes, as is the case in our five sites. Whereas conservationists concern themselves with the ecological functions of trees and promote their importance for biodiversity, habitat maintenance, carbon sequestering, the provision of NTFPs and socio-cultural functions, local people value trees also as construction material, tools, firewood and income-generating timber (including income for future generations). Although much forest loss in the tropics is today driven by population pressure and need for agricultural land, logging operations and plantation development in primary tropical forests are responsible for a significant amount of the deforestation taking place, and there is a large market for tropical timber species. Here the roles of precious timber species in Madagascar and Tanzania are examined.

Case 8: Valuable timber in Tanzania

Milicia excelsa (Moraceae), locally known as *mvule*, yields some of the most valuable timber in East Africa. In addition to its importance as a timber species, *M. excelsa* provides firewood and is used locally for magic: 'Ashes [from burnt leaves] are placed on the skin so that you cannot feel pain when you are beaten. You can even be beaten to death but you will feel no pain', Amina Njiku explained.

This tree was originally common throughout forests and farmland of the East Usambara Mountains, but its slow growth rate, high value as timber and the long and extensive history of logging in the area have, in combination, led to overharvesting. *M. excelsa* is now on Tanzania's list of nationally conserved trees. Special regulations govern its harvesting, whether the trees are found on private, public or reserved land. To cut a tree, the forest committee of the village government must grant permission and a permit from the district forest officer must be obtained. For the most part, local villagers seem to understand the importance of these regulations; however, illegal harvesting of *M. excelsa* is common and not concealed. Corruption is likely to be an alternative, easier and cheaper route to obtaining the appropriate paperwork required to cut down *M. excelsa* trees.

In addition to the strict governmental regulations surrounding the tree, there is also a rich collection of traditional beliefs that may have historically mediated its use and conservation. One such belief, which seems to have originated long before any environmental education efforts in the area, is the association of *mvule* with the creation of thunderstorms. In the past the tree was avoided as firewood because it was believed that it would attract lightning to the house. Another belief tells that some *mvule* trees have 'stones' inside, created when the tree was hit by lightning. A further Shambaa belief is that the trees house ghosts: '*Mvule* is the place to put ghosts. A platform is built under the tree and eggs and ripe bananas are placed on it. They put a piece of red or black cloth [on the platform] ... people sing this song: *Dakwea kwemzue nadiona dinga dienge swamanga swamanga* [It is climbing the *mvule*, it looks like a bird, like a dove]'. Ramadhani Salim of Misalai said, 'After 7pm in the Kisiwani area, people stop going near *mvule* trees because it is the time when ghosts come out'.

Case 9: Timber species in Madagascar

Because all forests in Madagascar are state property, logging activities require governmental permission. In remote areas, however, because of administrative difficulties, governmental regulations are subsidiary to customary laws, which give open access to timber. Virtually all forests that are less than two hours' walking distance from rivers or roads are now gone. Woodcutters from outside the community can enter forests to cut and sell timber, and the community can do nothing to stop them. Because the remaining forest is now a long distance from roads or rivers, individual woodcutters take only small quantities of high-value species. The rarest and most valued species are *Diospyros* (ebony), *Dalbergia* (rosewood) and *Capurodendron* (nanto) (Onjanantenaina, 2009). Open access is leading to increased conflict over natural resource governance, predominantly centred on precious timber species. With decreasing forest cover people are increasingly aware of their dependence on forest products, which are becoming more difficult and time-consuming to obtain. As a result, people's interest in defending their rights over natural resources is increasing.

Precious timber species may play a role in catalyzing novel forest governance practices in eastern Madagascar. Increasingly, in areas with sparse forest cover, people are interested in forming associations to safeguard rights over forest resources. Unfortunately, they lack knowledge of formal governance rules, and even if the planned associations can be established, obtaining community-based control over use of forest products will remain a huge challenge. Clearly, traditional governance practices for forest products are not adapted for sustainable use or conservation. Existing practices may have worked in areas with significant forest cover and low population density, but the current population density and market pressure on resources cannot be sustained without severe forest loss and increasing conflict.[1] Traditional practices are being replaced by locally initiated modern governance arrangements, but the majority of the population does not yet respect these systems. If local communities are to govern their own local forest resources, longstanding customary laws and traditional practices throughout the whole region will have to be broken.

Tree rights and conflict between different levels of government

People are particularly possessive of their precious timber species, and thus, at least in our Madagascar and Tanzania sites, these trees hold strong potential to catalyse community action towards changes in forest management. Local people are keen to establish community-based tree governance, but this may lead to conflict with higher-level governments, which are often reluctant to decentralize control over precious timber and the associated economic benefits.

In these cases, we see a sharp contrast between the very restrictive governance of precious timber in Tanzania and open access in Madagascar. In neither case, however, are local communities obtaining much benefit from harvesting, and in both cases local people have very limited control over their trees. Decentralized, fair, equitable and corruption-free management of valuable trees will need to be established before true community-based landscape management can be realized.

Subsistence Species

While all communities welcome development that increases their monetary income, it is possible to achieve food security without market integration. Communities living in and around tropical forests have traditionally relied on the local environment and natural resources for food and livelihood security, often with considerable success. However, as dynamics change, population densities climb and new markets for forest products emerge, the balance is altered and local people's ability to meet their daily subsistence needs from their local ecosystems can be threatened. Case studies from Cameroon, Indonesia, Tanzania and Laos highlight these changing dynamics.

Case 10: Eru in Cameroon

Gnetum africanum and *Gnetum buchholzianum* (Gnetaceae) are species of woody vines used as leafy vegetables, both known as *eru*. After bush mango, the harvest and sale of *eru* for people in the Takamanda National Park is the second most economically important NTFP (Sunderland et al, 2002; Chapter 8), and because, unlike other leafy vegetables, it is available all year round, it is also a very important dish in local diets, particularly in the dry season.

Men, women and children are all involved in *eru* collection, but it is generally women who harvest and men who trade. There are no restrictions on harvesting *eru* from the wild, and unlike most NTFPs, harvesting *eru* from other people's private farmland is generally tolerated. Reportedly, *eru* harvesting in forests is favoured because there it generally has more mature and stronger leaves, qualities sought after in the market. Harvesting involves picking the leaves from the vine, which would seem highly sustainable. But as market demand increases, unsustainable harvesting techniques, such as uprooting vines, are increasingly common. Because there is increasing market demand

Figure 7.2 *Truck loaded with* eru

Source: Stella Asaha

for this product, there is a risk that local resources will be overexploited and will no longer support local subsistence and dietary requirements.

Case 11: Wild fruits in Indonesian agroforests

In Indonesia, a rubber agroforest cycle begins with the clearing of a plot from the forest. Rubber seedlings are usually planted directly between the stumps and the fallen trees. During this phase of establishment, some useful species are conserved, especially if localized on the borders of the plot. Spontaneous regrowth is retained only for species considered useful by the farmer. Sometimes, species of economic interest, such as *petai* (*Parkia speciosa*) or *durian* (*Durio zibethinus*) are planted between rubber saplings.

Rubber agroforests are less profitable than monoculture plantations, in terms of cash return to land and labour, but they remain in production for a much longer period than oil palm or monoculture rubber plantations (Feintrenie et al, 2010). Moreover, they provide livelihood security in the face of environmental, social and economic change. For example, when the prices of rubber decreased on the international market in late 2008, secondary products from the agroforests provided a crucial alternative source of income (Feintrenie et al, 2009) (see also Colfer et al, 1988, for an earlier incarnation of a similar, neighbouring system's seasonality, diversity and benefits).

An average plot of rubber agroforest is about two hectares and, in addition to rubber, contains useful plants such as fruits (nuts, pods or fleshy fruits), timber and handicraft materials (rattan and bamboo). The most frequent valued trees are *durian*, *petai*, *kabau* (*Archidendron bubalinum*), *jengkol* (*Archidendron jiringa*), jackfruit (*Artocarpus integra*) and cat's-eye (*Dimocarpus longan*) (Lehébel-Péron et al, 2010).

Some of the useful species are marketed to generate income. Increasing cultivation of *petai* and *durian* illustrates farmers' interest in income diversification. Farmers have begun to densely plant these two species between rubber saplings. Some conserved species are also marketed, such as *manau*, a thick rattan. However, not all the useful species conserved in agroforests are marketable; some are useful purely for subsistence purposes. Products such as *jengkol*, *kabau* (used for their beans), cat's-eye and jackfruit are not often marketed because their prices are low, yields highly variable and pest damage high, and thus the return on labour is minimal. Although farmers pay less attention to these non-commercial species, and as a consequence they suffer higher losses due to pests, the density of cat's-eye (two trees per hectare), jackfruit (one tree per hectare) and *jengkol* (one tree per hectare) in rubber agroforest plots is not much lower than the density of marketed species such as *durian* (four trees per hectare) and rattan (one plant per hectare) (Lehébel-Péron et al, 2010), indicating the importance of these species to local farmers. By conserving these subsistence species, farmers are maintaining diverse agroforests that act to preserve fruit tree cultivars, including some of the last wild representatives of non-commercial species (Michon et al, 1986), as well as maintaining dietary diversity crucial for supplying adequate nutrient intake.

Case 12: Mchunga in Tanzania

Launaea cornuta is a highly valued leafy vegetable in the East Usambara Mountains: it is the region's most famous food. Local people call it by its Shambaa name, *mchunga*, which means 'bitter'. The species grows wild in fields and disturbed areas across much of Africa. Knowledge of its use as a leafy vegetable has been recorded in many tribes, but its consumption is limited outside coastal regions of northern Tanzania and southern Kenya (Maundu et al, 1999). Here, women are proud of their knowledge of correct preparation methods; lack of this knowledge may be the reason it is less commonly consumed in other areas.

As is often the case with culturally important foods, there are a number of taboos on the use of *mchunga*. One of the primary taboos is against pregnant women consuming it. Other less widespread taboos, usually against eating all bitter vegetables, are often prescribed by witchdoctors to prevent or treat illnesses or curses.

In comparative nutrient composition studies, *L. cornuta* has been shown to have higher levels of many micronutrients (including iron, zinc and beta-carotene) than many other leafy vegetables consumed in East Africa (Lyimo et al, 2003; Msuya et al, 2008). Despite strong cultural importance, excellent nutritional composition and widely reported medicinal properties, *mchunga* is praised primarily for its superior taste. Josephina Lukindo told us, '*Mchunga* is better than [other leafy vegetables]. We eat it because of its excellent flavour and not because it helps bodies to grow.'

Because *mchunga* is so common, there seems to be a low risk of overutilization. In Kwatango village, *L. cornuta* is plentiful enough that women reported sometimes harvesting it in groups. *Mchunga* is sometimes sold on the market. Mary Vincent explained that, unlike in Kwatango, in areas closest to markets one cannot pick *mchunga* from someone else's field. Women are aware that their harvesting methods may affect conservation. Mwanahamis Ayub told us, 'You must instruct your daughter how to harvest *mchunga* because some harvest by pulling up the whole plant'. Although there is no evidence of a need to conserve *L. cornuta* itself, the agricultural and fallow land it grows on is important for landscape-based management and biodiversity conservation in the East Usambara Mountains. The cultural and nutritional importance of this and other traditional food plants may provide impetus to local people to conserve the ecosystems in which they grow.

Case 13: Bamboo in Laos

Bamboos, locally known as *mai pong*, are species of major importance in the Lao site villages. Two small-diameter species, *Cephalostachyum virgatum* and *Oxytenanthera parvifolia*, and a larger-diameter species, *Dendrocalamus hamiltonii*, are considered to be important subsistence resources. Bamboo's value is explained by its abundance and versatility of use (including construction, tools, firewood, handicrafts and food, and the sale of shoots and bamboo worms) and restricted access to manufactured goods in remote villages.

Small-diameter species are widely distributed, abundant, quick-maturing (harvestable stems regenerate in one year) and widely available in fallow fields and close to inhabited areas. Growing in large patches along shady streams, the large-diameter species requires more than two years to produce usable stems. The large-diameter bamboo is also the preferable habitat of bamboo worms, which obtain high prices on the Chinese market. Local people note a decrease in the availability of this species over the past ten years.

Traditional, gendered segregation of labour is well illustrated by bamboo collection, as noted by the Gender Resource Information and Development Center (GRID, 2004). Men often travel more than an hour to reach collection sites for stems necessary for tools, handicrafts, construction and fences. From July to December, women, often with the help of children, collect shoots for consumption as they return from work in the fields. On one trip they can bring back between 3kg and 5kg, which is enough to provide two meals. Women are also in charge of collecting dead stems as firewood.

There is a significant market for numerous bamboo products in Laos, but because of poor road access, people in the project site receive little benefit from their commercialization. Bamboo handicrafts (rice containers, small tables and tools) are marketed only at the local level. Large bamboo stems and edible shoots are sold directly to traders along the road. Bamboo worms are sold to NTFP traders, who then sell them to Chinese buyers. Unlike other bamboo products, bamboo worms are under the control of the district Agriculture and Forestry Office, which sets quotas and receives trading fees.

As a result of bamboo's multiple uses, decreased availability would significantly threaten local people's subsistence. Bamboo also provides environmental services, such as erosion control in intensively used land, water retention and water source maintenance, as well as playing an important role in the reduction of fuelwood use. Enhanced incentives for improved management of this plant and the landscapes where it grows are much needed.

Case 14: Fish in Laos

For upland communities in northern Laos, fish is a vital resource for subsistence. A week-long household food survey in one project village showed that fish are a major part of the local diet and an important source of protein, since domesticated animals are rarely consumed. Fish are also a source of income: small fish are sold within villages and bigger fish enter the dry fish market. Local 'reserve' and 'conservation' areas, organized by villages in accordance with government conservation policy, make fish the only wild resource for which there are currently formal, local management and harvesting rules.

In the three project villages, two rivers and occasionally small streams are the local sources of fish. Villagers note a significant decline in fish availability, which they attribute to increased harvesting pressure because of human population growth and increasing use of modern fishing technology (plastic nets, masks, etc.). Explosives and chemicals have been banned since early 2000, when government officers first introduced official fish conservation strategies to villages. A significant decrease in the size of fish caught has also been reported, with large fish becoming increasingly rare. People report that they have had to diversify the types of fish consumed to meet their subsistence needs.

Fishing is primarily seen as an activity for men, especially for large, marketable fish, although women and children also fish. Fishing is a daily activity and the average daily catch ranges from 1kg to 2kg, about 70 per cent of which is for family consumption. Sometimes, groups of three to six men travel long distances to fish in more lucrative sites; they camp in the forest while they catch and dry fish before returning.

Traders come from nearby towns to purchase fish. The fish sold within the village usually bring less than half the profit of that sold to outside traders. Information on amounts of fish obtained from conservation areas is available from village organizations (youth and women's unions), which monitor and record the weight and prices (if sold) of fish caught in the areas they control. These groups have a strong interest in ensuring sustainable harvesting in their specific sites, and these management efforts may be a starting point for the spread of more formal, community-based management for other resources and

Figure 7.3 *Boys looking for fish*

Source: Amandine Boucard

possibly entire landscapes. Multiple villages all depend on the same rivers and this may help catalyse higher-level, inter-village management policies. The national policy on watershed protection could support efforts to establish broader community-based management practices and integrate them with national and district-level policies. All villagers express strong concern over future availability of this essential resource, and many propose further management regulations to ensure that fish stocks can be maintained or recovered. There would certainly be community support for the development of further local management of this resource.

Meeting basic local needs

Most of the species important for subsistence discussed above are obtained primarily from agroforests, agricultural land and fallow land. In the Madagascar site, it was also noted that subsistence species are generally sought in forests during the frequent periods of famine or when they are collected for sale (*eru* in Cameroon, fish and bamboo in Laos). Villagers in the Madagascar site note that, 'The products and species in forests exist during the whole year, and if people are suffering during crisis or famine, products can always be searched for, eaten or sold at any time' (villager, Manompana, Madagascar). Communities living in and around forests often face food security crises. For them, forests act as a reserve, providing security when households cannot obtain enough from agricultural and other livelihood activities. An emerging conclusion in conservation research is the need to include areas surrounding forest reserves – entire landscapes – in conservation practice and policy. The need to improve management of agroforests and agricultural land surrounding forest reserves is increasingly recognized as essential not only for the ecosystem but also for local people's livelihoods (see Chapter 2).

The importance of gender is highlighted again here: as in most instances, it is women who are responsible for the collection and provision of most food species, although they are often excluded from land-use and management decisions.

Of all the case studies presented here, local people seem to be most aware of changes in the availability of species that they rely on for subsistence. An excellent example of this is the strong concern expressed by local people in Laos over declining fish availability. Compared to species sold as NTFPs, there seems to be greater awareness of and more concern for sustainable harvesting techniques and practices of subsistence species. For example, compare the discourse around the medicinal NTFP *peuak meuak* with that of fish in the Lao site. Another example is the high level of awareness of sustainable harvesting methods for *mchunga* in Tanzania, even though no decrease in availability has been reported.

These trends support the growing body of work demonstrating that common resources can be, and in many cases already are, sustainably managed by local communities – most famously argued by Ostrom (1990) in her analysis of factors contributing to successful local management for common pool resources; also see Agrawal (2003). Moreover, subsistence species seem to hold potential for catalyzing novel management practices (such as in the case of fishing in Laos) and may prove key to modifying people's use and management of unprotected communal land, their own farms and entire landscapes.

Insights and Conclusions

The case studies from the five Landscape Mosaics project sites presented here help to reinforce many common beliefs and assumptions about forest and biodiversity governance, but challenge other widely held views.

Many species of megafauna are reported to have strong cultural and religious beliefs attached to them; however, the ability of these beliefs to help shape modern governance practices remains to be demonstrated. Other species represented in our case studies here are not associated with traditional beliefs; in particular, many commercialized NTFP species, newly important to people's livelihoods, lack any customary management practices or beliefs upon which new management practices could be built. In these situations, the existence or creation of strong governmental institutions, allowing for the rapid development and dissemination of new management practices, seems to hold significance (e.g. *msambu* in Tanzania).

Multidisciplinary and holistic approaches to community-based governance are increasingly promoted. Identifying existing local management strategies is seen as central to fostering approaches that build on local knowledge and institutions. The search for species with the capacity to enhance existing governance or motivate the development of novel practices has led to many interesting insights. From studying the cases presented, it appears that commercialization of NTFPs falls short of expectations. Although insecure land tenure can reduce local motivation to sustainably manage commercial species, the case studies here also suggest that, because local people quickly adapt to decreases in wild availability of many NTFPs, the importance of forests as a source of these products may be limited. Not only does the commercialization of NTFPs seem to lack anticipated benefits, but in cases where commercialized species are also vital for subsistence, it can also lead to reduced availability for local use (*eru* in Cameroon and fish in Laos are examples). The risks associated with NTFP commercialization, especially the effects on the livelihood and food security of the communities' poorest members, are laid out in detail by Belcher and colleagues (2005).

Many of the timber and subsistence species examined here hold the potential to stimulate novel local management regulations, possibly even leading to changes in landscape-wide practices. Local people seem particularly defensive of their precious timber. Whereas in Tanzania precious timber species are highly regulated, in Madagascar there is a complete lack of effective regulation. Yet in both sites timber species are proposed to have the potential to propel changes in management practices, and in both sites there is significant friction between various levels of government over devolution of rights.

As noted in Chapter 2, the landscape governance approach highlights the dual importance of social and natural environments. Location within the landscape is an important determinant of the interactions between people and their environment. The use, management and governance of each species vary within each of the project sites, sometimes with significant differences from one village to the next. In the case of *peuak meuak* in Laos, collection and sale vary greatly among villages because of road and market access. Likewise, people in Madagascar's villages with greater forest cover report less difficulty obtaining *Pandanus*. In Tanzania and Indonesia, locals in areas closer to reserves report greater conflict with local wildlife. Remoteness often means less market access, and greater likelihood that wild species from these areas will be used for subsistence (as seen in people's reliance on bamboo tools in Laos). In villages closer to roads, the market demand for wild species tends to be greater.

Location also influences what level of governance is involved and thus how a species is managed. The resource management of more remote villages is more likely to be based on traditional systems and less likely to conflict with higher levels of governance, as seen in the lack of regulation of precious timber species in Madagascar and fish in Laos. Recognizing the role of place will often mean that remote sites and more accessible sites will have different management issues.

Management regimes are affected by multiple levels of governance; this has a significant role in how wild species are used and managed. As we have seen, gorillas in Cameroon and large mammals in Indonesia are subject to national laws regulating their use and management, which at times conflict with the practices of villagers. More commonly, governance practices involve contestations within villages, among villages, and between multiple levels of government (*peuak meuak*, *Allanblackia*, *Pandanus*, fish, timber in Tanzania and Madagascar). This is less the case when the resource is located on lands considered private. Market demands, sometimes driven by distant global markets, visible on the landscape through trade, add a further dimension to the governance of landscapes.

When the entire landscape is considered, it seems that some types of species have greater potential to change local people's management of the land. Subsistence species presented here are virtually all obtained from agroforest, agricultural and fallow land, and may have the capacity to modify how local people manage land surrounding forest reserves. For example, the conservation of spontaneous subsistence species in Indonesia leads to more diverse agroforests. NTFPs are in some cases easily detached from the forests themselves, as people adapt to changing availability in the wild. Finally, although megafauna occur throughout different landscapes, their roles in local people's livelihoods change depending on the context: in agricultural landscapes these animals are pests, whereas in the forest they are bush meat. Conflicts between different land uses will continue to grow in years ahead, as population densities around the five Landscape Mosaic project sites increase still further. As local people seek to provide for their families and develop, balancing livelihoods and conservation in a participatory manner will be a complicated task. The case studies presented show that in most sites local people already participate or would like to participate more in the governance of their local resources and landscapes.

Note

1. Editor's note: see Keller (2008, 2009) for another view on what is an often acrimonious debate on the relationships among population, culture and natural resources/biodiversity.

References

Agrawal, A. (2003) 'Sustainable Governance of Common-pool Resources: Context, Methods and Politics', *Annual Review of Anthropology*, vol 32, pp243–262

Asaha, S., Balinga, M. P. and Egot, M. (2006) 'A Socially Differentiated View of the Significance of Non-timber Forest Products for Rural Livelihoods in Cameroon and Nigeria', Final Socio-economic Report for ARRP (R7636/R7636E)

Belcher, B., Ruiz-Perez, M. and Achdiawan, R. (2005) 'Global Patterns and Trends in the Use and Management of Commercial NTFPs: Implications for Livelihoods and Conservation', *World Development*, vol 33, no 9, pp1435–1452

Chamberlain, J. L., Cunningham, A. B. and Nasi, R. (2004) 'Diversity in Forest Management: Non-timber Forest Products and Bush Meat', *Renewable Resources Journal*, vol 22, no 2, pp11–19

Cocks, M. (2006) 'Biocultural Diversity: Moving Beyond the Realm of 'Indigenous' and 'Local' People', *Human Ecology*, vol 34, no 2, pp185–200

Colfer, C. J. P., Gill, D. and Agus, F. (1988) 'An Indigenous Agroforestry Model from West Sumatra: A Source of Insight for Scientists', *Agricultural Systems*, vol 26, pp191–209

FAO (Food and Agriculture Organization) (2006) 'Non-wood Forest Product Community-based Enterprise Development: A Way for Livelihood Improvement in Lao PDR', Working Paper 16, FAO, Rome

Feierman, S. (1974) *The Shambaa Kingdom,* University of Wisconsin Press, Madison, Wisconsin

Feintrenie, L. and Levang, P. (2009) 'Sumatra's Rubber Agroforests: Advent, Rise and Fall of a Sustainable Cropping System', *Small-Scale Forestry,* vol 8, no 3, pp323–335

Feintrenie, L., Lehébel,-Péron, A. and Pfund, J-L. (2009) 'Agrobiodiversity to Cope with the Crisis. A Case Study in Bungo District, Sumatra, Indonesia', Poster at Diversitas Conference, 13–16 October 2009, Cape Town, South Africa

Feintrenie, L., Chong, W. K., and Levang, P. (2010) 'Why Do Farmers Prefer Oil Palm? Lessons Learnt from Bungo District, Indonesia', *Small-Scale Forestry*, vol 9, no 3, pp379–396

Fisher, B. and Christopher, T. (2007) 'Poverty and Biodiversity: Measuring the Overlap of Human Poverty and the Biodiversity Hotspots', *Ecological Economics,* vol 62, no 1, pp93–101

GRID (2004) 'Gender, Forest Resources and Rural Livelihoods', Vientiane, Lao PDR: MAF/NAFES

Groves, J. L. and Maisels, F. (1999) 'Report on the Large Mammal Fauna of the Takamanda Forest Reserve, South West Province, Cameroon with Special Emphasis on the Gorilla Population', Report to WWF, Cameroon

Infield, M. (2001) 'Cultural Values: A Forgotten Strategy for Building Community Support for Protected Areas in Africa', *Conservation Biology*, vol 15, no 3, pp800–802

Kabeer, N. (2003) 'Gender Mainstreaming in Poverty Eradication and the Millennium Development Goals', IDRC, Ottawa

Keller, E. (2008) 'The Banana Plant and the Moon: Conservation and the Malagasy Ethos of Life in Masoala, Madagascar', *American Ethnologist,* vol 35, no 4, pp650–664

Keller, E. (2009) 'Who are "They"? Local Understandings of NGO and State Power in Masoala, Madagascar', *Tsantsa,* vol 4, pp76–85

Lehébel-Péron, A., Feintrenie, L. and Levang, P. (2010) 'Rubber Agroforest Profitability, The importance of secondary products', *Forests, Trees and Livelihoods*, vol 20, no 1, in press

Levang, P. and Sitorus, S. (2006) 'Rubber Smallholders Versus Acacia Growers: Socio-economic Survey of the Rambang Dangku Sub-district (South Sumatra)', IRD-CIFOR, Bogor, Indonesia

Lyimo, M., Mugula, J. K. and Temu, R. P. C. (2003) 'Identification and Nutrient Composition of Indigenous Vegetables of Tanzania', *Plant Foods for Human Nutrition,* vol 58, pp85–92

MAF (Ministry of Agriculture and Forestry) and STEA (Science Technology and Environment Agency) (2003) Biodiversity Country Report, Vientiane, Lao PDR

Maundu, P. M., Ngugi, G. W. and Kabuye, C. H. S. (1999) 'Traditional Food Plants of Kenya', Kenya Resource Centre for Indigenous Knowledge (KENRIK), Nairobi

Michon, G., Mary, F. and Bompard, J.-M. (1986) 'Multistoried Agroforestry Garden System in West Sumatra, Indonesia', *Agroforestry Systems,* vol 4, pp315–338

Msuya, J. M., Mamiro, P. and Weinberger, K. (2008) 'Iron, Zinc and β-carotene Nutrient Potential of Non-cultivated Indigenous Vegetables in Tanzania', Proceedings of the International Symposium on Underutilized Plants 2008, ISHS, *Acta Hort.* 806

NAFRI (National Agriculture and Forestry Research Institute) (2007) 'Non-timber Forest Products in the Lao PDR: A Manual of 100 Commercial and Traditional Products', NAFRI, Vientiane, Lao PDR

Nicholson, K., Ketphanh, S. and Sengdala, K. (2008) 'Review of Experiences in the Marketing, Production, Harvesting and Management of Agro-biodiversity and NTFP Products', Mission Report for SDC

Oates, J., Sunderland-Groves, J., Bergl, R., Dunn, A., Nicholas, A., Takang, E. et al (2007) 'Regional Action Plan for the Conservation of the Cross River Gorilla *(Gorilla gorilla diehli)*', Arlington, VA: IUCN/SSC Primate Specialist Group and Conservation International

Onjanantenaina, C. (2009) 'Etude de la Viabilité et de la Vulnérabilité des Habitats de Quelques Espèces Endémiques dans le Corridor Forestier de Manompana en Vue de la Conservation de la Biodiversité' [Study of the Viability and Vulnerability of the Habitats of Several Endemic Species in the Forest Corridor of Manompana, in View of the Conservation of Biodiversity], Travail d'Ingéniorats, Université d'Antananarivo, Antananarivo

Ostrom, E. (1990) *Governing the Commons: The Evolution of Institutions for Collective Action,* Cambridge University Press, New York

Rappaport, R. A. (1968) *Pigs for the Ancestors: Ritual in the Ecology of a New Guinea People,* Yale University Press, New Haven, Connecticut

Ruffo, C. K. (1989) 'Some Useful Plants of the East Usambaras', in A. C. Hamilton and R. Bensted-Smith (eds) *Forest Conservation in the East Usambara Mountains,* IUCN Tropical Forest Programme, World Conservation Union, Gland, and Forest Division, Ministry of Lands, Natural Resources and Tourism, Dar es Salaam

Sarmiento, E. E. and Oates, J. F. (2000) 'The Cross River Gorillas: A Distinct Subspecies, *Gorilla gorilla diehli* Matschie 1904', *American Museum Novitates,* vol 3304, pp1–55

Schmidt-Soltau, K. (2001) 'Human Activities in and around the Takamanda Forest Reserve', Unpublished Report for PROFA/GTZ, Mamfe

Sheikh, K. M. (2006) 'Involving Religious Leaders in Conservation Education in Western Karakorum, Pakistan', *Mountain Research and Development,* vol 26, no 4, pp319–322

Sunderland, T. C. H., Besong, S. and Ayeni, J. S. O. (2002) 'Distribution, Utilization and Sustainability of the Non-timber Forest Products of the Takamanda Forest Reserve, Cameroon', Consultancy Report for the Project 'Protection of the Forests around Akwaya (PROFA)', GTZ, Mamfe, Cameroon

Sunderland-Groves, J. L., Ekinde, A. and Mboh, H. (2009) 'Nesting Behavior of *Gorilla gorilla diehli* at Kagwene Mountain, Cameroon: Implications for Assessing Group Size and Density', *International Journal of Primatology,* vol 30, pp253–266

Tajoacha, A. (2008) 'A Market Chain Analysis of the Non-timber Forest Products (NTFPs) of the Takamanda Forest Area', MSc thesis, University of Dschang

Therville, C., Feintrenie, L. and Levang, P. (forthcoming) 'What Do Farmers Think about Forest Conversion to Plantations? Lessons Learnt from Bungo District (Jambi, Indonesia)', *Forests, Trees and Livelihoods*

Tress, B. and Tress, G. (2000) Second Draft for Recommendations for Interdisciplinary Landscape Research, Workshop 1, 'The Landscape – From Vision to Definition', in J. Brandt, B. Tress and G. Tress (eds) *Multifunctional Landscapes. Interdisciplinary Approaches to Landscape Research and Management,* International Conference on Multifunctional Landscapes, Centre for Landscape Research, University of Roskilde, Denmark, 18–21 October 2000

8
Governance and NTFP Chains in the Takamanda-Mone Landscape, Cameroon

Verina Ingram, Stella Asaha, Terry Sunderland and Alexander Tajoacha

Non-timber forest products (NTFPs)[1] provide a valuable contribution to rural and urban livelihoods (de Beer and McDermott, 1996, Nepstad and Schwartzman, 1992; Prance, 1992; Colfer, 1997; Prance, 1998; Shanley et al, 2002; Belcher and Schreckenberg, 2007; Shanley et al, 2008; Paumgarten and Shackleton, 2009). Particularly in areas that lack basic infrastructure and market access, these products provide food, materials and medicines, and are often culturally important (Sunderland et al, 2004). Their sale provides direct, and often the only, access to the cash economy (Arnold and Ruiz-Pérez, 1996; Ros-Tonen and Wiersum, 2005). In many instances, trade in NTFPs involves long and complex chains of beneficiaries (Belcher and Küsters, 2004).

As elsewhere in the tropics, many NTFPs are used in Cameroon, and trade in some species contributes significantly to both local and national incomes (Ambrose-Oji, 2003; Ndoye et al, 1997–98; Ndumbe, 2008), particularly for women, who are often the primary income-generators for the household (Ruiz-Pérez et al, 2002). In the Takamanda-Mone landscape, NTFPs are collected from forests of diverse status: protected areas, permanent forests, farmlands and timber concessions. They have also long played a vital role in terms of maintaining levels of subsistence and trade in rural livelihoods (Sunderland et al, 2003). The porous international border with Nigeria, coupled with significant markets for forest products there, results in considerable transboundary trade in both timber and non-timber resources (Malleson, 1998). Such trade can be seen in terms of 'value chains' that involve both direct actors, such as harvesters, processors, transporters, traders and final consumers, as well as indirect actors, such as regulators, service providers and support agencies. The continuum of regulation in the market chains, from open-access to highly regulated and institutional arrangements, influences the social, economic and environmental impacts of forest exploitation.

However, how is this trade regulated and who ultimately benefits? This chapter reviews the value chains of NTFPs harvested in Takamanda-Mone and explores ways in which to improve customary and statutory regulations to emphasize equity in revenue sharing and better resource management, in order to benefit both local livelihoods and sustainable land-use practices.

The Takamanda-Mone Landscape

The Southwest Region of Cameroon is an ecologically diverse area, with a coastal mangrove swathe along the Gulf of Guinea, the highest mountain in West-Central Africa (Mount Cameroon, 4095m), and extensive areas of humid tropical forests in the lowlands. The importance of the region's biodiversity is reflected in the fact that 11 per cent of Cameroon's 174 protected areas, including three national parks, Takamanda, Mount Cameroon and Korup, are located in this region, even though it covers just 5 per cent of the national territory. The fertile volcanic soils throughout much of the region have encouraged population growth and in-migration for commercial and subsistence agriculture, with an overall population density of 35 to 42 persons per km^2 (MINPAT, 2000). The Southwest has three urban centres, but 69 per cent of people live in rural areas, with lower population density (six to 12 persons per km^2). The lack of transport, communication, social infrastructure and access to markets has led to a heavy reliance on forest products for livelihoods and a strong cultural affinity to the forest and its resources (Malleson, 1998).

In Cameroon, *landscape* denotes a legal entity and a spatial delimitation known as a technical operations unit (TOU). This is a multiple land-use classification devised by the Republic of Cameroon at the request of external donors, primarily the Global Environment Facility of the World Bank. Created by Prime Ministerial decree, a TOU is defined as 'a delimited geographical area, based on ecological, socio-economic, cultural and political characteristics for the enhancement of integrated landscape management involving all stakeholders' (Republic of Cameroon, 2006). There are, to date, six TOUs distributed throughout Cameroon's humid forest zone. The Takamanda-Mone TOU was established in June 2007 and encompasses an area of almost 444,000ha. It is located in the northern part of the Southwest Region, with a small portion (1160ha) falling under the jurisdiction of the Northwest Region (Figure 8.1). The TOU includes a matrix of land uses and allocations, ranging from strict protection to production. These include a forest management unit

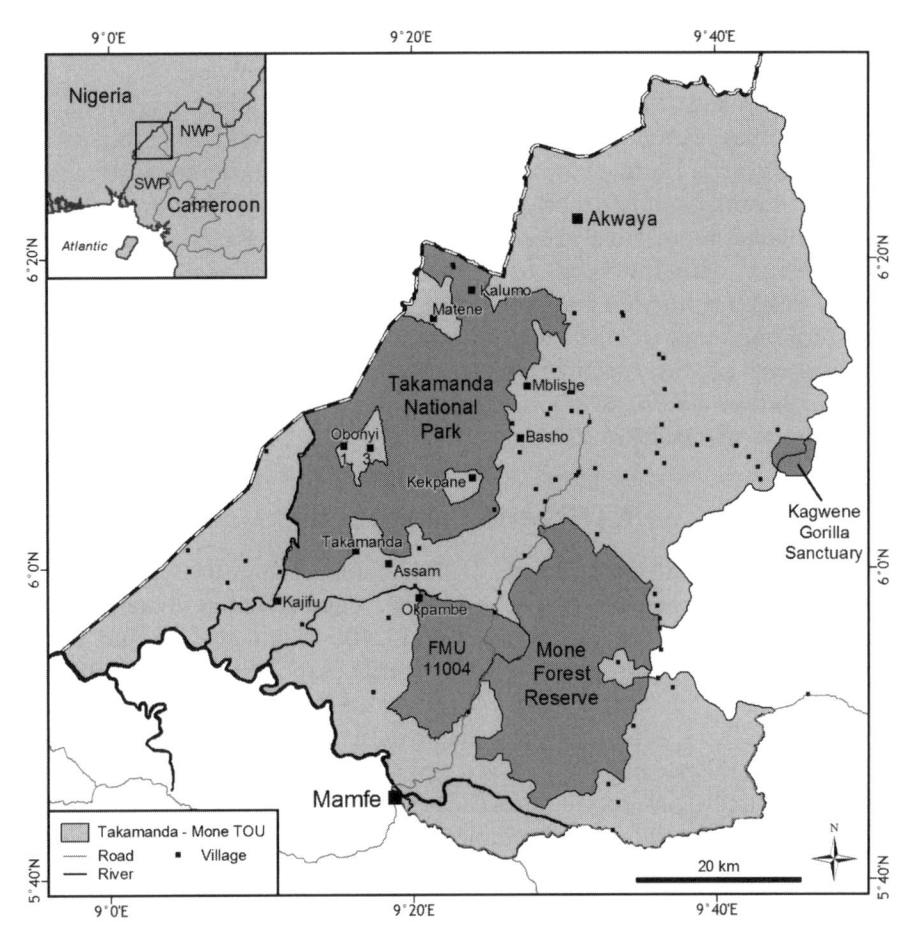

Figure 8.1 *Takamanda-Mone Technical Operations Unit, Southwest Region, Cameroon*

(i.e. logging concession); two protected areas, the Kagwene Gorilla Sanctuary (1900ha), established in April 2008,[2] and the Takamanda National Park[3] (69,599ha), formerly the Takamanda Forest Reserve, created in November 2008; and a production forest, the Mone Forest Reserve (55,872ha), introduced in 1941.

Recent developments include a Reducing Emissions from Deforestation and Forest Degradation (REDD) initiative, which aims to update the protection status of the Mone Forest Reserve. There are 64 villages in and around the Takamanda-Mone TOU, which, unlike elsewhere in Cameroon, has a relatively homogeneous population, with closely related ethnic groups linked by strong traditional ties (Sunderland-Groves et al, 2003a). The dominant tribe is Anyang, and the main spoken language is Denya. The majority of villagers, especially those living close to the Nigerian border, speak or understand the closely related Boki language, which is prevalent in Cross River State. Because of these ethnic links, Takamanda communities have a long-standing affinity, both ethnically and economically, with their Nigerian neighbours.

The Takamanda-Mone landscape is characterized by a high level of biodiversity (Comiskey et al, 2003) and significant levels of endemism across all taxonomic groups (Bergl et al, 2007). Of particular importance is the most endangered primate in Africa, the Cross River gorilla (*Gorilla gorilla diehli*), whose isolated populations are scattered across the landscape (Sunderland-Groves et al, 2003b; Bergl and Vigilant, 2007). Current landscape interventions are focused on establishing protected areas and maintaining ecological corridors to allow migration and gene flow for the Cross River gorilla, as well as other mammals, such as the Nigerian-Cameroon chimpanzee (*Pan troglodytes elliotii*), also restricted to this region, and the forest elephant (*Loxondonta africana cyclotis*).

NTFPs and Livelihoods

The rich biophysical environment of Takamanda-Mone provides natural resources and environmental services to local, regional and ultimately global beneficiaries. Cameroon harbours about 11,400 mammal, bird and plant species, of which 3000 are endemic to the region (Laporte and Justice, 2001). The uses of some 500 species have been recorded (Cunningham, 1993; FAO, 1997; Betti, 2002). About 50 per cent of these are for subsistence uses or traded locally; the remainder are traded more widely, with 25 per cent of these having national or international markets (Walter, 2001). This is a 'hidden economy': the value and scale of commercialization are unknown, even though some products have been traded across international borders for centuries (Falconer, 1990; Malleson, 1999).

Because of the strong 'extractivist' culture and heavy reliance on the forest for both subsistence use and cross-border trade, NTFPs generate an average 80 per cent of livelihood income in the region (Sunderland et al, 2003). Zapfack et al (2001) recorded more than 320 plants used in the area, as well as bush meat and fish, which contribute significantly to the regional economy. Seven plant-based forest products (covering eight genera and 11 species) were the most widely traded in 2001 (Table 8.1). A survey of harvesters, transporters and traders in Takamanda-Mone[4] indicates that the first five NTFPs listed in Table 8.1 continued to be the most heavily exploited NTFPs in 2008, for trade and domestic use, with at least 2500 people involved in their harvest and trade.

The route, consumers and volume of trade in these species has been heavily influenced by the inaccessibility of the landscape. Sunderland-Groves et al (2003a) emphasized that until the 1980s, most of the communities in Takamanda-Mone could be reached only by foot. Most products were therefore traded only locally and in nearby Nigeria. The development of logging roads in the early 1990s and more recent government-financed roads to Akwaya and Kajifu have improved access, with a corresponding increase in

Table 8.1 *Major NTFPs traded from Takamanda-Mone*

Common name	Local name	Scientific name	Life form	Ecological guild	Part used	Use
Bush mango	Kelua (Basho) Gluea (Anyang) Iweh (Ovande) Ogbono (Igbo) Bojep (Boki)	*Irvingia gabonensis* *Irvingia wombolu*	Forest tree to 30m	Closed-canopy forest; I. wombolu sometimes planted on farms	Seed Bark Timber	Condiments, medicine, cosmetics, construction
Eru	Gelu (Anyang) Ecole (Boki) Ukasi (Igbo) Ikokoh (Ovande) Ecole (Boki)	*Gnetum africanum* *G. buccholzianum*	Climber to 10m	Secondary forest and farm bush	Leaf	Vegetable, medicine
Njansang	Njansang (Oroko, common trade name in Cameroon) Ngoku (Basho) Itche (Becheve) Ngoge (Boki) Ngongeh (Anyang)	*Ricinodendron heudelotii*	Tree to 25m	Secondary forest and farm bush	Seed	Condiment
Bush pepper	Takuale (Basho) Kakwale (Ovande) Iyeyeh (Becheve) Ashoesie (Boki) Acachat (Anyang)	*Piper guineensis*	Climber to 10m	Secondary forest and farm bush; sometimes cultivated	Seed Leaf	Condiment
Cattle stick	Sanda (Hausa) Okah (Boki) Nyerem-mbe (Ovande) Essa (Nyang) Fesha (Basho)	*Carpolobia alba* *Carpolobia lutea*	Understorey shrub, grows to 10–12m	Closed-canopy forest	Stem	Walking and herding stick, construction
Chewing stick	Odeng (Boki) Pako (Yoruba) Egili (Ovande) Egili (Anyang) Feyili (Basho)	*Massularia acuminata*	Under-storey shrub to 12m	Closed-canopy forest	Stem	Dental hygiene, medicine, mortar stick
	Osun ojie (Boki) Okok (Efik) Aku ilu (Igbo)	*Garcinia mannii*	Tree to 20m	Closed-canopy forest	Stem	Dental hygiene
Bush onion	Eloweh (Ovande) Felou (Basho) Elonge (Becheve) Eloweh (Ovande) Elu (Anyang)	*Afrostyrax kamerunensis*	Understorey tree to 15m	Closed-canopy forest	Seed	Condiment, medicine

the harvest of many NTFPs by both local people and Nigerian traders. Over the past decade, such access, coupled with few formal controls, has led to a 'significant overexploitation of NTFPs' (Ebot, 2001, cited in Sunderland et al, 2003).

NTFP Value Chains in Takamanda-Mone

The trade in forest products can be viewed using a value chain approach that considers actors and activities both within and outside the landscape (Figure 8.2). These have direct impacts, through overexploitation leading to species mortality and changes in forest composition. However, they also have indirect effects, which can be seen, for example, in trade routes developed to export *Gnetum* spp. to Nigeria also stimulating imports of consumer products. Brief descriptions of the three major NTFP value chains follow.

Irvingia species

Two species of 'bush mango', *Irvingia gabonensis* and *Irvingia wombolu*, are found in the region. *I. gabonensis* fruits in the rainy season, from July to October; *I. wombolu* fruits in the dry season, from January to March. *I. gabonensis* is more common in closed-canopy forest and is sometimes found on farmland. *I. wombolu* is less common and less productive and hence brings higher prices during times of relative unavailability. Both species produce nutritious seed kernels, commonly used across Cameroon, Nigeria and most of coastal Central Africa as a condiment and a sauce thickener. The kernels are primarily sold whole, but also processed by grinding and crushing, and pressed into a cake for preservation and convenient use. *Irvingia* spp. are the most commonly harvested NTFPs in Takamanda-Mone, with the highest market value and largest household participation in harvesting and trading. In 2008, 46 tons were recorded from nine villages, with a total harvest estimated for the area of more than 1300 tons. Annual production varies widely, however; 2008 was a 'poor' year.

Harvesting *Irvingia* spp. from the forest is often a family affair, coinciding with school holidays. Often, fruits are collected, placed in heaps and allowed to rot for two to three weeks, then squeezed to remove the decayed mesocarp and reduce weight, and carried home to be cracked open with a stone hammer or machete. The fresh, creamy white cotyledons (kernels) are either sold immediately by individual households to passing traders or stored plastered on the sides of houses to dry further. Traders, in turn, sun-dry them for a few days and bag them for the market.

According to the Forest Law (1994), a permit is required for commercial harvest and trade in forest products, but not for personal consumption. Communities are entitled to collect forest products freely for their personal use, except for species protected by national legislation. On average, in 2008, each household harvested 70kg of *Irvingia* spp., of which 28 per cent was consumed

GOVERNANCE AND NTFP CHAINS IN THE TAKAMANDA-MONE LANDSCAPE | 189

BUSH MANGO
Quantity harvested: 655 tons

- Producers 540CFA/kg
  - 26.7% → Broker 990CFA/kg
    - 60% → Wholesalers 1450CFA/kg
    - 50% → Retailer
  - 13.3% → Wholesalers 1450CFA/kg
  - 22.2% → Retailer 3429CFA/kg
- Wholesalers 1450CFA/kg
  - 45.5% → Semi-wholesalers 5820CFA/kg
  - 9% → Retailer
- Semi-wholesalers 5820CFA/kg → 100% → Retailer 3429CFA/kg
- Retailer 3429CFA/kg → 100% → Consumers

ERU
Quantity harvested: 68 tons

- Producer 65CFA/kg
  - 11.1% → Brokers 55CFA/kg
  - 66.7% → Wholesalers 482CFA/kg
- Brokers 55CFA/kg → 100% → Wholesalers 482CFA/kg
- Wholesalers 482CFA/kg
  - 11.1 → Semi wholesalers 1750CFA/kg
  - 88,9% → Retailers 1560CFA/kg
- Semi wholesalers 1750CFA/kg → 100% → Retailers 1560CFA/kg
- Retailers 1560CFA/kg → 100% → Consumers

NJANGSANG
Quantity harvested: 5 tons

- Producer −1795CFA/kg
  - 75% → Wholesaler 2000CFA/kg
  - 25% → Retailer
- Wholesaler 2000CFA/kg
  - 71% → Retailer 470CFA/kg
  - 29% →
- Retailer 470CFA/kg → 100% → Consumer

BUSH PEPPER
Quantity harvested: 3 tons

- Producer −1909CFA/kg
  - 33% → Wholesaler 2460CFA/kg
  - 66.7% → Retailer
- Wholesaler 2460CFA/kg → 100% → Retailer 990CFA/kg
- Retailer 990CFA/kg → 100% → Consumer

CATTLE STICK
Quantity harvested: 168 tons

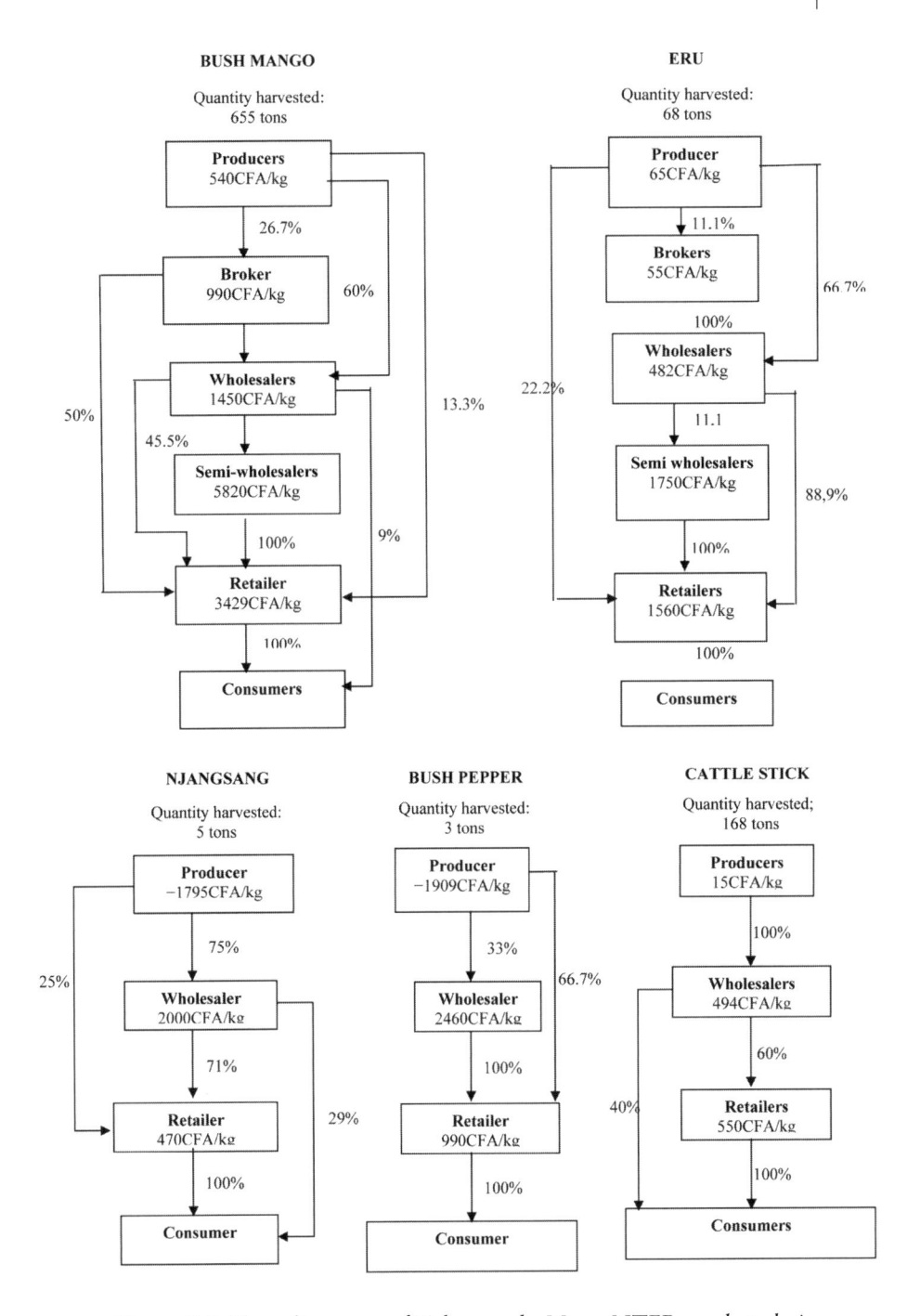

- Producers 15CFA/kg → 100% → Wholesalers 494CFA/kg
- Wholesalers 494CFA/kg
  - 60% → Retailers 550CFA/kg
  - 40% →
- Retailers 550CFA/kg → 100% → Consumers

Figure 8.2 *Flow diagrams of Takamanda-Mone NTFP market chains, values and profit margins, in CFA per kg, 2007*

and 58 per cent was sold (the rest perished). Despite the high proportion traded, *Irvingia* spp. kernels are not listed as a 'special forestry product'[5] by the Ministry of Forestry and Wildlife, which issues permits for trade and monitors exports of certain NTFPs.

In the Takamanda-Mone area, survey data indicated that more than 90 per cent of households are both collectors and consumers; there is little or no local trade. Consequently, there is no direct chain between producers and consumers; the chain has four or five different stages, as shown in Figure 8.2. More than 60 per cent of the *Irvingia* spp. harvest is sold to traders from Nigerian markets, and the rest to Cameroonian entrepreneurs. Because retailers are located far from producers and typically have little capital to pay for large quantities and their associated transport costs, wholesale quantities are typically small.

Gnetum species

The second most commonly harvested NTFP is a slender forest climber whose leaves are a staple vegetable. Although some harvesters and traders differentiate among species, many do not, and 'eru' is the collective term. *Gnetum buccholzianum* remains fresh for longer, without shedding its leaves, has a higher yield per vine, has larger leaves that are easier to slice, does not shrink after cooking and is considered to have a better taste; it is preferred by producers, traders, exporters and restaurant operators. In Takamanda-Mone, however, it is less common than *Gnetum africanum*, nearly 90 per cent of which is found in primary forest; this species has slightly narrower and elongated leaves. Four of the Landscape Mosaics villages produced 68 tons of *Gnetum* spp. in 2008, an average of 116kg per harvester, worth US$1297 per household that year.

In the forest, the leaves are collected by pairs or groups of up to five local women, youths or schoolchildren on holiday. The harvesters sell the majority of their product in the village, but also at rural markets and informal collection centres. Around 11 per cent of the collectors sell to brokers, 67 per cent to wholesalers and 22 per cent sell to retailers; none sell directly to consumers. Sales are opportunistic, related to the number of brokers visiting villages or in response to orders placed by buyers. Harvesters often feel they have no choice but to accept what they perceive as low prices offered by traders for their perishable leaves, even though the buyer may have travelled 20km, taking a whole day, to reach their remote village. A harvester's average profit margin is between US$0.48/kg and US$11/kg. Variations between *Gnetum* spp. and other NTFP margins are explained by the cost of production, with the distance to market being the most significant factor.

More than 70 per cent of female harvesters interviewed say *Gnetum* spp. are increasingly difficult to find, yet fewer than 10 per cent cultivate the species. Recently, the German Development Service, as part of its large-scale, long-term Programme for the Sustainable Management of Natural Resources,

Southwest Region, provided expertise on cultivation from cuttings and wild plants (the plant's long dormancy period and hard seeds make it difficult to propagate). Uptake has been low, however, and trial plots are often abandoned because farmers find that the method takes too long (the plants require two to five years to reach maturity), is too expensive (young plants need to be protected from livestock and thieves) and is not productive enough. Because *Gnetum* spp. are regarded as a forest product that anyone can collect, people remain reluctant to invest in cultivation, even as harvest opportunity costs rise with scarcity.

Buyers, referred to as 'buyam-sellams', are often the next link in the chain. They are predominantly Cameroonian or Nigerian men, and sometimes women, who are fairly well educated (more than 50 per cent have secondary-level education), an indicator of their business acumen. The majority specialize in *Gnetum* spp., although some combine it with farming or trade in other commodities and visit a circuit of villages by truck, bus or boat. Their biggest challenge is overcoming poor access via roads, which deteriorate significantly in the rainy season. This is coupled with what are euphemistically known as 'taxes' and checkpoints, where police and state authorities demand a range of financial settlements, depending on the size of the load, before the vehicle can pass. On average, a southwest trader earns US$1.5/kg, whereas a similar trader in southeast Nigeria earns US$4.12/kg. Some villages have local agents, known as 'buy and cuts', usually residents, who buy on behalf of wholesalers or buy to resell to wholesalers. When not sold immediately, *Gnetum* spp. are spread out in the shade for storage.

The fourth stage consists of transporters, both Cameroonians and Nigerians, who earn an average profit margin of between US$0.4/kg and US$0.51/kg, depending on the charges paid to the village councils and labour costs. There are also 'shredders', usually Cameroonian or Nigerian women in urban markets who slice and sell the shredded leaves directly to households and restaurant operators. About 70 per cent of shredders do this in combination with farming, and selling other foodstuffs, either daily in the local markets or on market days. Others involved in the *Gnetum* spp. trade are managers who negotiate purchases in nearby Nigerian markets of Calabar and Ikom on behalf of exporters, and traders in all the main markets in southwest Cameroon. In the markets are a host of casual labourers with descriptive names such as loaders, counters, tie-ers and waterers. Finally, the *Gnetum* harvest reaches the end consumers: 'mammies', who buy for their families, and restaurant operators.

In Nigeria, participants in the value chain include wholesaler-importer-buyers, their associations and retailers. No processing enterprises were identified here. Contrary to popular opinion in Cameroon, *Gnetum* spp. exported to Nigeria, with its large population, are mainly for consumption. Associations and unions act as oligopolies (few sellers, many buyers), regulating trade and competition so powerfully that individual traders are excluded.

These groups ensure that all members benefit from the trade equally. The associations are aligned with the federal government system, each state having its own registered association of *Gnetum* spp. dealers recognized by the Ministry of Agriculture. The retailer unions are female-dominated, whereas the transporters are male-dominated and also fulfil social, warehousing, security and financial functions. Because *Gnetum* spp. are part of a fast turnover market, with a large consumer population, most buying and selling between retailers, traders and transporters is by credit. High retail losses (perished and rejected stock) and consumers' refusal to pay a retailer's minimum reserve prices can mean that losses incurred by the retailer are eventually shared among the whole chain.

Gnetum spp. are classed as a special forestry product by the Ministry of Forestry and Wildlife, meaning a permit is needed for sale. None of the traders interviewed had such a permit and thus were operating illegally – hence the ability of the authorities to enforce systems of informal settlement, often known as 'dash'. More than 90 per cent of respondents in 2008 operated informally and/or were unaware of the special forestry product legislation. Surprisingly, given the open and well-documented trade going back at least a decade, permits for *Gnetum* spp. have rarely been issued annually, and when they are, the quantities are strikingly incommensurate with the historical volume of trade. For example, in 2008, a 50-ton quota and permit were given for the whole Southwest Region to one exploiter, yet an estimated 913 tons of *Gnetum* spp. was harvested from the Takamanda Forest Reserve in 2001 (Mdaihli et al, 2002) and at least 68 tons from four villages in 2007. The inconsistencies in rules and controls are largely due to general ignorance of the legal requirements for NTFP harvest and trade by producers, traders and government authorities, and the unmotivated and poorly organized government staff.

Ricinodendron heudelotii

This rapidly growing tree (known locally as *njangsang*), up to 30m high, is relatively common in secondary forest and farm fallows, where seeds germinate freely. Its seeds are harvested from April to June, with half collected from closed-canopy forest and the rest from secondary forest and farmland. Farmers select and nurture healthy seedlings when forest is cleared, and standing trees are often maintained on farms. The hard, trilobed fruits, which fall in the early part of the rainy season, are gathered into piles by about 68 per cent of households, mainly by women, and allowed to rot. The small, yellowish seeds are then removed and boiled. They are fermented for a week and boiled a second time to make them easier to crack open with a knife. The fragrant, oily seeds, a popular spice in many Cameroonian dishes, are then sun-dried and can be used immediately or conserved for a few years. The main traders of these seeds are women, who often sell a range of condiments on market days, directly to buyam-sellams (intermediaries and wholesalers). *Ricinodendron* seeds are a low-weight, small-volume, high-value commodity, with just over five tons harvested by five villages in 2007, on average 15kg a household. The means of transport include taxis, head loads and trucks. A large proportion (35 per cent) of this spice is consumed by the harvesters and a small quantity (9 per cent) perishes. Negative margins recorded are due to the high cost of labour involved in the processing of the product, which is often not factored into the selling price by harvesters.

Socio-economic Impacts

As the value chains for three NTFPs demonstrate, the route from forest to consumer across the Takamanda-Mone landscape is often long, complex and physically difficult. With the exception of *Irvingia* spp., which are in partial cultivation, a secure product supply is believed to be at the mercy of nature or god. Product seasonality, lack of access to market information and capital and 'dishonest' intermediaries are further significant problems that the surveyed harvesters confront in commercializing their products. Too few customers, yet insufficient quantities to meet variable market demand, as well as bad roads, are the principal problems named by traders. For transporters, bad roads, heavy taxes and illegal checkpoint settlements were seen as the major constraints.

Most NTFP harvesters sell their produce individually at the farm gate, counting on buyers to visit their village throughout the peak production periods. Although this is an easy and low-cost option for the harvester, it means villagers are price takers, accepting the price offered by buyers, and often have little knowledge of market trends beyond their village. The exchange is carried out in cash in Central African francs or Nigerian naira, with some trade by barter.[6] Most harvesters sell individually because buyers find small quantities affordable, the time between collection and sale is long, and the quantities sold vary from one harvester to another. These disparities – along with very weak producer profit margins – lead to variations in selling prices and break the chain of communication among producers, thus giving a comparative advantage to the well-travelled traders, who are more organized and better informed of market trends. Many of these traders enjoy the financial support of associations, which help set the purchase prices of NTFPs informally. Although the traders earn the highest profit margins along the NTFP chains (Figure 8.3), they also take the largest financial risks, as shown by the *Gnetum* spp. case, particularly with highly perishable products, and face the highest demands for bribes. Most involved in the value chains are self-employed and earn their income through physically intense labour, which results in a gender differentiation of tasks. The demanding nature of the work also means that poor health can have a devastating effect on income.

Most households depend on forest products (48 per cent) and farm products (41 per cent) for income; only 4 per cent of households are wage

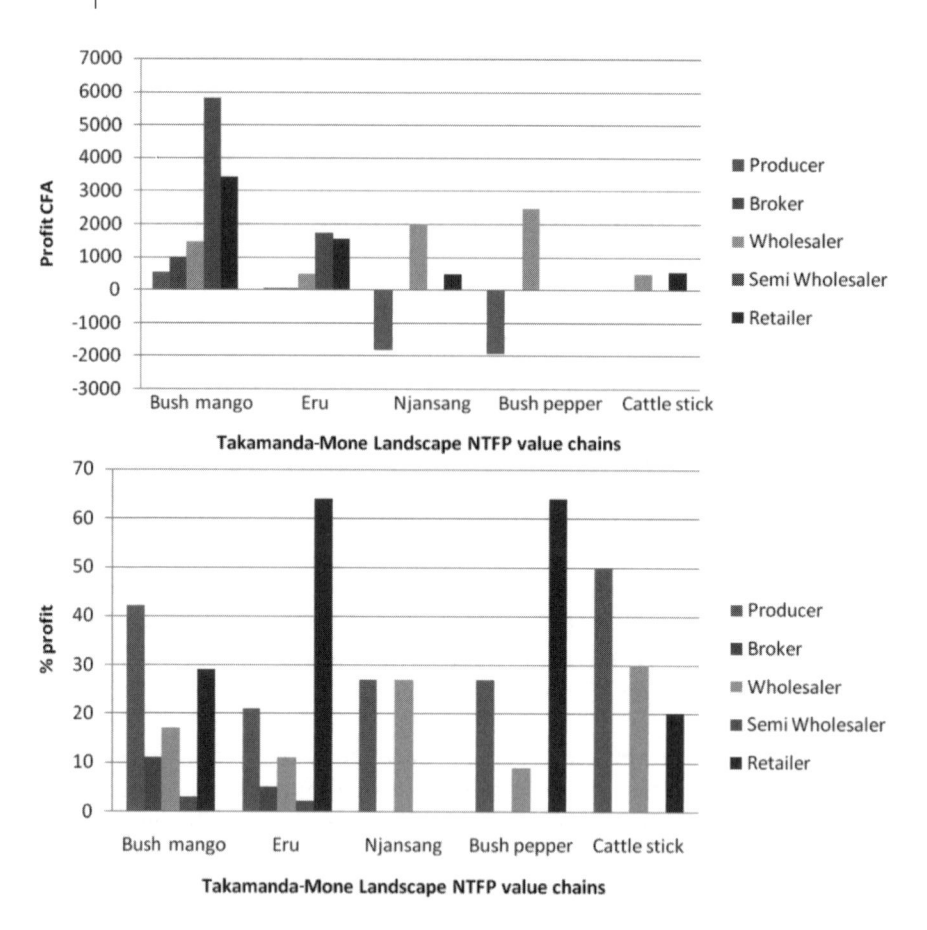

Figure 8.3 *Distribution of average profit and percentage profit margins in 2007 among surveyed NTFP market chains in Takamanda-Mone*

earners, mainly teachers and pastors. Nearly all households grow perennial cash crops, such as oil palm, cocoa and some coffee, and food crops, such as ground nuts, melon seeds, pepper, some rice, cassava, cocoyam, maize and plantain. Around 10 per cent of people depend partially or entirely on family remittances (Asaha and Fru, 2005). In general, more than 60 per cent of households report an annual income of less than US$234. This amounts to US$0.65 per day and indicates the level of (cash) poverty in the area. 25 per cent of *Irvingia* spp., *Gnetum* spp. and *Ricinodendron* harvesters earn between US$234 and US$469 annually. Except for the specialist cattle stick (*Carpolobia* spp.) harvesters, who have incomes ranging between US$46 and US$700, the majority of households – 'wealthy' and 'poor', men and women – harvest *Irvingia* spp. Many households also specialize in particular NTFPs. Some have made considerable incomes through the harvesting and trade of *Irvingia* spp. and *Gnetum* spp. The sales methods, seasonality of NTFPs and distance from

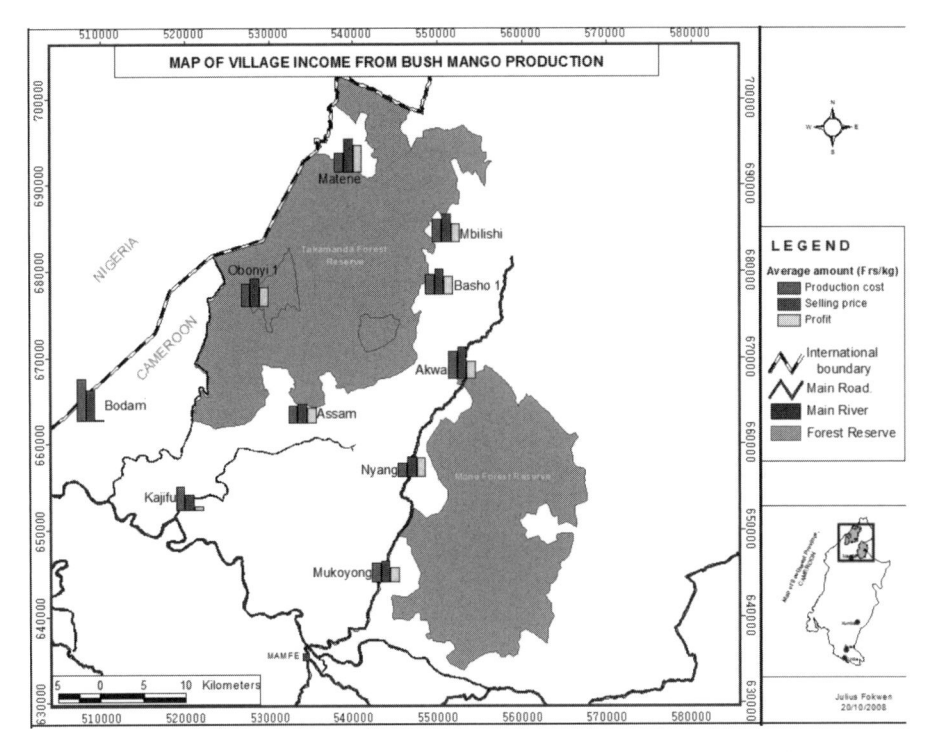

Figure 8.4 *Production cost, selling prices and profit margins from bush mango in Takamanda-Mone, 2008*

market create wide variations in the value of the annual contribution of NTFPs between villages, with average values for *Irvingia* spp. being US$204 to US$512, and for *Gnetum* spp., US$0 to US$105. *Ricinodendron* and *Piper guineensis* (bush pepper) are less profitable, generating an average US$37 to US$65 and US$28 to US$43, respectively (see Figure 8.3).

The variation among traders' margins, as highlighted in Figures 8.4 and 8.5, can be explained by remoteness, since the distance from production village to end market means higher transport costs. For example, an *Irvingia* spp. kernel trader from the southwest makes an average profit margin of US$1.8/kg, whereas a similar Nigerian trader makes US$10.2/kg. A Cameroonian *Gnetum* spp. trader earns on average US$1.5/kg, and a Nigerian earns US$4.12/kg. Transporters earn an average profit margin of between US$1.28/kg and US$-0.51/kg for *Irvingia* spp., US$-0.01/kg and US$-0.21/kg for cattle stick, and US$0.49/kg to US$-0.51/kg for *Gnetum* spp., depending on the charges paid to the village councils and the labour cost incurred.

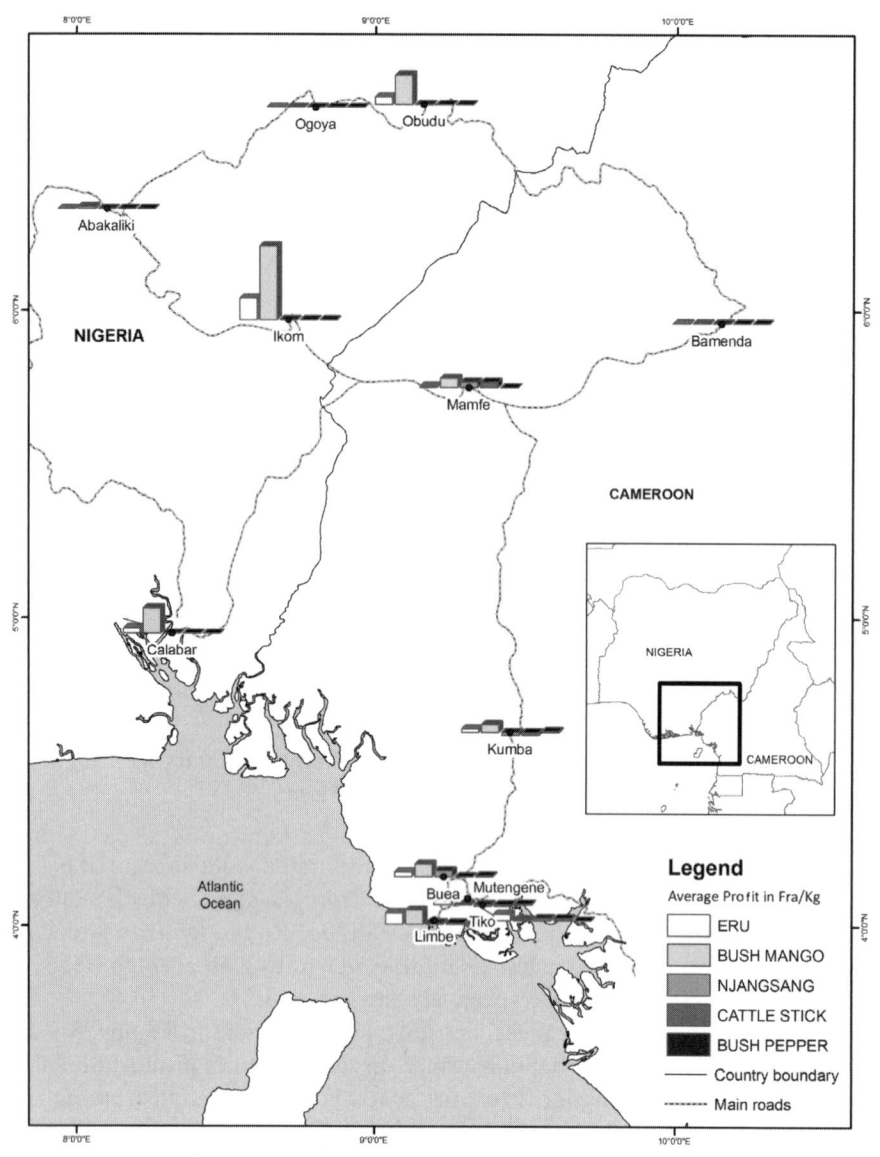

Figure 8.5 *Average profit of NTFPs per trader in surveyed Takamanda-Mone markets, 2008*

Governance of NTFPs?

Governance here means the process of policy-making, implementing and monitoring the allocation of forest land, resources and products. At a landscape level, it can be assessed in terms of (1) interactions within and between the state, civil society and private sector; (2) institutions for economic, political and social affairs, their policies and the principles guiding their interactions; and (3) cultural values.

Interactions: NTFP value chains

Value chains provide a good framework for assessing the interactions between state, civil society and the private sector. These include economically active individuals, organizations, small enterprises and associations.

How value chains are governed has critical implications for how incomes and benefits are distributed. Authority, power relations and organizations largely determine whether a chain will alleviate poverty (Gereffi et al, 2004; ILO, 2006). Governance arrangements also influence the social, economic and environmental effects of forest exploitation (Ingram and Bongers, 2009). Value chains are not closed systems but parts of larger institutional frameworks. They receive external inputs, such as knowledge from technical research institutes and extension services, and they are influenced by advocacy groups including trade associations and NGOs, as well as by government or international policy. They are also affected by social institutions, such as the level of organization of producers and traditional hierarchical relations. These institutional frameworks may provide effective channels that strengthen value chains or create barriers that block exchanges along the chain, particularly harming the harvesters and those at the beginning of the chain. A third aspect, the structure, complexity and spatial dispersion of inputs and outputs of the interrelated value-adding activities (including processing and marketing), is also an important variable.

Highly governed chains can reduce production costs, increase quality, and provide information to improve skills, production flows and the distribution of gains along the chain. This is an opportunity for policy initiatives and technical assistance (Humphrey and Schmitz, 2001). Thus the type of value chain governance can fundamentally determine the success of intervention strategies. ILO (2006) has highlighted how typical governance structures in chains affect governance. Many chains have a dominant player who determines the overall character of the chain and seeks to govern it. In the *Gnetum* spp. market, the well-organized Nigerian retailer and wholesaler unions play this role, acting as oligopolies that set and enforce terms under which other actors in the chain operate. When such extensive control is exercised over parts of the value chain, the relationship is hierarchical. Understanding this powerful role and the relationships between leading institutions and local producers – and the opportunities and constraints that result from entering such relationships – is the key to change.

The cattle stick chain involves many interactions, but the relationships among actors are unequal. Using terminology from ILO (2006), such a network can be seen in terms of the following: (1) *modular relationships*, where suppliers offer services or products, sometimes in a package (e.g. processing *Irvingia* spp.); (2) *relational relationships* that are often complex interactions between buyers and sellers, and create mutual dependence and asset specificity (e.g. ethnic ties and social relations between cattle stick suppliers and a few buyers in a fixed market); and (3) *captive relationships* that are typical of situations where small suppliers are dependent on larger buyers (e.g. suppliers who prefer to sell to other buyers face significant transaction costs and are therefore captive). *Irvingia* spp. and particularly *Gnetum* spp. harvesters, who remain in the villages, are dependent on the larger buyam-sellams for transport and sales, and if not sold quickly, the total harvest of *Gnetum* spp. can be lost. Many chains, however, are not governed by powerful leading enterprises or economic operators. For *Irvingia* spp., *Ricinodendron heudelotii* and bush pepper, there is little exchange of information and learning from others through buying and selling transactions. This type of value chain governance consists of *market-based relationships*, where the conditions of exchanging goods and services are negotiated daily, on the basis of the market price.

Institutions and policies: multiple and overlapping layers

The forest and its non-timber products in Takamanda-Mone are governed by an array of institutions and their policies. These include government organs, mainly the Ministry of Forestry and Wildlife, which regulates forest exploitation and protected areas; customary institutions, such as traditional councils with usufruct laws and customs; and forest-based businesses, such as timber companies, small enterprises and associations of traders, which have rules, codes of conduct and bylaws.

Government institutions

In the past two decades, Cameroon has developed one of the most advanced institutional and legal frameworks with supporting instruments to promote sustainable and participatory forest management in Central Africa (Topa et al, 2009). The institutional framework principally consists of the Ministry of Forestry and Wildlife and the Ministry of the Environment and Nature Protection, and is implemented through the Forestry and Environment Sector Programme (2003), developed by the government of Cameroon with international development partners. Other ministries, such as Agriculture and Rural Development, Lands and Land Tenure, Territorial Administration and Decentralization (responsible for the National Zoning Plan, which specifies land use), Commerce and Industrial Development, Trade and Small Scale Enterprise, Economy and Finance, Tourism, National and Higher Education, are also directly or indirectly involved. Forest policy emphasizes an integrated conservation and development approach, intended to ensure sustainable management of forest ecosystems. The Land Law (1974) of Cameroon distinguishes between private land that has been registered by individuals or entities and national domain land. Within the TOU, the land belongs to the state (Table 8.2); there are no examples of formal, private ownership of land. The difference between the law and common perception – that communities and individuals own their farmlands and forests – has been the subject of much unresolved debate (Assembe-Mvondo 2009).

Table 8.2 Legal classifications of Cameroon forests

	Permanent forests		Non-permanent forests		
	State forests	Council forests	Communal forests	Community forests	Private forests
Defining characteristics	Areas protected for wildlife: national parks, game reserve, hunting areas; state game ranches, wildlife sanctuaries, buffer zones, state zoological gardens. Forest reserves proper: integral ecological reserves, production forests, protection forests, recreation forests, teaching and research forests, plant life sanctuaries, botanical gardens, forest plantations	Forests classified on behalf of local council or forests planted by local council	Forests that are partly neither permanent forest estate nor private forest; do not include orchards, agricultural plantation, fallow land or wooded land adjoining agrosylvopastoral facility; and managed by forest and wildlife services. Derived products belong to state, except if management agreement had been signed with local community	Forest subject to management agreement signed between state services and local community; have single managed plan approved by state services. Derived products belong to concerned community; agreement may be terminated if law or agreement is violated, but population maintains logging rights	Forest planted by natural persons or corporate bodies on land they acquired legally; management plan drawn up in collaboration with state services. Certain products found in natural forest on private land remain state property (i.e. ebony, ivory, certain animal, plant and medicinal species of particular interest)
Governing property regime	De jure state property	De facto state property	De facto state property	De jure commons	De jure private property

Management of forest resources is conducted through a programme called the Sustainable Management of Natural Resources, Southwest Region, whose members are often in the field. The Ministry of Forestry and Wildlife implements this programme with financial support from the German Development Bank, technical conservation assistance from the Wildlife Conservation Society and the World Wide Fund for Nature, and support for development and the sustainable use of natural resources from German Technical Cooperation and the German Development Service.

Physically, however, representatives of many government institutions are all but absent in the landscape, on both sides of the border. Staff occasionally can be found in the towns of Mamfe and Mundemba, but generally it is rare to find officers at sub-divisional posts, where roads are few. Remoteness and the lack of adequately trained government staff, basic infrastructure and financial and logistical support hinder the implementation of much forestry legislation in both Cameroon and Nigeria. Staff are demoralized and disorganized, and taking bribes, or 'private settlements', has become standard practice. Confusion exists in the field about which government institutions are responsible for regulating NTFPs.

The Ministry of Forestry and Wildlife has established TOUs for the decentralization of decision-making. A TOU is headed by a conservator; administratively, it falls under a regional delegation and technically under the Forest Directorate. In theory, partnerships with local populations are established through rural forestry or village committees. Financial and administrative power and authority have not yet been fully transferred (Oyono, 2005), but the Southwest, with a high ratio of protected areas to land area, may be making more headway than most in the cultural change from a central to local government mentality. Village development committees have been set up in all the villages in the TOU zone to provide an interface between the population, timber concession holder and other partners. To date, however, there is no explicit framework for the TOU management or planning, although a minimal budget for administrative management and field functions, such as patrols, has been allocated by the government. Since its designation in 2008, attention has focused on the upgrade of the Takamanda Forest Reserve to a national park, rather than planning landscape-level issues.

The Takamanda reserve was established in 1934, but no management plan was ever developed, as is the case for the majority of Cameroon's protected areas. Donors, development organizations and international NGOs have played a strong role in setting up the TOU and protected areas – the change to a national park is one result. As a result there are a number of other institutional layers. The enhanced status for Takamanda means that a conservator should be appointed and guards posted to monitor and enforce a management plan, once such a plan has been written and adopted. The village management committees and community forests (adjoining the park) would also introduce new co-management arrangements. All these new institutions could strengthen governance arrangements and devolve power and authority, but may also increase complexity, cause coordination problems and create new opportunities for (mis)appropriation of power, authority and resources, both financial and forest assets. Experience from TOUs that have been set up in protected areas elsewhere in Cameroon (notably the Bamenda Highlands Forest Project) suggests that these multiple layers may not ensure long-term conservation protection or balance livelihoods needs beyond the project's life cycle (Abbot et al, 2001; WHINCONET, 2005).

The Ministry of Forestry and Wildlife's list of special forestry products[7] is seemingly arbitrary, containing exotic, endemic, endangered and commercial species that require a non-inventoried, quota-based permit for any trade. *Gnetum* spp. appear on this list, but none of the other NTFPs do, and most traders and transporters working in the Southwest do not have permits – both according to our survey and evident from the inconsistency between the number of permits and the quantities traded. Those few traders with permits were equally likely to be targeted by government and forestry officials seeking bribes to allow continued passage (Ndumbe et al, 2009; Ndoye and Awono, 2009). Even though large quantities of *Gnetum* spp. have for decades been exported daily to Nigeria (Clark and Sunderland, 1998) from both the Takamanda and other areas of the Southwest, Littoral and Centre regions, this trade is not reflected in the ministry's database for tracking special forestry products. The European diaspora markets, once small (Tabuna, 1998) and now probably larger, are also not reflected in official documents.

Table 10.2 shows Cameroon's low governance indicators and Table 10.3, its high corruption ranking, reflecting how bribery is part of everyday life.[8] Such corruption creates a level of informal 'governance' that affects how resources are exploited and trade conducted. It also adds irregular, uncontrollable costs and delays with few tangible benefits and no addition of value to the forest products.

Customary structures

At village level, the traditional council is considered the highest customary authority, headed by the chief. The council's influence on village people has traditionally been strong, as it is considered the traditional owner of resources and legitimate customary ruler over land use and resources; it controls access to NTFPs, particularly by outside parties. The council's role is to enforce customary law and order in the village and also serve as an administrative link between the village and the local administration. It sits with the village chief to deliberate issues of importance in the village. Although women are represented in the village council, they do not always have the power to oppose unfavourable decisions about income-generating resources. For example, in one of the villages the women are asked to pay a token fee to the council before harvesting *Gnetum* spp. for sale, whereas the men are free to hunt and sell without restrictions.

Many villagers belong to the powerful and respected sacred societies of the Anyang, Boki, Bayangi and Banyang tribes, the *ékpe* and the *makwo*.[9] These men-only societies originated with the decline of the Calabar kingdom in the 18th and 19th centuries, in the Cross River basin (Hacket, 1989). Each settlement has its own society 'house' called *ékpe* (leopard), which is the highest indigenous authority in that village. The *ékpe* lodges work alongside traditional councils to rule local communities while also managing regional and long-distance trade (Zapfack et al, 2001; Hacket 1989). Our surveys indicate that these cultural practices remain strong but are deteriorating among youths, who are becoming increasingly disaffected, linked to the societies' weakening with outmigration. The youths in some communities believe that they are being deprived of their rights by the stringent controls on the use of resources, especially timber.

In Boki villages, laws pass through the council, which then brings the matter to the whole village. The *ékpe* society upholds such laws; with a vigilante group acting as police. In Anyang villages, the most serious laws pass through the council to a spiritual shrine (*makwo*), then to the general village through the *makwo*. The *makwo* will then be the general overseer of the edicts. Both institutions solve cases of conflict (including access to resources and tenure) among community members and reinforce traditional council decisions. Traditional council decisions are, however, often criticized by community members who do not benefit from certain revenues; such institutional arrangements are a cause of internal conflict within many forest-edge communities where expropriation of forest resources by 'society elites' is common.

Tenure and ownership, under such customary rules, dictate that *Irvingia* spp., *Gnetum* spp., bush pepper vines and cattle stick bushes within the forest, are not owned by individuals or families, and access to the resource is generally on a first-come, first-served basis. About 99 per cent of NTFP harvesters did not acquire prior authorization from any regulatory authority before entering the forest to collect NTFPs. Only strangers from Nigeria coming into the forest reserve in Cameroon require permission, which they usually obtain from the village chiefs or the traditional councils (Mdaihli et al, 2002). Trees planted or maintained on farmland are owned by the landowner, and others are not allowed access without permission. As *Irvingia* spp. have increased in value, some people have begun to clear land around these trees in the forest; others claim trees hosting particularly abundant *Gnetum* spp. vines, in order to establish long-term collecting rights (Ndumbe et al, 2009). This extension of tenure through clearance usually relates to farmland, but resources from retained trees are also considered to be owned by the family that cleared the land.

When harvested by outsiders, natural resources such as NTFPs, particularly those of high value (see Table 8.1), are controlled and 'taxed' by the traditional authorities. Many communities require both outside and indigenous *Irvingia* spp. buyers to register in the village before they are permitted to purchase any material. There is no standard rate: in Kajifu a buyer pays US$11 for the whole season (US$7 for indigenes), but in Mbilishi, the cost is US$4.7 and an unspecified amount of palm wine for the council. Fees increased in

2001 by up to 50 per cent (Sunderland et al, 2002). Mdaihli et al (2002) reported that in 2001, neighbouring villagers could collect resources freely 20 per cent of the time, but paid compensation to the traditional council in 80 per cent of cases, and 25 per cent of Nigerians paid the traditional council and occasionally the chief or others directly. Sunderland et al (2002) note that for certain products, particularly those generating high incomes, there are well-defined and effective means of controlling access and ensuring benefits. For others, there are few, and their exploitation is characterized by inequity and poor levels of revenue generation, as discussed below.

For *Irvingia* spp., the most valuable resource, the primary customary control is that non-indigenes are discouraged from harvesting. Encroachment from Nigerian communities leads to considerable resource conflict between indigenous forest users and itinerant harvesters from the Nigerian side of the border. Because of such conflicts, the Mbilishe people began planting both species of *Irvingia*. Harvest is on a first-come, first-served basis, but in practice, the same families tend to harvest in the same area each year, based on the location of their bush houses; there are tacit acknowledgements of resource 'ownership' within most communities. Other traditional controls include prohibitions on felling individual *Irvingia* spp. trees under any circumstances, climbing trees and harvesting unripe fruit; the fruit may be harvested only after it is ripe and has fallen to the ground.

Many villages also have rules for the harvest and sale of *Gnetum* spp. Harvesting norms stipulate that only the leaves should be plucked and the stem must not be uprooted; observance of these rules contributes to regrowth. Although generally adhered to, Kajifu, Takamanda and Obonyi villages have reported problems controlling more destructive methods, such as tree felling. Additionally, in many communities women often prefer to harvest *Gnetum* spp. because of their profitability, to the neglect of farmwork. Outsiders are not permitted to enter the forest to harvest *Gnetum* spp.; benefit accrues to villages, as the resource 'owners'. Buyers and dealers from outside are also expected to register and pay taxes to the council; the tax varies from US$0.03 to US$0.40 per head load (about 1kg of leaves) for the transporter, to US$4.50 in Obonyi; they are also expected to accept a restricted purchasing period of four or five days per month.

Because cattle stick is not traded locally, villagers are often ignorant about its real value. This is reflected in the lax control on access: outside harvesters usually pay village councils for unlimited harvest rights, which cost around US$11 and two crates of beer for the Mfakwe council, and US$16 in Mbu.

Communities often have formal associations known as common initiative groups. These social structures are legally recognized by the government to promote economic development at the village level. Several groups are starting to form, largely promoted by external projects to assist in marketing schemes and obtaining funds. Informal associations known as *njangi* are also common; these are neighbourhood, youth, women's or men's groups that often have cultural and financial functions. None of them regulate forest use, but the credit and savings aspect is critical for the support of families and their NTFP-based activities.

Village forest management committees are recent institutions that combine customary with administrative rights. Logging companies are responsible for managing forest concessions, or forest management units, which are allocated under provisional three-year agreements. During this period, the concession-holder (Hall et al, 2000) prepares a 30-year management plan. Following its approval by the Ministry of Forestry and Wildlife, a renewable 15-year management plan on harvesting is developed. The local concession, FMU11004, was allocated to the company Seficam, and is exploited by Transformation Reef Cameroun under a temporary title. As of early 2010, the management plan was being prepared but was not yet final. Management plans are obliged to take into account areas that are important for communities' cultural or livelihood value, such as where NTFPs are harvested. The draft recommendations in the management plan stipulate that communities are conferred the right to use a 3km strip along the road for their agricultural activities within the forest management unit. Villages located inside the unit have the right to benefit from the logging company's improvements of local infrastructure (schools, health centres or roads).

Local communities also have the right to benefit from the 2007 forest royalties, 40 per cent of which should have been used by the Akwaya sub-division to conduct development projects for the 41 villages inside the forest management unit, and 10 per cent should have directly reached the communities since 2006. However, no development projects have been implemented to date and royalties have yet to benefit communities. Expectations are high that the company will build roads and employ villagers. Many communities are also mistakenly under the impression that they will receive cash, even though royalties are paid only to the council. Village heads have contested claims that these obligations have been fulfilled. The government-run newspaper, The Cameroon Tribune, reported on 23 June 2009 that the Forestry Administration convened a meeting of traditional rulers, the elites of 41 villages and Transformation Reef Cameroun to encourage them to 'bury their differences and turn a new leaf in forest management'. This 'Information and Sensitization Meeting on Forest Exploitation' was the first of its kind, creating village forest management committees to solve what was termed a 'misunderstanding among various actors in forest management' and to implement the participatory intention of the law. The local population and the logging company chosen by the state were named as custodians of the forest, with the government laying down rules, preparing the log book (*cahier de charges*), choosing the logging company, collecting taxes and supposedly giving forest rebates to beneficiary communities.

Irvingia spp. and *Gnetum* spp. traders' unions in Nigeria provide support to traders in times of need but do not set quotas or formal prices, which are market determined (Asaha et al, 2005). In response to domination of the *Irvingia* spp. trade by Igbo buyers, in early 2000 indigenes of Matene purchased *Irvingia* spp. within their communities for sale in Nigeria, in an attempt at greater 'indigenization' and resource control. Surveys in 2008 found that this practice was difficult to maintain. The revenues go to the village community fund, which is controlled by the village council, and are expected to contribute to community activities. This system is more effective in some areas than in others, depending on a village's culture of cooperation and benefit sharing. The benefits accrued from *Irvingia* spp. are more visible (e.g. complete construction of a village hall in Matene in 2002) than those of other NTFPs, despite high incomes generated by forest products (Sunderland et al, 2003).

Cultural values

Cultural values relating to the forest and its use are strong in the region, as could be expected in a landscape so heavily dominated by forest (Sheil and Wunder, 2002; Bengston, 1994). Consciously or not, the cultural framework encourages forest conservation, with secret areas established for the worship of various gods and as meeting grounds for the *ékpe* and *makwo* traditional societies. As described above, a range of rules relating to access, participation and use continue to exist.

Cultural values can be understood in terms of how current legal structures interface with customary management arrangements. The 1994 Forest Law defines user rights in the permanent and non-permanent forest domain and provides for customary user rights in permanent forest land, with the local population given the right to harvest NTFPs, wildlife and fish (except protected species) for their personal use. The current classification of national forest takes into account indigenous populations, who should retain their normal rights of use. However, these rights may be limited if they conflict with the aims of the protected area or a concession; if local user rights are restricted, the law provides for compensation.

During the gazettement of the Takamanda Forest Reserve in 1934, traditional rights were granted to the local populations to use the forest for their subsistence-based livelihoods and to pass legally through the area. Indeed, the route through the forest reserve is the basis of a strong cross-border trading pattern. Agriculture, hunting, fishing and gathering NTFPs were then, as now, widespread among the main ethnic groups present in the TOU, the Anyang and the Becheve.

The change in status from forest reserve to national park will, at the very least, imply a change in user and access rights. Validation of amendments in the proposed management plan is expected soon, and with this, a restriction of user rights. One of the challenges in managing Takamanda National Park is to elaborate management strategies that support conservation goals while securing continued, regulated access of local communities to natural resources. During the elaboration of the management plan, the resolutions on usufruct rights that had been approved during the previous consultations with local communities appear to have been disregarded. Until now, the restrictions proposed in the draft management plan are the result of a single consultation workshop with communities in December 2007. The current draft plan lists several restrictions, including a total ban on hunting and agricultural practices within the park, abandonment of current farms, removal of temporary and permanent camps, and a phasing out of the harvesting of NTFPs. The draft plan might provide for the relocation of the communities inside the park, although the required compensation costs make it unlikely. Notably, remote sensing data from 2008 Landscape Mosaics studies indicates that these communities have little impact on forest clearing, and boundaries are generally respected.

Sustainability of NTFPs

The *Irvingia* spp. trade has largely been perceived as ecologically sustainable (Ainge and Brown, 2001). The species are widespread in West and Central Africa, as well as in Takamanda, and not considered either threatened or endangered. Because of their many values, the trees are unlikely to be cleared from farm and fallow lands. However, across Cameroon the natural habitat for *Irvingia* spp., humid lowland forest, is being cleared for agricultural land and its products are often overexploited. In the Takamanda-Mone landscape, the current low levels of forest clearance are not yet a cause for concern, but since about 2000, rates of deforestation have increased. In light of that, and the increasing demand from growing urban populations in Cameroon and Nigeria, attention will need to be given to the status of *Irvingia* spp. Both countries have high average annual population growth rates – 2.7 and 2.8 per cent, respectively (UNDP/ARPEN, 2006) – and growing urban populations that cherish traditional dishes. The loss of wild *Irvingia* spp. trees would reduce the diversity of the gene pool from which trees with desirable market characteristics may be domesticated.

For *Gnetum* spp., however, concerns about unsustainable harvest practices and high levels of demand, particularly from Nigeria (Blackmore and Nkefor, 1998; Fuashi, 1997; Tanda, 2009; Nde-Shiembo, 1999; Shiembo, 1998), coupled with the very low levels of domestication (Blackmore and Nkefor, 1998; Tanda, 2009), are legitimate; the trade can be considered environmentally unsustainable. One indicator of unsustainable harvests is that 97 per cent of the producer population sampled reported that the distances they travelled to collect *Gnetum* spp. had increased in the past decade (Ndumbe et al, 2009). The same percentage of respondents observed that in the forest areas around their villages, *Gnetum* spp. had become less abundant; only 2.7 per cent had not yet observed any changes. Sixty-eight percent of respondents attributed the change to forest clearance for farmlands, and 25 per cent attributed it to the creation of palm plantations. This indicates that the rate of harvesting is above the natural regeneration rate, a conclusion confirmed by the fact that some villagers no longer harvest the leaves because of resource scarcity. This parallels the experiences in the Centre, East and Littoral regions (Awono et al, 2002; Blackmore and Nkefor, 1998; Fondoun and Manga, 2000).

For *Ricinodendron* seed and bush pepper, the small volume of harvest and the non-destructive harvesting technique, combined with few specific reports of decreased supply, imply that there are no sustainability issues for these species. Still, the lack of resource control, destructive harvest methods, and sequential harvest of cattle stick by all the study villages, indicates that harvest rates could be well above sustainable levels. Despite the vegetative and NTFP survey data from the Takamanda area (Sunderland et al, 2003), without a full resource inventory, it is impossible to quantify this impression.

Landscapes are dynamic, and the pattern and location of changes are rarely random. The recent major changes in land-use status of the Takamanda-Mone landscape – from customary, village-managed forests to forest reserves, and now a national park and a logging concession – have been purposeful. Although forest cover remains high, at 94 per cent of the total area, around 3 per cent of forest cover has been lost since 1980, increasingly replaced since 2001 by non-forest land cover and growing settlements along the major river and road routes, particularly in elevated areas. The drivers that determine the extent, pattern and location of such changes have themselves changed rapidly with infrastructure (new logging and public roads), policies (renewed support for the Cameroonian cocoa sector), economies (from local to global trade) and increasingly extreme climatic events. Some changes are due to long-term cycles within land-use systems, such as shifting cultivation, and some are 'permanent'. The land-use system in Takamanda-Mone is obviously location specific, and reflects local culture, history and responses to external factors, such as increasing urbanization and population growth, particularly in Nigeria, and demand for traditional NTFPs.

Improving Governance of NTFP Chains

The strong 'extractivist' culture and heavy reliance on forest products for subsistence and to supply cross-border trade mean that NTFPs contribute between 20 and 80 per cent of incomes in the nine villages surveyed in the Takamanda-Mone landscape. These products are also critical for at least 2500 people interlinked in the value chains in locations across Cameroon and into Nigeria. These NTFPs also provide significant livelihood benefits for the households and individuals involved in their trade and harvest. They contribute on average more than half of cash incomes, are consumed as food and in many cases provide additional benefits, from medicines to fuelwood. As the products move from the forest to the final consumer, they pass through an array, often simultaneously, of governance arrangements. Initially, these are traditional customary arrangements that focus on tenure, access and exploitation. The government regulations that exist at this level to control and monitor extraction are largely unenforced. The new governance arrangements for Takamanda National Park will have to provide better enforcement, plus alternative income-generating activities to compensate for the unexpected effects on livelihoods and conservation. Further along the chain, customary rights and rules become mixed with administrative regulations aiming to control quantities and trade. Here, informal arrangements and corruption add costs but do little to regulate products or competition.

Better customary and legal governance

The governance of forest products in the remote Takamanda-Mone landscape is formally the domain of the government, but in the effective absence of government intervention in the field, governance is largely by customary rules and institutions and by commercial actors, particularly Nigerian traders. Any increased government and donor intervention that focuses largely on the government actors and their rules and policies, and fails to capture and address the dynamics of actual governance arrangements on the ground for NTFP chains needs to recognize this. Therefore, regulatory changes affecting customary access and control (e.g. in the national park or timber concession) that do not have support from traditional structures and customary authorities are unlikely to work in this region, where a homogenous, traditional culture predominates.

Additional regulation of NTFPs, in this case, will probably be ineffective for improving either the conservation or livelihood profile of these species (Laird et al, 2010) unless government structures are given power and resources to fulfil their responsibilities and remedy the lack of transparency and accountability. The crucial usufruct rights to protect and manage resources must be recognized and strengthened, and responsibility to ensure that they continue to be maintained in the landscape should be devolved to the local level. Customary regulations are already the arrangements best recognized by many local inhabitants and are arguably the most effective in controlling harvesting practices, resource rights and the allocation of benefits within the community.

Informing the traditional governors and users of the long-term sustainability impacts of changing trade and increased demand could help them govern their resources better. Because the consequences of higher and unsustainable harvest are not often directly apparent, more awareness and better management options could help communities improve both livelihoods and conservation.

Mechanisms to counter corruption could also improve profits and resource management. The village councils, traditional authorities and government officials along the chain often siphon off fees for granting permission to outsiders, to collect, transport and retail goods. These create unpredictable levels of bribes, causing delays, stock losses and higher consumer prices.

Trade

The low level of vertical integration by producers into trading and retailing is both a result and a cause of the lack of market information. Increased integration could offer, for suitably entrepreneurial individuals or communities, improvements in prices and resource control. Sustainable trade – particularly if promoted by support actors – as a route out of poverty, appears largely unachievable unless it is accompanied by a programme of domestication. As we saw in the cattle stick and *Gnetum* spp. chains, the more valuable a resource becomes, the more likely it is to be overexploited until its local populations and, ultimately, trade are threatened. Many actors at the end of the chain, such as traders, are mobile, but harvesters are limited to the forest around their village. Their livelihoods are most likely to be harmed by the local loss of NTFPs unless the species are cultivated.

Improved equity

The differences in income at either end of the market chain are striking. The desire to 'make trade fairer' by bypassing the buyam-sellams and retailers and enabling harvesters to sell directly to consumers is unrealistic, however. Given their remoteness and lack of infrastructure, villagers cannot take this route to improve profits and need to recognize the high-risk and essential role of traders and retailers, who transport perishable products across difficult terrain, fending off multiple requests for 'dash'.

Equity along the chain could be improved, both for harvesters (particularly for high-value products like *Gnetum* spp., cattle stick and *Irvingia* spp.) and for traders, by providing information on tariffs, regulations, market prices and product availability. This could significantly improve governance arrangements, especially in hierarchical and network relationships. The high level of imperfect competition is largely related to the remoteness, lack of communication and the tendency of villagers to be 'price takers', with little access to information on prices in other areas. Greater equity for producers is critically linked to improving investment and access to capital and credit, which is essential to kick-start such information systems.

Trade is overseen primarily by statutory laws, although when these are inapplicable (for *Gnetum* spp. particularly, the laws are inconsistent and ambiguous), they create further opportunities for corruption. The overlapping customary, regulatory and protected area management systems to regulate the harvest and trade of NTFPs in the cross-border region are either inconsistent or insufficiently enforced.

More effective resource management

Extensive long-term data on the Takamanda-Mone landscape provide the basis for continued monitoring of indicators at the landscape and market chain level to ensure that livelihoods and conservation are better integrated in the landscape. Cameroonians can look at how Nigerians have organized products and services at all levels of the chain, improving prices, equitable distribution of profits, access to credit, security and market information. Improved resource management – whether by increasing quality through domestication, controlling access to wild resources or practising sustainable harvesting techniques – is absolutely essential if the market chain is to continue to provide for both the Takamanda-Mone communities and the dependent actors along the chain.

Payments for environmental services under schemes such as Reducing Emissions from Deforestation and Forest Degradation (REDD) are an opportunity to support both forest conservation in the protected areas and socio-economic development of communities. REDD may indirectly compensate for the loss in NTFP incomes and provide an alternative source of forest-based revenue for communities.

This remote landscape and the increasing trade in its valuable forest resources illustrate the dilemmas of collaborative governance on a landscape and product level. There is unfortunately no one magic bullet, nor is it feasible to pin all hopes on one strategy. Given the multiple actors and complexity of the governance arrangements, an array of recommendations, developed through collaborative processes, is needed.

Notes

1. 'Non-timber forest products are biological resources other than timber which are harvested from either natural or managed forests. Examples include fruits, nuts, latexes, resins, gums, medicinal plants, spices, wildlife and wildlife products, dyes, ornamental plants and raw materials such as bamboo and rattan' (Peters, 1994, p1).
2. Ministerial Decree 2008/1064/PM, 3 April 2008.
3. Ministerial Decree No 2008/2751/PM, 21 November 2008.
4. The survey was conducted between April and July 2008, using pre-tested questionnaires, of 55 (8 per cent of total) collectors in nine villages (29 per cent of total villages) in the Takamanda and Mone reserves and in the corridor. Following a rapid market survey, interviews were held with 66 traders (10 per cent of total) in seven Southwest markets and 43 traders (9.2 per cent) in five markets in Cross River State, Nigeria, selected according to the volume of NTFP commercial activities, number of sellers and market period. Twenty-seven (11 per cent of total) NTFP transporters from the local to main markets were interviewed. Focus group meetings were held in each village and local weights and measures calibrated. Data was analysed in XLSTAT.
5. Cameroon's 1994 Forestry, Wildlife and Fisheries Law, Section 9 (2), refers to 'certain forest products, such as ebony, ivory, wild animals, as well as certain animal, plant and medicinal species or those which are of particular interest and shall be classified as special'. Criteria or definitions of the terms 'certain', 'interest' and 'special' are not given. A 2006 decree defined 13 such products, including '*Gnetum* (Eru)'.
6. Exchange rate US$1 = 425.73 Central African CFA franc (CFA), as of 1 August 2008.
7. Décision No 336/D/MINFoF, 06 Juillet 2006 Fixant la liste des Produits Forestiers Spéciaux présentant un intérêt particulier au Cameroun. [Decision No. 336/MINFoF, 6 July 2006, Fixing the List of Special Forest Products of Particular Interest to Cameroon.]
8. Cameroon was ranked 141 of 161 countries in 2008 and was in the second-lowest category, with 2.2 on the 2009 corruption perception index; an overall score of 4 is considered 'very corrupt' www.transparency.org (also discussed in Chapter 10).
9. These are also the names for community halls in Boki ethnic areas, and *ékpe* is the Basho name for *Terminalia ivorensis*, a forest tree.

References

Abbot, J. I. O., Thomas, D. H. L., Gardner, A. A., Neba, S. E. and Khen, M. W. (2001) 'Understanding the Links Between Conservation and Development in the Bamenda Highlands, Cameroon', *World Development,* vol 29, no 7, pp1115–1136

Ainge, L., and Brown, N. (2001) 'Irvingia Gabonensis and Irvingia Wombolu', in *A State of Knowledge Report,* Central African Regional Program for the Environment, Oxford

Ambrose-Oji, B. (2003) 'The Contribution of NTFPs to the Livelihoods of the "Forest Poor": Evidence from the Tropical Forest Zone of Southwest Cameroon', *International Forestry Review,* vol 5, no 2, pp106–117

Arnold, J. E. M. and Ruiz-Pérez, M. (1996) 'Framing the Issues Relating to Non-timber Forest Products Research', in J. E. M. Arnold and M. Ruiz-Perez (eds) *Current Issues in Non-timber Forest Products Research,* CIFOR, Bogor, pp1–18

Asaha, S. and Fru, M. (2005) 'Socio-economic Survey of Livelihood Activities of the Communities in and around the Mone Forest Reserve and the Mbulu Forest Area, South West Province of Cameroon', in A Consultancy Report for the Wildlife Conservation Society/Cameroon-Nigeria Transboundary Project (WCS/CNTP), R. A. P. F. Forests, WCS, Limbe

Assembe-Mvondo, S. (2009) 'Sustainable Forest Management Practice in Central African States and Customary Law', *International Journal of Sustainable Development and World Ecology,* vol 16, no 4, pp217–227

Awono, A., Ngono, D. L., Ndoye, O., Tieguhong, J., Eyebe, A. and Mahop, M. T. (2002) 'Etude sur la Commercialisation de Quatre Produits Forestiers Non-Ligneux dans la Zone Forestière du Cameroun: *Gnetum* spp., *Ricinodendron heudelotii, Irvingia* spp., *Prunus african*a' [Study of the Commercialisation of Four Non-Timber Forest products in the Forest Zone of Cameroon], FAO, Rome

Belcher, B. and Küsters, K. (2004) 'Non-timber Forest Product Commercialisation: Development and Conservation Lessons', in K. Küsters and B. Belcher (eds) *Forest Products, Livelihoods and Conservation: Case Studies of Non-timber Forest Product Systems,* vol 1, Asia, CIFOR, Bogor, pp1–22

Belcher, B. and Schreckenberg, K. (2007) 'Commercialisation of Non-timber Forest Products: A Reality Check', *Development Policy Review,* vol 25, pp355–377

Bengston, D. N. (1994) 'Changing Forest Values and Ecosystem Management', *Society and Natural Resources,* vol 7, pp515–533

Bergl, R. A., Oates, J. F. and Fotso, R. (2007) 'Distribution and Protected Area Coverage of Endemic Taxa in West Africa's Biafran Forests and Highlands', *Biological Conservation,* vol 134, pp195–208

Bergl, R. A. and Vigilant, L. (2007) 'Genetic Analysis Reveals Population Structure and Recent Migration Within the Highly Fragmented Range of the Cross River Gorilla *(Gorilla gorilla diehli)*', *Molecular Ecology,* vol 16, pp501–516

Betti, J. L. 2002. 'Medicinal Plants Sold in the Yaoundé Market, Cameroon', *Africa Study Monographs,* vol 23, no 2, pp47–64

Blackmore, P. and Nkefor, J. T. (1998) *The Transfer of the Eru (Gnetum africanum, G. buchholzianum) Domestication Model to Village-based Farmers on and around Mount Cameroon,* L. B. Garden, Limbe, Cameroon

Clark, L. E. and Sunderland, T. C. H. (1998) 'A Regional Market Survey of the Non-wood Forest Products Traded in Central Africa', in T. C. Sunderland, L. E. Clark and P. Vantomme (eds) *Non-wood Forest Products in Central Africa; Current*

Research Issues and Prospects for Conservation Development, International Expert Meeting on Non-wood Forest Products in Central Africa, FAO, Limbe, Cameroon

Colfer, C. J. P. (1997) *Beyond Slash and Burn: Building on Indigenous Management of Borneo's Tropical Rainforests,* New York Botanical Garden, New York

Comiskey, J. A., Sunderland, T. C. H. and Sunderland-Groves, J. L. (eds) (2003) *Takamanda: The Biodiversity of an African Rainforest,* Smithsonian Institution's Monitoring and Assessment of Biodiversity Program, Washington, DC

Cunningham, A. B. (1993) 'African Medicinal Plants: Setting Priorities at the Interface Between Conservation and Primary Health Care', *People and Plants,* Working Paper 1, UNESCO, Paris

de Beer, J. H. and McDermott, M. J. (1996) 'The Economic Value of Non-timber Forest Products in Southeast Asia', Netherlands Committee for IUCN, Amsterdam

Falconer, J. (1990) 'The Major Significance of "Minor" Forest Products: The Local Use and Value of Forests in the West African Humid Forest Zone', *Forests, Trees and People,* Community Forestry Note 6, FAO, Rome

FAO (Food and Agriculture Organization) (1997) 'Medicinal Plants for Forest Conservation and Health Care: Non-wood Forest Products 11', FAO, Rome

Fondoun, J. M. and Manga, T. T. (2000) 'Farmers' Indigenous Practices for Conserving *Garcinia kola* and *Gnetum africanum* in Southern Cameroon', *Agroforestry Systems,* vol 48, pp289–302

Fuashi, N. A. (1997) 'Production and Marketing of NTFPs in the Korup Project Area Cameroon, and Cross Border Trade with Nigeria', Paper Presented at International Workshop on the Domestic Market Potential of Non-timber Products (NTTPs), Eyumojock, Cameroon, 24 January 1997

Gereffi, G., Humphrey, J. and Sturgeon, T. (2004) 'The Governance of Global Value Chains', *Review of International Political Economy,* vol 12, pp78–104

Hacket, R. I. J. (1989) *Religion in Calabar: The Religious Life and History of a Nigerian Town,* Mouton de Gruyter, Berlin

Hall, J., O'Brien, E. and Sinclair, F. (2000) *Prunus africana, A Monograph,* Publication 18, School of Agricultural and Forest Sciences, University of Wales, Bangor

Humphrey, J. and Schmitz, H. (2001) 'Governance in Global Value Chains', *IDS Bulletin,* vol 32, no 3, p17

ILO (2006) *A Guide for Value Chain Analysis and Upgrading,* International Labour Organisation, Geneva

Ingram, V. and Bongers, G. (2009) 'Valuation of Non-timber Forest Product Chains in the Congo Basin: A Methodology for Valuation', FAO-CIFOR-SNV-World Agroforestry Center-COMIFAC, Yaoundé, Cameroon

Laird, S., Ingram, V., Awono, A., Ndoye, O., Sunderland, T., Lisinge, E. and Nkuinkeu, R. (2010) 'Bringing Together Customary and Statutory Systems: The Struggle to Develop a Legal and Policy Framework for NTFPs in Cameroon', in S. A. Laird, R. McLain and R. P. Wynberg (eds), *Wild Product Governance: Finding Policies that Work for Non-timber Forest Products,* Earthscan, London

Laporte, N. and Justice, C. (2001) 'Monitoring of Forest Cover in Central Africa: Why, What, How and When to Monitor', Central African Regional Program for the Environment Briefing Sheet 13, CARPE/USAID

Malleson, R. (1998) *Community management of Non-Wood Forest resources: a case study from the Korup forest, Cameroon,* International Expert Meeting on Non-Wood Forest Products in Central Africa: Non-Wood Forest Products in Central

Africa; Current research issues and prospects for conservation development, Limbe, Cameroon, FAO

Malleson, R. (1999) 'Community Management of Non-wood Forest Resources: A Case Study from the Korup Forest, Cameroon', in T. C. H. Sunderland, L. E. Clark and P. Vantomme (eds) *Non-wood Forest Products of Central Africa: Current Research Issues and Prospects for Conservation and Development,* FAO, Rome, pp117–122

Mdaihli, M., Schmidt-Soltau, K. and Ayeni, J. S. O. (2002) 'Socio-economic Baseline Survey of the Villages in and around the Takamanda Forest Reserve', PROFOR, Mamfe, Cameroon,

MINPAT, 2000, Etudes Socio-economiques Regionales au Cameroun, Province du Sud-Ouest [Regional Socio-economic Studies on Cameroon, South-West Region], Projet PNUD-OPS, CMR/98/005/01/99, Yaoundé, Cameroon

Nde-Shiembo, P. (ed) (1999) 'The Sustainability of Eru (*Gnetum africanum* and *Gnetum buchholzianum*): Over-exploited Non-wood Forest Products from the Forests of Central Africa', in T. C. H. Sunderland, L. E. Clark and P. Vantomme (eds) *Non-wood Forest Products of Central Africa: Current Research Issues and Prospects for Conservation and Development,* International Expert Meeting on Non-wood Forest Products in Central Africa, FAO, Limbe, Cameroon, 10–15 May 1998

Ndoye, O., Ruiz-Pérez, M. and Eyebe, A. (1997/98) 'The Markets of Non-timber Forest Products in the Humid Forest Zone of Cameroon', Paper 22c, Rural Development Forestry Network, London

Ndoye, O. and Awono, A. (2010) 'Regulatory Policies and *Gnetum* spp. Trade in Cameroon', in S. A. Laird, R. McLain and R. P. Wynberg (eds) *Wild Product Governance: Finding Policies That Work for Non-timber Forest Products,* Earthscan, London

Ndumbe, L. N. (2008) 'An Assessment of the Economic Potential of *Gnetum* spp. in the Humid Lowlands of Cameroon', Department of Geology and Environmental Science, University of Buea, Cameroon

Ndumbe, L., Ingram, V. and Awono, A. (2009) 'Baseline Study on *Gnetum* spp. in the Southwest and Littoral Regions of Cameroon', FAO-CIFOR-SNV-World Agroforestry Center- COMIFAC, Yaoundé, Cameroon

Nepstad, D. C. and Schwartzman, S. (eds) (1992) 'Non-timber Products from Tropical Forests: Evaluation of a Conservation and Development Strategy', *Advances in Economic Botany*, vol 9

Oyono, P. R. (2005) 'Profiling Local-level Outcomes of Environmental Decentralizations: The Case of Cameroon's Forests in the Congo Basin', *Journal of Environment and Development,* vol 14, no 2, pp1–21

Paumgarten, F. and Shackleton, C. (2009) 'Wealth Differentiation in Household Use and Trade in Non-timber Forest Products in South Africa', *Ecological Economics*, vol 68, pp2950–2959

Peters, C. M. (1994) 'Sustainable Harvest of Non-timber Plant Resources in Tropical Moist Forest: An Ecological Primer', Biodiversity Support Program, World Wildlife Fund, Washington, DC

Prance, G. T. (1992) 'Rainforest Harvest: An Overview', in S. Counsell and T. Rice (eds) *The Rainforest Harvest: Sustainable Strategies for Saving the Tropical Forests,* Friends of the Earth Trust, London, pp21–25

Prance, G. T. (1998) 'Indigenous Non-timber Benefits from Tropical Rain Forest', in F. B. Goldsmith (ed) *Tropical Rain Forest: A Wider Perspective*, Chapman and Hall, London, pp21–42

Republic of Cameroon (2006) 'A Management Plan for the Proposed Takamanda National Park 2006–2010'

Ros-Tonen, M. A. F. and K. F. Wiersum (2005) 'The Scope for Improving Rural Livelihoods Through Non-timber Forest Products: An Evolving Research Agenda', *Forest Trees and Livelihoods*, vol 15, pp129–148

Ruiz-Pérez, M., Ndoye, O., Eyebe, A. and Ngono, D. L. (2002) 'A Gender Analysis of Forest Product Markets in Cameroon', *Africa Today*, vol 49, no 3, pp97–126

Shanley, P., Luz, L. and Swingland, I. R. (2002) 'The Promise of a Distant Market: A Survey of Belém's Trade in Non-timber Forest Products', *Biodiversity Conservation*, vol 11, pp615–636

Shanley P., Pierce, A., Laird, S. and Robinson, D. (2008) *Beyond Timber: Certification and Management of Non-Timber Forest Products*, Center for International Forestry Research (CIFOR), Bogor, Indonesia

Sheil, D. and Wunder, S. (2002) 'The Value of Tropical Forest to Local Communities: Complications, Caveats, and Cautions', *Conservation Ecology*, vol 6, no 2

Shiembo, P. N. (1998) 'The Sustainability of Eru (*Gnetum africanum* and *Gnetum buchholzianum*): Over-exploited Non-wood Forest Product from the Forests of Central Africa', in T. C. H. Sunderland, L. E. Clark and P. Vantomme (eds) *Non-wood Forest Products of Central Africa: Current Research Issues and Prospects for Conservation and Development*, International Expert Meeting on Non-wood Forest Products in Central Africa, FAO, Limbe, Cameroon, 10–15 May 1998

Sunderland, T. C. H., Besong, S. and Ayeni J. S. O. (2002) 'Distribution, Utilization and Sustainability of Non-timber Forest Products from Takamanda Forest Reserve', Cameroon, Consultancy Report: Project for Protection of Forest around Akwaya (PROFA), Mamfe, Cameroon

Sunderland, T. C. H., Besong, S. and Ayeni, J. S. O. (2003) 'Distribution, Utilization, and Sustainability of Non-timber Forest Products from Takamanda Forest Reserve, Cameroon', in J. A. Comiskey, T. C. H. Sunderland and J. L. Sunderland-Groves (eds) *Takamanda: The Biodiversity of an African Rainforest*, Monitoring and Assessment of Biodiversity Program, Smithsonian Institution, Washington, DC

Sunderland-Groves, J. L., Sunderland, T. C. H., Comiskey, J. A., Ayeni, J. S. O. and Mdaihli, M. (2003a) 'The Takamanda Forest Reserve, Cameroon', in J. A. Comiskey, T. C. H. Sunderland and J. L. Sunderland-Groves (eds) *Takamanda: The Biodiversity of an African Rainforest*, Monitoring and Assessment of Biodiversity Program, Smithsonian Institution, Washington, DC, vol 8, pp1–7

Sunderland-Groves, J. L., Maisels, F. and Ekinde, A. (2003b) 'Surveys of the Cross River Gorilla and Chimpanzee Populations in Takamanda Forest Reserve, Cameroon', in J. A. Comiskey, T. C. H. Sunderland and J. L. Sunderland-Groves (eds) *Takamanda: The Biodiversity of an African Rainforest*, Monitoring and Assessment of Biodiversity Program, Smithsonian Institution, Washington, DC, vol 8, pp129–140

Tabuna, H. (1998) 'The Markets for Central African Non-wood Products in Europe', in T. C. H. Sunderland, L. E. Clark and P. Vantomme (eds) *Non-wood Forest Products of Central Africa: Current Research Issues and Prospects for Conservation and Development*, International Expert Meeting on Non-wood Forest Products in Central Africa, FAO, Limbe, Cameroon, 10–15 May 1998

Tanda, G. (2009) 'Domestication of Eru (*Gnetum* spp.) in Southwestern Cameroon: Case Study of CENDEP's Field-based Eru Domestication Intervention', Paper presented at Natural Products of Western and Central Africa: Naturally African Sub-Regional Consultative Workshop, Nairobi, 27–29 October 2009

Topa, G., Karsenty, A., Megavand, C. and Debroux, L. (2009) *The Rainforests of Cameroon: Experience and Evidence from a Decade of Reform*, World Bank, Washington, DC

UNDP/ARPEN (2006) Plan d'Action National de Lutte Contre la Désertification (PAN/LCD) [National Plan of Action for the Struggle Against Desertification], vols 1 and 2, CCD/MINEP et du Programme APREN/PNUD, Yaoundé

Walter, S. (2001) 'Non-wood Forest Products in Africa: A Regional and National Overview/Les Produits Forestiers Non-Ligneux en Afrique: Un Aperçu Régional et National', Working Paper/Document de Travail, E.-F. P. Programme, European Commission (EC) and Food and Agriculture Organization (FAO), Rome

WHINCONET (2005) 'Report on the Illegal Harvesting of *Prunus africana* in the Kilum-Ijim Forests of Oku and Fundong', Western Highlands Nature Conservation Network, Bamenda, Cameroon

Zapfack, L., Ayeni, J. S. O., Besong, S. and Mdaihli, M. (2001) 'Ethnobotanical Survey of the Takamanda Forest Reserve', Consultancy Report, Project for Protection of Forest around Akwaya (PROFA), MINEF-GTZ, Mamfe, Cameroon

9

A Dozen Indicators for Assessing Governance in Forested Landscapes

Carol J. Pierce Colfer and Laurène Feintrenie

This chapter is more directly pragmatic than the rest of this book. Its aim is to provide an evolving tool that will enable users – readers, project planners and implementers, local and regional governments and (with help) communities – to conduct a comparatively quick assessment of local-level governance, typically within a village or district. Such tools may be particularly important for assessors of schemes participating in the United Nations' programme of Reducing Emissions from Deforestation and Forest Degradation (REDD), a popular approach to mitigating climate change. Assessors for payments for environmental services (PES) projects, an idea with a longer history, may also find it useful.

The importance and shortcomings of governance (as defined in Appendix 1.1) in forested areas of the world have been widely acknowledged. Efforts to analyse and address such problems are increasing. Below, we briefly outline some of these before moving to the tool itself and its partial application in our Indonesian site.

The ways day-to-day governance plays out in a given landscape are very dependent on historical and cultural contexts (Capistrano and Colfer, 2005). Disciplines, beyond governance *per se*, that provide insights on governance realities in landscapes include natural resource management, ethnography, collective action and participation. Here, rather than accepting, for instance, predefined principles of governance, such as those provided by the World Bank (2009) or the United Nations Development Programme (UNDP, 1997) or by Jesse Ribot (2004, 2008), we have kept diverse literature sources in mind while striving to develop a useful assessment tool. The elements in the tool have primarily been derived inductively, building on what we have seen in the field; our goal has been to develop a tool specific enough to recognize the diversity one encounters at the landscape level, yet also capable of identifying widely valued features.

We have envisioned two likely uses for a governance assessment tool. Its primary use, in our minds, is diagnostic. In such a case, each governance factor (Table 9.1) should be carefully evaluated and considered, in terms of areas in need of improvement. The second use, more controversial, would be as a scorecard, for use in evaluation. In this case, one can compute a score, averaging the individual factor scores. The advantage of such an approach is the possibility of comparing governance across sites, measuring changes within a site, or (most controversial) linking the governance score to a benefit (such as participation in a REDD or PES scheme), which would mean that low-scoring communities would be ineligible for participation, but would learn in which areas they should focus their efforts to improve.

What makes a governance assessment scorecard controversial is that definitions of 'good governance' are strongly influenced by the power and prestige of the globe's West and North. Whereas some (particularly in the 'developed' world) consider democracy and transparency to be universally valued and an obviously desirable end-state for humanity, there are others for whom these concepts do not ring so true. We acknowledge the ethnocentrism of the tool. Our assumption that gender equity is desirable is not shared by elite males in many African societies, for instance; nor are conservative, rural, high-caste Brahmins likely to welcome the effort to solicit the views and desires of the lower castes. There are many other examples that could illustrate the tool's ethnocentrism. We therefore feel more comfortable with the tool's use as a diagnostic aid rather than as a scorecard. Governance issues cannot be delinked from human values, which differ from group to group; we cannot escape elements of subjectivity in such assessments.

Theoretical Background

In this section, we briefly cover some of the literature that has led to the selection of issues that the tool measures. Our effort in producing this tool has been to make extensive use of Occam's razor (the assumption that the simplest solution is usually the correct one) and make the tool as brief as possible, while addressing the central and crucial issues. The literature cited below has informed our efforts.

Stakeholders and their relationships

The importance of identifying relevant stakeholders in each area has been thoroughly addressed in the literature. The 'Who Counts?' Matrix (Colfer et al, 1999) was an early attempt to highlight this issue in the context of assessing the sustainability of timber management at the forest management unit level. A more nuanced and interesting discussion of why stakeholder identification is important is available in Wollenberg et al (2005). A central purpose is to ensure that the voices of all those affected are heard in any decision-making process. This improves our analytical capabilities and thereby increases the chances of success with any given plan. It also increases the opportunities for the equitable participation of all concerned.

Uphoff (1996) and Krishna (2002) made particularly compelling cases for looking at the issues of trust, social capital and collective action as they affected governance. The relevance of their ideas was reinforced in research undertaken as part of the Center for International Forestry Research's (CIFOR's) Adaptive Collaborative Management (ACM) programme (see Colfer, 2005). Conflict emerged as a related issue in this study.

Komarudin and colleagues (2008) have analysed the governance relationships between local communities and district-level governments in two districts in Sumatra, Indonesia (including our Landscape Mosaics site), with particular attention to collective action, trust, social capital and elite capture.

Corruption and elite capture issues have been widely reported but rarely thoroughly analysed because of the sensitivity of the subject and the power of many practitioners (see World Bank, 2003, for an exception to this rule). Dudley (2000) used system dynamics modelling to examine corruption and the social interactions that maintain it in Indonesia. The Indonesian case study by Komarudin et al (2008) gives several specific instances of the use of collective action to counter elite capture in places near our Indonesian sites (also later in this chapter).

Levels of conflict have also been important issues in local governance. Conflict was one of seven important factors in the above-mentioned 30-site, 11-country comparison of CIFOR's ACM research. A thorough examination of conflict was conducted by Yasmi (2007), who studied our Indonesian landscape.

Access to resources

Land tenure issues are central in developing country governance and to a lesser extent in almost all forested areas. In the countries studied in this volume, forests technically belong to the government. Cronkleton et al (2010) make use of a 'tenure box', which shows the bundles of forest rights and their allocations to various stakeholders in a useful, visual way (Barry and Meinzen-Dick, 2008). Rights and resources issues are examined thoroughly, with their governance implications, in a new book edited by Larson et al (2010). A recent World Resources Institute (WRI, 2008) report and a companion volume by

Sunderlin (2008) provide an excellent global overview of some forest dwellers' lack of rights to local resources and encouraging news on other fronts. Both documents also provide suggestions for improvement, with particular reference to the potential dangers and opportunities posed by climate change abatement efforts now underway.

Structure and function of governance

Particularly in remote tropical forested areas, customary governance may be the most important governance at play. Marfo et al (2010) have recently analysed the difficulties and potential for collaboration between traditional and formal systems of governance, using examples from Burkina Faso, Ghana, Guatemala and Indonesia – relevant issues for the Landscape Mosaics sites that form the core of this book (see Chapters 5 and 6).

But formal governance structures are increasingly extending their reach, sometimes with unintended adverse effects (Edmunds and Wollenberg, 2003). Many forest governance problems have been analysed in the context of global efforts to decentralize; see the collections by Colfer et al (2008) for Asia, German et al (2009) for Africa, and Ribot and Larson (2005) and Colfer and Capistrano (2005) for global overviews. Ribot has written extensively on the importance of downward accountability for improving governance (e.g. Ribot, 2004, 2008; Ribot et al, 2008). His West and Central African experience has led him to distrust traditional or pseudo-traditional governance mechanisms, favouring more conventional, 'western' style democracy. Others have critiqued his view, seeing clear advantages to many traditional systems (see Diaw, 2005) and the recurrent shortcomings of attempts to import western-style democracy. Oyono (2004, 2005a, 2005b) and colleagues (Oyono et al, 2008) have written compellingly about the problems with decentralization in Cameroon. The recent edited book by Diaw et al (2009) includes chapters on Cameroon's mixed experience with governmental efforts to distribute national assets and largesse more equitably among the citizenry.

Nobel Laureate Elinor Ostrom (1990) has noted the importance (and difficulty) of ensuring mutual monitoring of rules and regulations, particularly in areas governed as commons. The usefulness of collective action in ensuring good governance has been shown repeatedly (see Krishna, 2002, for an excellent Indian example).

Efforts to enhance governance

An important element of improved governance is citizen participation. Luyet (2005) provides an appealing method for strengthening citizen participation (synthesized in Luyet et al, 2008), providing practical suggestions for identifying, characterizing, encouraging and evaluating public participation. See Ojha (2008) for a symbolic, analytical assessment of ways people and governance interact. Pokharel et al (2008) have developed a technique for coaching leaders and user group members in good governance in Nepal. Some of the guidance provided in McDougall et al (2009), also prepared for the Nepal situation, is relevant for improving governance more widely.

Komarudin's team in Indonesia, which included district officials, produced a dozen governance policy briefs on topics related to local and district-level governance, based on their 2004–2006 research (available from CIFOR, Bogor) in collaboration with district-level officials. The purposes included both improved analysis of local conditions and capacity building for local officials. Several manuals for improving governance came out of this research as well. Chhettri (2009) documents a recent attempt to scale up 'good governance' from the user group level to village development committees in Nepal.

In terms of addressing governance problems directly, the WRI has created a governance assessment tool for use at the national level (Brito et al, 2009). Like ourselves, the authors anticipate that such an evaluation tool may prove important because of the current interest in REDD, a use also anticipated by the World Bank (2009). Evidence of good governance may prove to be a condition for receiving programme funds. The WRI tool is complex, however, with 94 governance indicators to be assessed, and is intended to be applied at a broad scale. We anticipate that a simpler tool will be needed at the landscape level. With this need in mind, we have produced and tested the tool presented below. The issues addressed provide a succinct if broad view of good governance.

Governance Assessment Tool

Using the tool involves three to five assessors working as a multidisciplinary team for one to two weeks in the field. This team should include at least a researcher, a local government official and a trader, industry representative or conservationist, depending on local context, with representation of both genders to strengthen possibilities for equitable access to information. Skills in ethnographic methods and rural development and awareness of gender and diversity issues are particularly useful in such evaluations. Each team member should assess the features below independently. Towards the end of the exercise, the assessors share and discuss their findings and reach agreement on appropriate scores.

During the assessment process, special care must be taken to go beyond community and district elites to include women, youth, disadvantaged castes or ethnic groups, and any locally marginalized groups who use, have knowledge of and/or depend on local resources. Fear of the powerful can be a constraint to communication: interviews or communication platforms must be structured to give marginalized people the freedom to speak up without endangering themselves (at the hands of politically or economically powerful actors). This may require separate meetings (as took place in Indonesia). Care must be taken to protect people's anonymity when conclusions are drawn and recorded.

Simple lack of self-confidence or inexperience in public speaking can be a potent constraint, particularly for women (see Rantala et al, 2009 for

Table 9.1 Governance assessment tool (GAT)

Section	Scored observation	Scoring
1. Stakeholders and their relationships (4 scores) A first step is to list the stakeholders, including both the powerful and the less powerful (e.g. marginalized ethnic or tribal groups, low castes, women, 'the poor,' etc.) to ensure completeness. For each stakeholder, compute a score and average, yielding composite scores for #3 and #4.		
	#1 Do mechanisms exist to obtain the input of marginalized groups in decision-making that affects them?	1 (no); 5 (yes)
	#2 List the mechanisms and evaluate the functionality of each	1 (not functioning well); 5 (functioning effectively)
	#3 Evaluate actual access for each stakeholder category to forest lands and products	1 (not at all); 5 (consistently, well)
	#4 Evaluate the voice of each stakeholder category in decision-making affecting them	1 (not at all); 5 (consistently, well)
2. Formal access to resources (3 scores)	#5 Do formal land-use categories in state law reflect the types of land uses common in the field?	1 (no); 5 (very well)
	#6 Are people's rights with regard to land and forests clearly defined? a) Are rights of access to forest lands clear? b) Are rights to withdraw products from forest lands clear? c) Are rights to manage forest lands clearly defined? d) Are rights to exclude others from forest lands clearly defined? e) Are rights to alienate forest lands clear?	1 (no); 5 (yes) (average items a–e)
	#7 Are people's rights outlined above secure (whether by law or by custom)?	1 (no); 5 (yes)

A DOZEN INDICATORS FOR ASSESSING GOVERNANCE IN FORESTED LANDSCAPES | 223

3. *Structure and function of governance (5 scores)* See page 224 for explanations about possible improvements to items asterisked below.

#8 Are the different formal and customary governance structures functioning?

    a) Are the positive aspects of customary systems used effectively in day-to-day management?*   1 (no); 5 (yes, very effectively)

    b) Does the formal governance structure support or complement decisions rendered by the customary actors?   1 (no); 5 (yes, consistently and well)

    c) Do the governance structures (formal and informal) function effectively (consider issues of corruption, funding, personnel, training, motivation, experience, recognized authority)?   1 (no); 5 (yes)

#9 Are effective sanctions applied when natural resource rules and regulations are ignored?

    a) at the community level?   1 (no); 5 (yes)
    b) at the district or landscape level?   1 (no); 5 (yes)
    c) at higher levels?   1 (no); 5 (yes)

#10 Are conflicts effectively managed?

    a) Is the level of conflict acceptable at the community level?   1 (no); 5 (yes)
    b) Are there effective means to resolve conflicts?   1 (no); 5 (yes)
    c) Is the level of conflict between communities acceptable?   1 (no); 5 (yes)
    d) Are there effective means to resolve these conflicts?   1 (no); 5 (yes)
    e) Is the level of conflicts acceptable between communities and higher levels of governance?*   1 (no); 5 (yes)
    f) Are there effective means to resolve these conflicts?*   1 (no); 5 (yes)

#11 Are there high levels of trust among people in the landscape?

    a) Among community members?   1 (no); 5 (yes)
    b) Among communities in the landscape?   1 (no); 5 (yes)
    c) Between communities and higher levels of government?*   1 (no); 5 (yes)
    d) Among government officials?*   1 (no); 5 (yes)

#12 Are there good links between communities and outside resources (information, funding)?

    a) Do community members have access to external resources?
        i. Through higher levels of government?   1 (no); 5 (yes)
        ii. Through industry?   1 (no); 5 (yes)
        iii. Through external projects?   1 (no); 5 (yes)
        iv. Through their own networks?   1 (no); 5 (yes)

    b) Do governmental offices coordinate effectively among themselves, for the common good?   1 (no); 5 (yes)

Tanzania). The work of McDougall et al (2009) in Nepal provides hints on overcoming such constraints, as does Fajber (2005). Singh (2009) offers convincing ethnographic evidence of the degree to which 'participation' in some Laotian contexts is largely a 'performance', staged by powerful actors (in this case government and donors); Cooke and Kothari (2002) also document some of the dangers of 'participation' done carelessly. An assessment of local governance that inadvertently harmed local people would clearly be undesirable.

The tool (Table 9.1) has three sections: stakeholders and their relationships, access to resources, and structure and function of governance. The scores for 12 indicators are computed, based on a variable number of observations. Following our test, one of the indicators (2) was eliminated. We reproduce it here for purposes of summarizing the data, and because it may be more important in another context.

In general, we found the tool useful, as discussed in greater detail below. Several ideas for improvements have come to us, after the fact; these are listed and identified with an asterisk (*) in the scorecard.

- For Indicator 8, item (a), we mention only the positive aspects of customary systems. One of our anonymous reviewers reasonably suggested that we might also want to look at negative aspects of customary systems.
- In Indicator 10, items (e) and (f), we question conflicts between communities and higher levels of governance. We suggest noting conflicts with other outside actors, such as conservation NGOs or industry.
- For Indicator 11, items (c) and (d) regarding levels of trust, we would now add other outside actors.

Finally, an unresolved issue is the question of weighting the different sections, scores and elements. As currently fashioned, the 12 scores are equally weighted. This issue warrants contextual attention from future users of the assessment tool.

Jambi: Test Case

The governance tool criteria listed in Table 9.1 inspired a questionnaire used to conduct closed interviews in the Indonesian site in Bungo District, Jambi Province (see Chapter 3). One scientist (Laurène Feintrenie) and one field assistant assessed people's perceptions of regional and global governance, with a specific focus on forest management. Each criterion was given a value between 1 and 5 by the respondents. A score of 1 meant strong discontent about the issue raised, and at the other extreme, a score of 5 meant that the respondent was very happy with the present situation and considered that no improvement was needed. The questionnaire was used in structured interviews with representatives of each stakeholder group. We asked respondents to evaluate the governance system, its structure, its implementation and its efficiency.

Individual interviews were conducted for civil servants, conservationists and heads of villages, whereas villagers were interviewed in groups, with strict separation between men and women. We interviewed nine representatives of four public agencies: the Regional Planning and Development Agency (BAPPEDA, Badan Perencanaan Pembangunan Daerah), the Plantations Department (Dinas Perkebunan), the Forestry Department (Dinas Kehutanan), the Transmigration Department (Dinas Transmigrasi) and the National Land Agency (Balai Pertanahan Nasional). We also interviewed four people working for research centres (World Agroforestry Center, Center for International Forestry Research and an independent consultant), and 138 villagers (Feintrenie and Levang, forthcoming). In a previous study, Therville et al (forthcoming) classified the villages of Bungo District according to their type of agriculture, their distance from the forest, and their access to roads and markets. We built on this research and selected the same 12 sample villages to represent the district's diversity (Feintrenie, forthcoming).

We adapted the questionnaire to apply the governance tool in a context we knew well. We developed two parts: One part, dealing with respondents' perceptions of what was happening in their village or region, focused on GAT questions 1, 5, 8–12. The second part, aimed at estimating the differences in rights and participation in decision-making among categories of people at the village level, addressed GAT questions 3/6, 4, and 7. This categorization was based on prior research in the district; in another situation there may be a need for prior discussion with key informants to categorize villagers (indeed other strategies besides questionnaires are also reasonable).

In this field test, respondents were unable to distinguish between Indicators 3 and 6. Actual access and rights are not easy to separate, and our questions about rights were typically answered about access.

Indicator 2, on the mechanisms available for obtaining the input of marginalized groups in decision-making that affects them, was deleted because of the length of the interviews. Feintrenie informally interviewed the village heads, officials and villagers on this subject to round out her understanding, consistent with the governance tool's intent.

The questionnaire was well received by officials and researchers, providing a good stimulus for discussion. Indeed, the respondents often sought to explain or to justify their scores by giving insightful examples. The questions were adapted and simplified in the interviews with villagers, and examples were given immediately to illustrate the most difficult questions. Nevertheless, overall respondents did not find it too difficult. In some cases (in the most remote and isolated villages) people seemed worried about presenting a bad image of their village or getting a 'bad mark' if they were negative about their governance situation. We had to explain and reassure them that this was not the case (another potential problem if the tool is used as a genuine scorecard).

The respondents seemed more comfortable in groups (from three to ten people); the ideal group size being five or six (consistent with findings specifying seven as an ideal size (see www.leadingleaders.net/articles/entry/does_the_size_of_your_group_matter)). We found that such groups generated useful discussion, with each respondent arguing his or her own score and often giving examples. For women, the group work was useful in overcoming their shyness and encouraging them to give their personal points of view. Men in groups were more likely to say when they didn't understand the question, whereas individually, they tended to give an intermediate value without really thinking about the question. We made it clear that group members could give several scores for the same question. They liked the fact that they didn't have to reach an agreement but could express different feelings and opinions.

Some interpretations by the respondents need clarification here. The questions about 'rights' (*hak*), for instance, were understood as 'allowed or not' (*boleh, tidak boleh*). The notion of trust needed more thought than the other concepts, especially for 'trust among people' and 'between communities'. People found it difficult to differentiate between customary and formal rights. Indeed, in Jambi, the customary leaders are generally now included in the village council (*badan desa*), as the religious and customary heads. Thus customary rules have been, since Indonesia's decentralization process began, well respected by the formal government in this district.

The process went more smoothly when we began with simple questions that didn't need too much reflection and were likely to elicit 'positive' answers. Asking questions that people could answer easily and with good marks appeared to reassure them. On the other hand, when beginning with questions about conflicts or judgments of the government's actions, the respondents expressed worries and were reluctant to continue answering; after a few simple questions about the village that got people interested in the questionnaire and comfortable with the enumerator, the same questions were easily answered.

We also found the instrument useful in eliciting the history of the village; such information emerged spontaneously during the discussions. Conflicts, help received from the government, projects from NGOs, installation of a factory: all these were presented by one or another respondent to explain a score. Thus this questionnaire also makes a useful first entry to discover the recent history of a village and the present problems people are facing. It would, for instance, be a good tool to use before initiating a participatory action research project.

Tables 9.2 and 9.3 provide the synthesized results from our use of the governance tool. Table 9.2 presents our findings on people's perception of their own governance-related realities.

Table 9.3 provides responses about the roles of groups of people in the community, by category (villagers, village men, village women, civil servants and researchers).

The main conclusions of the study were that local people, as well as newcomers, feel secure in their access to land, forest and forest products (Indicators 3, 6 and 7). Even though some insecurity may exist that is not fully perceived by people, the general perception is that the district policy and

Table 9.2 *Synthesis of results on people's perceptions of governance (from first part of questionnaire)*

Indicator of governance	All respondents	Villagers	Villagers (men)	Villagers (women)	Civil servants	Researchers
1. Efficiency of mechanisms of participation in decision-making	**3.5**	3.7	3.9	3.5	3.4	3.3
3,6. Rights to access forest and to use forest land and products	**3.4**	3.5	3.6	3.3	3.1	2.5
4. Actual voice of each group of interest in decision-making process	**4.2**	4.3	4.4	4.1	4.0	2.2
5. Formal land categories' conformity to actual land use	**3.1**	3.1	3.1	3.1	2.3	3.8
7. Security of rights to access forest and use forest products	**4.1**	4.2	4.4	4.0	3.6	2.4
8. Efficiency of formal and customary governance	**4.0**	4.2	4.3	4.0	3.6	3.8
9. Enforcement of rules and regulations	**4.0**	4.2	4.4	4.0	3.7	4.0
10. Acceptability of the level of conflicts	**4.0**	4.2	4.5	3.9	3.9	3.3
11. Level of trust among stakeholders	**3.7**	3.9	4.0	3.8	3.8	3.2
12. People's access to external networks	**3.2**	3.4	3.5	3.2	3.6	2.4

actions are consistent with local needs. Nevertheless, tenure security and access to resources are complex issues that should be assessed in a broader sample of situations, especially in villages with a history of land conflicts. In future, we would hope to expand our queries beyond categories of villagers to other categories of stakeholders, including outsiders. Such interviews would be very useful for understanding conflicts.

Former conflicts related to forest conversion or land given to agribusiness companies by the government are a major factor in present dissatisfaction and generate some feeling of insecurity among villagers. Conflicts are not numerous in the district (Indicator 10), but the unclear land tenure (Indicator 5) is a major difficulty in places with former or unresolved conflicts. People expressed

Table 9.3 *Synthesis results on rights (6) and access (3) of stakeholders (from second part of questionnaire)*

Group	All respondents	Villagers	Villagers (men)	Villagers (women)	Civil servants	Researchers
Local people	3.7	3.7	3.8	3.7	3.4	2.3
Migrants	2.7	2.7	2.9	2.4	2.4	2.1
Village council	3.7	3.7	3.8	3.6	3.8	2.9
Civil servants	3.5	3.6	3.7	3.5	2.5	2.8
Brokers, traders	3.5	3.6	3.6	3.5	3.0	2.4
Women	3.4	3.5	3.6	3.5	2.5	2.0
Men	3.6	3.7	3.8	3.6	2.7	2.3
Average for all categories of people	3.4	3.5	3.6	3.3	3.1	2.5

satisfaction with the present political strategy oriented towards economic development (Indicators 8 and 9). This trend is clearly linked to their interest in improving livelihoods and increasing incomes to meet such needs as educational costs for children (Feintrenie and Levang, forthcoming). For more details, see Chapter 3.

Because of insufficient time to test this method, we could not organize interviews with representatives of the agroindustrial, mining and forestry sectors, important stakeholders in issues of natural resources, land and forest management. Ideally, a complementary survey should be conducted among them, and in a broader sample of villages. The results summarized in the tables are thus incomplete and only aim at illustrating the type of information one can get from using such a tool.

Conclusions

In sum, we have found this abbreviated assessment useful in evaluating governance at the village, district and landscape levels. We would like to see its wider use and further evaluation and improvement. We find this particularly important because so many new actors are moving towards incentive programmes that may assume a kind of 'good governance' not always visible in the field, the district or the nation-state.

References

Barry, D. and Meinzen-Dick, R. (2008) 'The Invisible Map: Community Tenure Rights', *International Association for the Study of the Commons*, Cheltenham, England

Brito, B., Micol, L., Davis, C., Nakhooda, S., Daviet, F. and Thuault, A. (2009) *The Governance of Forests Toolkit (Version 1): A Draft Framework of Indicators for Assessing Governance of the Forest Sector*, WRI, Washington, DC

Capistrano, D. and Colfer, C. J. P. (2005) 'Decentralization: Issues, Lessons and Reflections', in C. J. P. Colfer and D. Capistrano (eds) *The Politics of*

Decentralization: Forests, Power and People, Earthscan, London, pp296–313

Chhettri, B. (2009) 'Workshop Report: Review and Next Steps of NSCFP Supported Village Level Development Planning Pilots', Chetana Kendra, Banepa, Nepal, Swiss Community Forestry Project

Colfer, C. J. P. (2005) *The Complex Forest: Communities, Uncertainty, and Adaptive Collaborative Management,* Resources for the Future/CIFOR, Washington, DC

Colfer, C. J. P. and Capistrano, D. (eds) (2005) *The Politics of Decentralization: Forests, Power and People,* Earthscan, London

Colfer, C. J. P., Dahal, G. R. and Capistrano, D. (eds) (2008) *Lessons from Forest Decentralization: Money, Justice and the Quest for Good Governance in Asia-Pacific,* Earthscan/CIFOR, London

Colfer, C. J. P., with Prabhu, R., Gunter, M., Mcdougall, C., Porro, N. M. and Porro, R. (1999) *Who Counts Most? Assessing Human Well-Being in Sustainable Forest Management,* CIFOR, Bogor

Cooke, B. and Kothari, U. (eds) (2002) *Participation: The New Tyranny?,* Zed Books, London

Cronkleton, P., Barry, D., Pulhin, J. and Saigal, S. (2010) 'The Devolution of Management Rights and the Comanagement of Community Forests', in A. Larson, D. Barry, G. R. Dahal and C. J. P. Colfer (eds) *Forests for People: Community Rights and Forest Tenure Reform,* Earthscan/CIFOR, London, pp43–68

Diaw, M. C. (2005) 'Modern Economic Theory and the Challenge of Embedded Tenure Institutions: African Attempts to Reform Local Forest Policies', in S. Kant and R. A. Berry (eds) *Sustainability, Institutions, and Natural Resources: Institutions for Sustainable Forest Management,* Springer Publishers, The Netherlands, pp44–82

Diaw, M. C., Prabhu, R. and Aseh, T. (eds) (2009) *In Search of Common Ground: Adaptive Collaborative Management of Forests in Cameroon,* CIFOR, Bogor

Dudley, R. G. (2000) 'The Rotten Mango: the Effect of Corruption on International Development Projects, Part 1: Building a System Dynamics Basis for Examining Corruption', Report for System Dynamics Society, Bergen, Norway

Edmunds, D. and Wollenberg, E. (eds) (2003) *Local Forest Management,* Earthscan, London

Fajber, E. (2005) 'Participatory Research and Development in Natural Resource Management: Towards Social and Gender Equity', in J. Gonsalves, T. Becker, A. Braun, D. Campilan, H. De Chavez, E. Fajber, M. Kapiriri, J. Rivaca-Caminade and R. Vernoy (eds) *Participatory Research and Development for Sustainable Agriculture and Natural Resource Management: A Sourcebook,* vol 1: Understanding Participatory Research and Development, International Potato Center-Users' Perspectives with Agricultural Research and Development, and International Development Research Centre, Laguna, Philippines and Ottawa, pp51–57

Feintrenie, L. and Levang, P. (forthcoming) 'The Alarming Fate of Jambi's Forests: A General Consensus on their Conversion', *Forests, Trees and Livelihoods*

German, L., Karsenty, A. and Tiani, A.-M. (eds) (2009) *Governing Africa's Forests in a Globalized World,* Earthscan/CIFOR, London

Komarudin, H., Siagian, Y. L., Colfer, C. J. P., with Neldysavrino, Yentirizal, Syamsuddin and Irawan, D. (2008) *Collective Action to Secure Property Rights for the Poor: A Case Study in Jambi Province, Indonesia,* Collective Action and Property Rights System-Wide Initiative, Washington, DC

Krishna, A. (2002) *Active Social Capital,* Columbia University Press, New York

Larson, A., Barry, D., Dahal, G. R. and Colfer, C. J. P. (eds) (2010) *Forests for People: Community Rights and Forest Tenure Reform,* Earthscan/CIFOR, London

Luyet, V. (2005) *Bases Methodologiques de la Participation Lors de Projets ayant des Impacts sur le Paysage, cas d'Application: La Plaine du Rhone Valaisanne,* [Methodological Bases for Participation in Projects with Impacts on the Countryside, Test Case: The Rhone Valaisanne] École Polytechnique Fédérale de Lausanne

Luyet, V., Schlaepfer, R., Carter, J. and Parlange, M. B. (2008) 'A Framework to Implement Public Participation in Environmental Projects: Lessons from Experience in the West and East, and Practical Tools', Intercooperation, Berne (unpublished report)

Marfo, E., Colfer, C. J. P., Kante, B. and Elias, S. (2010) 'From Discourse to Policy: The Practical Interface of Statutory and Customary Land and Forest Rights', in A. Larson, D. Barry, G. R. Dahal and C. J. P. Colfer (eds) *Forests for People: Community Rights and Forest Tenure Reform,* Earthscan/CIFOR, London

McDougall, C., Pandit, B. H., Banjade, M. R., Paudel, K. P., Ojha, H., Maharjan, M., Rana, S., Bhattarai, T. and Dangol, S. (2009) *Facilitating Forests of Learning: Enabling an Adaptive Collaborative Approach in Community Forest User Groups, A Guidebook,* CIFOR, Bogor

Ojha, H. R. (2008) *Reframing Governance: Understanding Deliberative Politics in Nepal's Terai Forestry,* Adroit Publishers, New Delhi

Ostrom, E. (1990) *Governing the Commons: The Evolution of Institutions for Collective Action,* Cambridge University Press, Cambridge

Oyono, P. R. (2004) 'One Step Forward, Two Steps Back? Paradoxes of Natural Resources Management Decentralisation in Cameroon', *Journal of Modern African Studies,* vol 421, pp91–111

Oyono, P. R. (2005a) 'From Diversity to Exclusion for Forest Minorities in Cameroon', in C. J. P. Colfer (ed) *The Equitable Forest: Diversity, Community, and Resource Management,* Resources for the Future/CIFOR, Washington, DC, pp113–130

Oyono, P. R. (2005b) 'Profiling Local-level Outcomes of Environmental Decentralizations: The Case of Cameroon's Forests in the Congo Basin', *Journal of Environment and Development,* vol 14, no 2, pp1–21

Oyono, P. R., Kombo, S. S. and Biyong, M. B. (2008) 'New Niches of Community Rights to Forests in Cameroon: Cumulative Effects on Livelihoods and Local Forms of Vulnerability', Report for CIFOR (RRI), Bogor

Pokharel, B., Branney, P. W. and Carter, J. (2008) 'Governance Coaching: A Tool for Promoting Good Governance in Community Forest User Groups', Report for Nepal Swiss Community Forestry Project, Issue 5, Kathmandu

Rantala, S., Lyimo, E., Powell, B., Kitalyi, A. and Vihemäki, H. (2009) 'Natural Resource Governance and Stakeholders in the East Usambara Mountains: Governance Report', Report for CIFOR/ICRAF, Bogor

Ribot, J. C. (2004) 'Waiting for Democracy: The Politics of Choice in Natural Resource Decentralization', Report for WRI, Washington, DC

Ribot, J. C. (2008) *Building Local Democracy through Natural Resource Interventions: An Environmentalist's Responsibility,* WRI, Washington, DC

Ribot, J. C. and Larson, A. M. (eds) (2005) *Democratic Decentralisation through a Natural Resource Lens,* Routledge, London

Ribot, J. C., Chhatre, A. and Lankina, T. (2008) 'Introduction: Institutional Choice and Recognition in the Formation and Consolidation of Local Democracy', *Conservation and Society,* vol 6, no 1, pp1–11

Singh, S. (2009) 'World Bank-directed Development? Negotiating Participation in the Nam Theun 2 Hydropower Project in Laos', *Development and Change,* vol 40, no 3, pp487–507

Sunderlin, W. D. (2008) 'From Exclusion to Ownership? Challenges and Opportunities in Advancing Forest Tenure Reform', Report for Rights and Resources Initiative, Washington, DC

Therville, C., Feintrenie, L. and Levang, P. (forthcoming) 'What do farmers think about forest conversion to plantations? Lessons learnt from the Bongo dsitrict (Jambi, Indonesia)', *Forests, Trees and Livelihoods*

UNDP (United Nations Development Programme) (1997) 'Governance for Sustainable Human Development: A UNDP Policy Document', UNDP Governance Policy Paper, vol 97

Uphoff, N. (1996) *Learning from Gal Oya: Possibilities for Participatory Development and Post-Newtonian Social Science,* Intermediate Technology Publications, London

Wollenberg, E., Anderson, J. and López, C. (2005) *Though All Things Differ: Pluralism as a Basis for Cooperation in Forests,* CIFOR, Bogor

World Bank (2003) 'Combating Corruption in Indonesia: Enhancing Accountability for Development', Report for World Bank, East Asia Poverty Reduction and Economic Management Unit, Jakarta

World Bank (2009) 'Roots for Good Forest Outcomes: An Analytical Framework for Governance Reforms', Report for Agriculture and Rural Development Department, World Bank, Washington, DC

WRI (World Resources Institute) (2008) 'Seeing People Through the Trees: Scaling Up Efforts to Advance Rights and Address Poverty, Conflict and Climate Change', Report for WRI, Washington, DC

Yasmi, Y. (2007) 'Institutionalization of Conflict Capability in the Management of Natural Resources: Theoretical Perspectives and Empirical Experience in Indonesia', PhD thesis, Forest and Nature Conservation Policy Group, Wageningen University

10

Minefields in Collaborative Governance

Carol J. Pierce Colfer, with Etienne Andriamampandry, Stella Asaha, Imam Basuki, Amandine Boucard, Laurène Feintrenie, Verina Ingram, Michelle Roberts, Terry Sunderland and Zora Lea Urech

In this chapter we raise some red flags. Most fundamentally, after analysing our research findings, some of us are left wondering whether our effort to strengthen collaborative governance represents a betrayal of the people whose lives we have sought to improve. We also recognize that many powerful actors within tropical governments manifest little concern for the people and landscapes in which we have worked, and that local people have developed self-protective mechanisms that serve to distance themselves from states.

Many studies of governance emphasize the desirability of representational democracy, downward accountability, and the rule of law. This literature typically builds on the experience of the West or North of the globe, with hopes of transferring such experience, often as whole cloth, to developing countries. Success in this endeavour has been rare.

In our work, we have assumed the existence of multiple relevant scales, the desirability of collaboration among stakeholders and hierarchical levels and the need for mutually supportive institutions of governance. But this dual emphasis on top-down and bottom-up approaches has led to more questions about the viability of the most common conceptual models of governance. We, like many others, have found governance at the 'higher' (more central) levels to be marked by serious shortcomings, and patterns at village level to be so variable they defy succinct description.

In searching for a framework within which to look at and better understand the ubiquitous failures of central governance and the diversity of local patterns within tropical forests, we turned to a 2009 work by James C. Scott, *The Art of Not Being Governed*. Although focused on mainland Southeast Asia, his historical analysis has implications that touch on all our sites. He takes a historical view of the hinterland region he terms Zomia[1] and characterizes two kinds of societies: those in lowland, rice-growing areas where Southeast Asian states have formed, and those in remote hinterlands, far from state control.[2]

Using extensive evidence from Zomia, he builds his central thesis: that, rather than being primitive groups 'left behind' in the quest for civilization – a common interpretation that makes complete sense from a state (or 'lowland') perspective – upland peoples (like the forest peoples of our sites) have, over time, fled to the hills precisely to avoid the taxes, *corvée* (unpaid labour) and military service that being so governed has typically entailed. He makes a very convincing case for this 'upside-down' and proactive interpretation, focusing on the aspects of hill areas that have served to stymie those who would govern them:

- swidden agriculture and 'escape crops' (diverse, movable, less easily taxed);
- multiple ethnicities with porous boundaries (hard to define, count, make legible to the state);[3]
- acephalous social systems (without easy administrative 'handles' by which a state could exert control);
- non-literacy (allowing societies to quickly redefine their histories and change political affiliations as circumstances dictate); and
- value systems within which flexibility and autonomy, sometimes egalitarianism, rank high.

Scott's story applies most precisely to our site in Laos, which lies squarely within Zomia, but his observations are nearly as applicable in Indonesia. Furthermore, parallels – many of the characteristics that serve to protect the people of Zomia's hills and forests from state control – apply to our other sites as well. In all our sites there are culturally different groups that control national politics. The centres of power in Cameroon, as in Zomia, are in the lowlands (Yaoundé, Douala) in culturally different (francophone, with a different ethnic array) parts of the country. In Madagascar, the national power base is historically in the uplands, and coastal regions (where Manompana is located) are comparatively disempowered. In Tanzania, our most remote sites are also in the lowlands, although conditions in this country are least consistent with Scott's theory: the East Usambaras are clearly more integrated into the national picture than our other landscapes. The central feature, in all these situations, is remoteness (rather than altitude) and associated characteristics, such as those listed above.

Scott's narrative starkly demonstrates one of the difficulties we have had in interpreting our results. On the one hand, most policy analysts see governance from macro perspectives and in a positive light, seeking to strengthen it while correcting any shortcomings. Those who have looked at macro-scale governance from local perspectives (anthropologists, some geographers and sociologists) typically see a strikingly different picture, one in which the motivations of government actors become far more suspect. In our work, we have strived to bring these two perspectives together, both in our analyses and on the ground. The clash of perspectives is clearest in this chapter. We begin by looking at the policies on decentralization and on corruption – both to remind ourselves of the weaknesses of existing governance and to share our cross-site findings on the legal systems. Such weaknesses are cogent reminders of the reasons hinterland peoples might want to distance themselves from formal government. We then proceed in subsequent sections to look at certain features implied by Scott's characterization of hinterlands, examining their governance implications.

If we accept Scott's story – a story he argues persuasively and with abundant evidence – our efforts at collaborative governance become a simple aid to a harmful state's agenda (see Ferguson, 1994, for another sceptical, Africa-based interpretation of governance, or Ojha, 2008, on Nepal). We become a mere extension of the 'long arm of the state' (with its political interference, taxes, corruption, and military and labour demands). Such an interpretation is also implicit in some work of scholars looking at the effects of decentralization (see Contreras' 2003 collection; Edmunds and Wollenberg 2003; Elias and Wittman 2005).

However, as Scott also points out, the story he tells is mainly a historical one, written about the millennia when running to the hills (or swamps or seas) was an option, when there was land available to which people could turn, to escape the downsides of state rule. That era has been nearing an end in recent decades (Rights and Resources Initiative, 2010). With a global population of more than six billion people and the capability to reach even the most remote areas with Kalashnikovs, smart bombs and tax collectors, the option to 'escape' becomes less and less probable. Besides the downsides of statehood (or what state actors see as 'civilization'), there are mobile phones, cash, education and health care – inducements that can further alter forest people's perceptions and decisions about joining those 'being governed'. Decentralization processes have brought state controls closer to forest peoples' daily lives; and the recent global preoccupation with Reducing Emissions from Deforestation and Forest Degradation (REDD), REDD+[4] and payments for environmental services (PES) suggests that more state intervention, whether bringing welcome benefits or disturbing intrusions, is likely.

Our effort, partially described in this book, has been fundamentally to begin a process that would alter the relationship between those whose landscapes and livelihoods have evolved in an interdependent if antagonistic relationship with states, into something more equitable and sustainable. We recognize and value many of the 'wild' qualities of these forest and human systems that have evolved in remote areas, far from state control. We have pondered the dilemma of how to maintain the wonders of biological and cultural diversity while facilitating access to the benefits of the 'modern'; of participation in a state. How can the flexibility, autonomy, creativity and comfort of local systems be maintained in interaction with the inflexible, hierarchical, universalistic bureaucracies that typify states?

We envisioned local systems meshing more smoothly with the broader-scale systems of which they form a part. Hinterland systems can probably no longer function truly autonomously and separately – at least not given current global realities. But the position of hinterland peoples needs strengthening, in interaction with actors operating on larger scales. This need becomes increasingly critical as global efforts to tackle climate change begin to impinge on remote forested landscapes. We struggled, in these communities and landscapes, with how to do this.

In this chapter, we look first at our sites from a state perspective. The idea of governance is, after all, a lens with a state bias. We begin with two widespread realities, decentralization and corruption, to demonstrate the potential dangers of trusting in states. We then briefly examine each country's legal framework, relating to land, forests and forest products, to demonstrate the typical lack of legal support for local systems governing natural resources (despite such *de facto* governance in many areas). With the features of Scott's 'wild' forest systems in mind, we then turn to a review of policies on swidden agriculture, resettlement and, imagining solutions, participation policies.

In the penultimate section, we briefly turn our lens from the top to the bottom, from the state to the hinterlands, and consider one example of the complexity encountered there: gender and other locally relevant social inequities. We conclude by reiterating the need – exacerbated by the global response to climate change – to move forwards in fashioning equitable arrangements that marry the positive elements of local realities with the needs and potential contributions of actors on broader scales.

Realities of Governance in Five Countries

Decentralization

Forest decentralization has been very popular with governments in recent years, but reviews have been mixed (Colfer and Capistrano, 2005; Ribot and Larson, 2005; Colfer et al, 2008; German et al, 2009). Governance scholars have seen decentralization in its ideal form as opening doors to greater hinterland autonomy and self-determination, but the ideal form has been largely a chimera in tropical forests.[5]

A genuine interest by the state in devolving power to local communities would seem to be an important element in any kind of collaborative governance. Cameroon, Indonesia, Madagascar and Tanzania have laws designed to decentralize governance (Table 10.1), marred by the same failings described for

Table 10.1 *National laws relating to decentralization in five countries*

Country	National laws with decentralization implications	Sources
Cameroon	1994 Forest Law recognizes council forests, community forests, community-managed hunting zones and annual forest fees from timber concessions	Assembe, 2009a; Zoa, 2009; Sandker et al, 2009; Oyono, 2004; Logo, 2010
Indonesia	Law 22/1999 on regional governance; Law 25/1999 on fiscal balancing between central and regional governments	Barr et al, 2006; Resosudarmo and Dermawan, 2002
Laos	Strong central control, moderated by Land and Forest Allocation Programme (Prime Minister Order No 3, 1996) and (potentially) the creation of village clusters (2009)	Sunderlin, 2006; Fujita and Phengsopha, 2008; Milloy and Payne, 1997; Oberndorf, 2009
Madagascar	Malagasy Constitution and Laws No 93-005, 1994 and No 94-039, 1995; Law No 96-025 of 30 September 1996 (GELOSE); decree 2001–122 (GCF) of 14 February 2001	Randrianatoandro, 2009; Ranjatson, 2009
Tanzania	Local Government Act, 1982; Village Land Act, 1999; 2002 Forest Act	Blomley et al, 2009; Wily and Dewees, 2001; Rantala et al, 2009

other tropical forested countries; Laos has moved only slightly away from its extreme centralized approach.

Figure 10.1 summarizes the length of time each country has been decentralized, and the levels of implementation and envisaged change in each research country.

Although considered a regional leader in forest decentralization, with legal mechanisms that devolve significant rights to various forest actors,[6] Cameroon has experienced extensive problems in implementation (see Cerutti and Fométe, 2008; Oyono, 2004, 2005b; Oyono et al, 2008), including confusing and conflicting laws, slow and corrupt implementation, disregard for customary systems and excessive bureaucratic requirements.

Indonesia, which was extraordinarily centralized under Soeharto's long reign, took drastic measures in 2001 when it implemented Law 22, devolving much authority to districts, bypassing provinces and resulting in dramatic interdistrict variation. The district-level concern for fund raising (true of all districts; Kadjatmiko, 2008) has also had important effects on the landscape (see Chapter 3). Many have remarked on Indonesia's 'decentralization of corruption', as district heads followed Jakarta's historical lead, handing out smaller-scale concessions and favours to their own evolving network of cronies. One of the important shifts in Bungo District, with more immediate social than ecological impacts, has been the attempt to resuscitate the traditional mode of local governance, called Rio (Hasan et al, 2008).

Laos represents the one country among our sites that continues to espouse strong central authority (Sunderlin, 2006). Recently, some additional authority

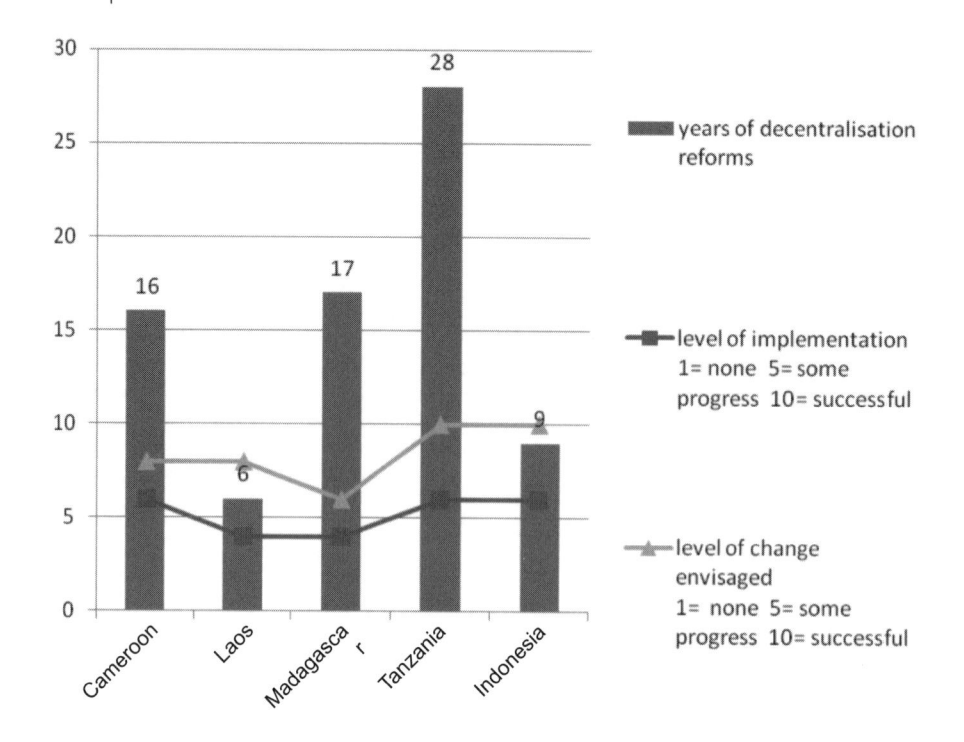

[Figure: Bar and line chart showing years of decentralisation reforms (Cameroon 16, Laos 6, Madagascar 17, Tanzania 28, Indonesia 9), with lines for level of implementation and level of change envisaged across the five countries.]

Figure 10.1 *Years and levels of decentralization in Landscape Mosaics countries*

has been given to local governments to enforce national forestry policies and regulations (Phongoudome and Sirivong, 2007), in a process described as 'deconcentration', wherein responsibilities are shifted downward, without corresponding authority, and accountability remains upward (Ribot, 2008; see also Morris et al, 2004, on Lao government structure). A recent innovation has been the creation of the *kumban*, or village cluster, designed to extend government policies and development programmes (Foppes, 2008).

Laotian policies on land and forest allocation, which formally allocate lands to forest-dwelling communities, were considered among the most progressive in Southeast Asia in the 1990s (Fujita and Phengsopha, 2008), but may have created more problems than solutions (cf Milloy and Payne, 1997; Sunderlin, 2006; Oberndorf, 2009, on 'village forestry' for further discussion of the strong central control maintained in Laos). One challenge has been the lack of central government motivation to support the development of regional implementing capacities. As with so many governmental efforts, funding to fully implement laws and staffing at the district level is inadequate, centralistic and paternalistic bureaucratic attitudes interfere with participatory efforts, and corrupt practices can further adversely affect implementation.

According to Madagascar's constitution, the decentralization policy is envisioned as a plan of action, deliberate, coordinated and designed to encourage equitable and harmonious development in the nation. The policy's stated goal is to provide a rational organization of the national territory and an institutional framework for citizens' effective participation in the management of public affairs and in the poles of economic growth (Randrianantoandro, 2009; see also Ranjatson 2008, 2009, for fuller description of the various laws affecting decentralized forest management, as well as their incompatibilities). Madagascar's difficult decentralization experience, with legal incompatibilities, uncertain areas of authority, corruption, burdensome bureaucracy and mobile bureaucrats, is hardly unique.

Tanzanian decentralization has proceeded most successfully among our sites (cf Wily and Dewees, 2001; Markelova and Swallow, 2008; Blomley et al, 2009; Vihemäki, 2009), benefiting from a longstanding governmental commitment. One important piece of legislation, the Village Land Act (1999), gave village governments legal responsibility for land in the 'village area', across all land (and forests) claimed and used by the villagers, sometimes covering tens of thousands of hectares (see Chapter 5). The 2002 Forest Act built on this legal framework to give communities legal rights to own, manage or co-manage forests under a wide range of conditions.

Still, suspicions remain about government officials' motivations to promote decentralized natural resource governance, and our team expressed concerns about the financial viability of district-level implementation. Our efforts to institute more effective cross-sectoral collaboration via multistakeholder action groups at the district level, for instance, proved impossible in the absence of district financing.

Although all five sites are involved in efforts to decentralize to some degree, there are none in which the process is moving smoothly. Indeed, if one looks at the literature on forest decentralization globally, one finds the same pattern. We must ask ourselves, how far can governments be trusted to enforce their laws, designed to empower their own people?

Corruption

Allegations of malfeasance have been made in all five countries, although systematic substantiating evidence is unavailable. Table 10.2 (also displayed as Figure 10.2) provides some widely accepted indicators relating to governance, as they apply to the countries where we work.

Transparency International has surveyed national perceptions of corruption in 180 countries (Table 10.3). On its index scale of zero to ten, none of our sites scored higher than 3.4. Perceptions of corruption increased between 2008 and 2009 in Indonesia, Madagascar and Tanzania; they stayed roughly the same in Laos and declined slightly in Cameroon. In this section, we draw heavily on the work of others, since an explicit focus on local or national corruption could have adversely affected our other work.

Corrupt practices have been reported in all our landscapes as well, although systematic cross-site comparisons have been impossible. Genuine

Table 10.2 *Governance indicators in five countries, 1996–2008*

Country	Rule of law Rank	Score	Control of corruption Rank	Score	Political stability Rank	Score	Government effectiveness Rank	Score	Voice and accountability Rank	Score
Cameroon	17.2	−0.99	18.8	−0.90	27.8	−0.53	19.9	−0.80	18.3	−1.02
Indonesia	28.7	−0.66	31.4	−0.64	15.8	−1.00	47.4	−0.29	44.2	−0.14
Laos	20.1	−0.90	5.8	−1.23	43.5	−0.01	17.5	−0.84	6.3	−1.71
Madagascar	40.2	−0.46	55.1	−0.10	30.1	−0.42	33.2	−0.59	43.8	−0.16
Tanzania	47.8	−0.28	36.2	−0.51	45.0	+0.01	39.3	−0.45	45.2	−0.09

Rank = percentile rank (0–100)
Score = governance score (−2.5 to 2.5)
Note: The governance indicators presented here aggregate the views on the quality of governance provided by a large number of enterprise, citizen and expert survey respondents in industrial and developing countries. These data are gathered from survey institutes, think tanks, NGOs and international organizations. The system of Worldwide Governance Indicators (WGI) does not reflect the official views of the World Bank, its executive directors, or the countries they represent. WGI is not used by the World Bank Group to allocate resources.
Source: Kaufmann D., Kraay, A. and Mastruzzi, M. (2009), Governance Matters VIII: Governance Indicators for 1996–2008

collaborative governance will require serious efforts to address such practices, which both encourage forest peoples to distance themselves from states and formal governance, and to adopt similar practices themselves.

In Cameroon, corruption allegations have been longstanding and persistent (see Oyono et al, 2008; Assembe, 2009b). In 1999, Cameroon was estimated to be the most corrupt country on earth by Transparency International. Researchers have convincingly analysed the country's losses from illegal logging (Auzel et al, 2004), including the international scale of corruption (Roda, 2009). On the positive side, Cameroon has been actively involved in the European Union's Forest Law Enforcement, Governance and Trade (FLEGT) process (Gasana and Samyn, 2009). Despite progress in dealing with illegal forest activities since 2001, serious difficulties remain, often linked to corrupt governance practices (Cerutti and Fomété, 2008; Karsenty, 2009). The dependence on external actors and the conditions attached to aid have meant that there is little ownership of reform efforts by government actors.

On a more local scale, the people of Takamanda-Mone are subject to informal 'taxation' along trade routes by forest officers, police and others, and to ambiguous rules, which provide additional opportunities for corruption (e.g., in the case of bush mango; Chapter 8; Sunderland et al, 2010). Community leaders give outsiders rights to harvest non-timber forest products (NTFPs), causing intra-community conflict. Women and people under 40 have had difficulty being heard in decision-making forums (Oyono et al, 2008), and there is evidence of corruption in many of the governmentally organized management groups connected with community and communal forestry (Oyono, 2005b).

Indonesia was the third most corrupt country on Transparency International's list in 1999 but moved to 130th in 2008 and improved further in 2009 (see World Bank, 2003). Complex national systems have been put in

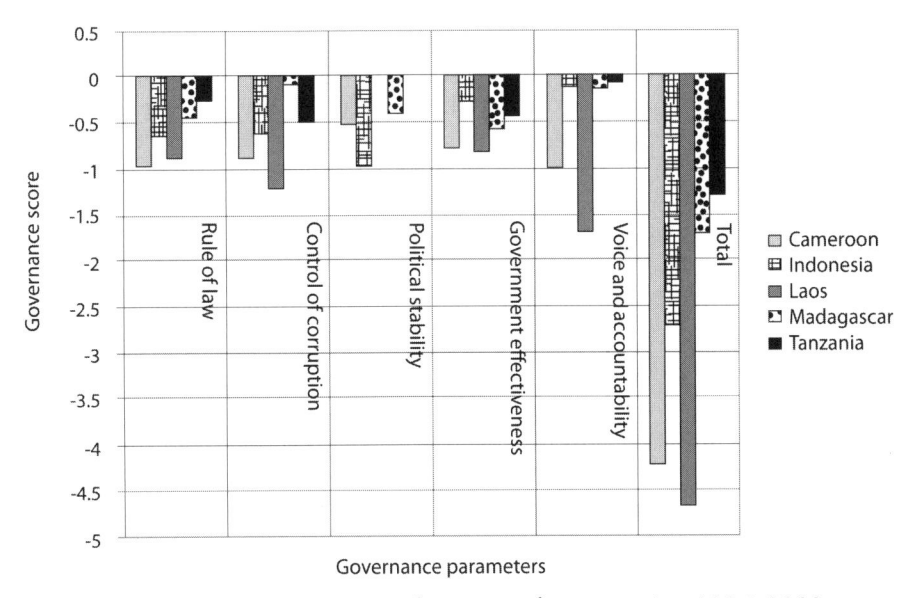

Figure 10.2 *Governance indicators in five countries, 1996–2008*

place to address illegal logging (Wells, 2008), including active leadership in FLEGT (Gasana and Samyn, 2009). Indonesia appears to have greater governmental buy-in than Cameroon or Laos. But corruption has been a pervasive part of governmental culture for decades.[7] Despite significant reforms (e.g. an anticorruption law and establishment of both a corruption court and the Corruption Eradication Commission in 2003), fighting corruption has been an uphill battle. Two system dynamics models showing how corruption has worked in Indonesia are available (Dudley, 2000, 2002): the 2000 model shows the difficulty of stopping it under a variety of scenarios; the 2002 model shows its role in the rampant logging that characterized the turn of the millennium there. Between 2004 and 2006, Komarudin et al (2008) conducted participatory action research with district-level officials in Bungo, as did Indriatmoko. Their analyses and personal communications indicate that irregularities remain widespread.[8]

Table 10.3 *Perception of corruption rankings and indices in five countries, 2008 and 2009*

Country	Rank 2008	Index 2008	Rank 2009	Index 2009
Cameroon	141	2.3	146	2.2
Indonesia	130	2.6	111	2.8
Laos	155	2.0	158	2.0
Madagascar	87	3.4	99	3.0
Tanzania	108	3.0	126	2.6

Note: an index of 10 indicates the least perceived corruption, 0 the most.
Source: Transparency International, www.infoplease.com/world/statistics/2008-transparency-international-corruption-perceptions.html; www.transparency.org/policy_research/surveys_indices/cpi/2009

Laos, which has the worst corruption rating of our five countries, has 'rampant corruption and illegal logging', according to Sunderlin (2006), a conclusion also supported by Yen et al (in preparation, 2009), and a non-transparent quota system for log sales that 'distorts the market and industry structure [and] creates incentives for corruption' (Morris et al, 2004, p41). The holistic, cultural interpretation of corruption in this region provided by Stuart-Fox (2006) explains some of the interacting forces that keep corruption at high levels there.

Laos' Land and Forest Allocation (LFA) policy has been a significant force in terms of landscape management in the country, but Fujita and Phengsopha (2008) have found gaps between theory and practice:

> *Those with more power – determined by years of residence, proximity to resources, links with political leaders, and ability to mobilize capital and labour – within the village are often able to use LFA to their advantage and make legal claims to lands and resources, while socially marginalized people have limited opportunities to negotiate their access to productive lands.* (Fujita and Phengsopha, 2008, p123)

> *Anybody with the right connections with political power at the district, provincial or national level can override local resource management agreements. Under the generally weak institutional and regulatory body of local authorities, claims for productive resources continue to be influenced and defined by existing social relationships.* (Fujita and Phengsopha, 2008, p127)

These authors also note the need for monitoring of resource tenure and resource conditions to strengthen village organizations' and local authorities' accountability, in support of better resource management and human rights protection. However, both local people and government staff in our research sites expressed concerns about the lack of funds and expertise for the resource monitoring without project support.

Despite Madagascar's history of governmental corruption (Horning, 2009), conditions may have been exacerbated during the political turbulence of 2008 and 2009: '[C]orruption is a daily part of life in Madagascar. It is standard practice to pay bribes for the transportation of goods, a new permit, traffic violations, and even a high school diploma', according to the United States Agency for International Development (USAID, 2009). Airport posters warn travellers against corrupt practices, potent reminders of its ubiquity. Many people believe that officials are corrupt; for example, the belief that officials illegally grant permits to loggers to obtain rare timber from nearby national parks is widespread. Still, within our five-country data set, Madagascar is perceived as the least corrupt (Table 10.2).

There is ample historical evidence of longstanding corruption in Tanzania (Vihemäki, 2009). The benefits, to local communities and the government, from the community-based forest management programme, are significantly reduced by corruption and patronage networks (Blomley et al, 2009). Opportunities for corrupt behaviour have been enhanced by improved infrastructure (especially roads and bridges) and market demand, mainly from China, since 2004, involving everything from 'petty corruption to highly organized patronage and personal involvement' to 'self-dealing, nepotism and cronyism' (Milledge, 2009, pp289–290).

On a more local level, the contradictory advice from some officials may be deliberate (Vihemäki, 2005), since rules and regulations that can be variously interpreted leave room for some actors to adjust the system to their own benefit (Rantala et al, 2009). Such corruption weakens people's trust and willingness to participate in natural resource management.

One element of corruption that has affected Indonesia, Laos and Tanzania is its social acceptability. When deemed natural and normal, corruption is more difficult to control (see Dudley, 2004 and Milledge, 2009, for Indonesia and Tanzania, respectively). In Laos, people display passivity about the corrupt practices of government officials and their cronies, perhaps because of their lack of power to control such governance (see also Wescott, 2003). Pervasive corruption reduces people's faith in the functioning of their systems, warps systematic processes designed to improve local conditions, and at least historically, has driven forest peoples further from the influence of the state.

Legal Context of Landownership and Land-use Planning

In the governance of forest resources, control of land is a central issue for both states and forest communities (see Larson et al, 2010; Sunderlin, 2008). In all five countries, a vast majority of rural land technically belongs to the government. In Indonesia (Barr et al, 2006; Moeliono et al, 2009), Laos (Morris et al, 2004; Fujita and Phengsopha, 2008; Fitriana et al, 2009), Tanzania (Wily, 1997; Rantala et al, 2009) and Madagascar (Randrianatoandro, 2009), all unregistered (untitled) land is technically owned and formally managed by the state. But other arrangements coexist in all our sites:

- competing systems of ownership, with traditional or local tenure systems taking a variety of forms and being variously recognized;
- management by external actors (conservation groups, timber companies, plantations); and
- new systems.

Formal land-use planning is underway in Indonesia (Irawan et al, 2008; Komarudin et al, forthcoming), Laos (Morris et al, 2004; Fujita and

Phengsopha, 2008; Fitriana et al, 2009) and Tanzania (Rantala et al, 2009), and our teams have engaged seriously with these processes.

Here, we strive to clarify relevant governmental forest policies that affect the hinterland peoples with whom we have worked. We describe some pertinent elements of the legal contexts within which peoples and their forest use are currently governed. Governments, typically ignoring customary uses and governance, impose laws that purport to grant people more rights but in reality deprive them of their customary rights. Some have argued that people's legal powerlessness makes a mockery of our attempts to foster collaborative governance – a conclusion we are reluctant to draw. But the recurrent clashes between governmental and customary policy yield ambiguity that interferes with good governance at all levels.

Cameroon

Although local communities in remote areas such as Takamanda-Mone have continued informally to govern local landscapes and manage their resources to a large degree, this freedom has been largely due to inaccessibility. The central government claims ownership and control of 86 per cent of the country as the formal forest estate (Egbe, 1997; Oyono, 2005a). This means that the government has the legal right to determine the uses of the land.

Several Cameroonian laws are ostensibly designed to grant communities more rights to manage their resources. The Forestry Law (1994) includes the possibility for councils to own forests (and communities to manage them). Council forests have functioned in many communities as another mechanism to take away land (Assembe, 2009a). Such initiatives in the Southwest Region are only starting to emerge, with donor support. Takamanda-Mone has no examples of formal, private ownership of land. Land is technically under the management of the Ministry of Forests and Wildlife (supported by donors and NGOs). Timber companies are responsible for managing their concessions. No formal land-use planning has occurred, but the government, supported by projects, has been active in zoning lands for different uses. Day-to-day subsistence use remains in the hands of local people. Recognition of this in the regulatory framework contributes to the ambiguities in the legal system, resulting in considerable associated conflict (Van Vliet et al, 2009). A serious attempt was made in the early 2000s to implement an adaptive management process involving multiple stakeholders (Comiskey and Dallmeier, 2003). The fact that the current formal managers of the park have no interest in such an approach demonstrates the degree to which people's well-being depends on the whims of others.

The park was previously a 'forest reserve' designed for timber production, a designation that technically prohibited agriculture, but allowed subsistence hunting and fishing, collection of NTFPs, oil palm and palm wine, as well as fuelwood and poles, all for domestic uses. The decision to convert the area to a national park (begun in 2004, finalized in 2008) implies stricter regulations.

These could potentially be moderated by a people-friendly management plan, something yet to be completed – but again, insofar as state actors can access the region, the fate of the people and the landscape depends on outsiders.

Cameroon's Ministry of Forestry and Fauna has established a department ostensibly responsible for NTFP development, promotion and revenue generation, though its mandate remains unclear (Chapter 8; Laird et al, 2010). Locally, hunting and fishing have been important (Sunderland-Groves and Maisels, 2003; Mdaihli et al, 2003) and largely unregulated (Mdaihli et al, 2002), but recently built roads and other infrastructure suggest that this is likely to change. The state's presence is likely to be more seriously felt in the near future.

Indonesia

The Indonesian constitution, Article 33(3), states that 'the land and the waters as well as the natural riches therein are to be controlled by the state to be exploited to the greatest benefit of the people' (Arnold, 2008, p82). For decades, some 90 per cent of Indonesia's Outer Islands were classified as part of the national forest estate (Fay and Sirait, 2001). Law 41/1999, on forestry, confirms state recognition of customary law and community. But as elsewhere, the state maintains the legal rights to own and manage most forests, and the more remote the area, the more likely are the people to govern themselves according to their customary practices. Accessibility is increasing in Bungo District, and the role of government in people's daily lives has correspondingly grown. This trend is generally well perceived by local communities, but there are those who fear local people's innocence about possible consequences of the agreements they sign with businesses or government agencies (Feintrenie and Levang, 2010).

In recent years, there has been considerable wrangling between the central government, on the one hand, and the district and provincial governments, on the other, about control of forests – an issue that remains in dispute (Barr et al, 2006; Kadjatmiko 2008; Wollenberg et al, 2009). Some of this conflict has revolved around district land-use plans (Komarudin et al, 2008; Chapter 3). Law 24/1992, on land-use planning,[9] requires that each level of government incorporate input from lower levels. Although sub-district and village spatial plans are intended to be developed by local people and to recognize traditional land use, people's involvement has so far been nonexistent or superficial (see Komarudin et al, 2008).

District land-use planners struggle with national ambiguity about legal authority over forest land (Adnan et al, 2008), often wishing to recategorize lands as 'areas for other uses' (*areal penggunaan lain*, which allow non-forest uses). District officials seek greater community agricultural use, district revenues and often, less socially responsible goals. As elsewhere, lack of clarity about land tenure complicates land-use planning and generates conflicts (see Chapter 3). Most people farm lands the Ministry of Forestry considers part of the national forestry estate. They generally feel confident about their customary ownership of this land (see Chapter 9), but the Ministry of Forestry refuses to consider formal recognition.

During the late 20th century, the informal governmental recognition of the planting of tree crops as indicative of 'traditional' ownership encouraged villagers to plant trees to strengthen their land claims, which were traditionally granted to those who cleared forest. Certainly in Bungo, people's interest grew in planting rubber agroforests in upland fields planted first to rice (Feintrenie and Levang, 2009). Agroforests have traditionally been complex, including numerous useful fruits, fibres and timber, and have typically been linked with subsistence rice cultivation, which is decreasing in importance; Lehébel-Péron (2008) describes the current, somewhat reduced, complexity of these systems.

There are nationwide regulations on tree use, particularly endangered species, although such regulations are marked by ambiguity (e.g. Kusumanto, 2001; Colfer's field experiences). Several national government programmes are designed to address local people's land and forest use,[10] including HKm (Hutan Kemasyarakatan, Community Forests), PIR (Perkebunan Inti Rakyat, Nucleus Estate and Smallholder [plantation] systems), HTR (Hutan Tanaman Rakyat, People's Planted Forest) and Hutan Desa (Village Forest, which in 2009 was inaugurated with much pomp and circumstance in our most remote site, Lubuk Beringin). But important legal controls by the central government have, to date, been maintained in all such programmes. With regard to indigenous populations, the programmes involve the central government's *conveying* (reduced) rights, rather than *recognizing* existing rights.

Laos

Laotian government landownership is clearly written into the constitution (Fujita and Phengsopha, 2008). 'The Forestry Law essentially designates all of the land area in Lao PDR as forest land, with the exception of established paddy' (Morris et al, 2004, p17); cf Scott's (2009) emphasis on paddy rice as a 'state-friendly crop'.

Land-use planning is tightly linked to the Land and Forest Allocation policy, ostensibly designed to allocate and recognize people's rights to use and to manage forest resources. Fujita and Phengsopha (2008) describe the actual effects and identify the strengths and weaknesses of the policy's implementation.[11] Adverse effects include reduction in swidden lands, with resulting shortening of fallows, land degradation, increases in secondary forests and reductions in NTFP access (Fujita and Phengsopha, 2008), along with other adverse impacts on people.[12] Management responsibility has been shifted under LFA, from the Department of Forestry to the Department of Agriculture and Forestry Office, and to village cluster organizations. A decline in the programme's budget since 2000 has rendered implementation less effective.

Traditionally, forest clearing has granted tenure, which remained during fallow periods. But legally, the Land Law (1997) has created two main types of tenure: temporary and permanent. In theory, the government allocates three plots of land per family for a three-year trial period; if the conditions are met and the laws respected, a permanent land title is provided (Morris et al, 2004). Intended to provide more secure land rights, this policy has actually decreased security while reducing the sustainability of many swidden systems. It also accentuates what Cramb et al (2009) have called a ratchet effect by speeding up a shift to smallholder cash crops; Scott (2009) finds that these 'state accessible products' decrease food security.

In our sites, although LFA has to some extent been implemented, identification of the size and locations of actual land holdings by individual families remains obscure. The process has also been intimately linked with resettlement (Chapter 4; and below), with widely recognized adverse effects (e.g. Baird and Shoemaker, 2005). More accessible areas have been predictably deforested by swiddeners resettled into more densely populated settlements, and people's insecurity about their futures has increased.

Madagascar

> ... Malagasy legislation dictates ... that private, nontitled land belongs to the state ... Furthermore, the state can use private titled lands according to its needs' (Randrianatoandro, 2009, p18).

In Manompana, the KoloAla policy allows devolving forest management rights and responsibilities to associations formed from local clusters of villages – a process that is just beginning (see Chapter 6). Forest zoning is underway at national and regional levels, and donors have initiated landscape, eco-regional initiatives (e.g. USAID, 2009; Freudenberger, 2010). Nevertheless, formal land-use planning at the landscape level remains a future possibility. At the local level, the KoloAla process includes initial mapping and zoning for each association territory and landscape, in this case, a forest corridor (Randrianatoandro, 2009) – a process in which field teams were involved.

Forestland belonging to the state that is converted to agricultural land becomes traditional family land and is generally so recognized in project-led land classifications. GELOSE (La Gestion Locale Sécurisée, Local Securised Management) was the first law to promote the transfer of natural resource management authority from the state to local communities. This act was followed by a decree known as GCF (La Gestion Contractualisée des Forêts, Contract Management of the Forests), which transferred only forest management rights. Both policies have been important components of forest decentralization and influenced, for a period, the basic principles of Madagascar's environmental policy. However, community forestry has yielded mixed results (Toillier et al, 2008; Montagne et al, 2007), and forest managers have faced pressing national changes. In 2000, the state forest service, already struggling to control logging, stopped delivering new permits, resulting in a period of generalized illegal logging. In 2003, the Durban Declaration, in which the president vowed to expand the nation's protected areas from 1.7 million to 6 million hectares, led to revisions in conservation and production forestry planning. A study by Rabenandrasana (2007) stressed the need for ensuring the country's timber supply and was an important precursor to the KoloAla approach for production forestry with local communities (see Chapter 6).

Madagascar's law recognizes two basic categories of trees; species in one category cannot be exploited, and those in the other are strictly regulated. Trees with local or cultural significance are simply 'illegible' to the state. Officially strictly controlled, Palissandre (*Dalbergia* sp), a beautiful wood, is popular for new buildings. Its ready availability in any wood market reflects Madagascar's general problems with law enforcement in an area so dramatically remote. Most commercial activities, such as hunting, technically require licences, but even more than in our other sites, remoteness makes compliance with external regulations unusual and improbable. Practically speaking, these people govern themselves.

Tanzania

The president of the nation is a trustee of all land on behalf of the citizens and has the authority to revoke land rights and take land for public use. The important 1999 Village Land Act has been described above. As in several of our sites, anything on Tanzania's 'forest estate' belongs to the state, and forested areas are likely to be considered as part of this land. Certainly there is comparatively strong governmental control of the protected areas near our field sites, Shambangeda and Misalai.

In southern Tanzania, more than 85 per cent of forests are on such 'open' lands (Milledge, 2009). In our field sites, 'the district officials seem to be mostly preoccupied with implementing the government control over tree harvesting' (Rantala et al, 2009, p36). Both local communities and districts see this as bureaucratic intrusion – particularly a required permit to harvest on-farm trees. There remains significant confusion over rights to trees, harvesting rules and regulations in villages.

When our team began work in the East Usambaras, a World Wide Fund for Nature (WWF) project was already supporting village land-use planning. We worked closely with the consultant and district-level personnel to enhance the quality of implementation. The boundaries of all three study villages had been recently demarcated, entitling them to apply for the certificates that would strengthen their management rights. We worked to enhance the depth of people's participation through visioning, pathway analysis, and participatory action research. Such multistakeholder engagement is consistent with the intent of Tanzania's decentralized laws and policies, but the team expressed its concerns about whether the process, led by outsiders, could be sustained; the costs were borne by the project, not the district or local communities.

There are two official ways that communities can be involved in tree management, through community-based forest management or via joint forest management (both described by Blomley et al, 2009). The first grants communities fuller rights and control; the second involves sharing of benefits between communities and the state. The mixing of ethnic groups, deriving partially from the Ujama'a policy of the 1960s,[13] has meant a different meaning for 'customary' law than holds in our other sites (discussed in Chapter 5). Still, there are contestations between government policy and local practice.

Strong vs weak governments

Legally, central governments in all five countries claim strong rights to control land and other natural resources. Remoteness and illegibility have so far granted communities much more autonomy than the laws would imply, but this situation is fast changing. Because of the potential for forestry to mitigate climate change, the constriction of their self-determination is likely to accelerate.

Land-use planning and related processes in these five landscapes have typically been among the most effective means to stimulate cross-sectoral and cross-scale discussions about how to use natural resources. They could be a useful mechanism for better collaborative governance, given more time to facilitate such processes (or as a more effective state wedge for further state penetration into people's daily lives). Our experience encourages us to seek collaborative governance (as probably the best available option for these peoples) while watching out for governmental minefields.

The forest management roles of NGOs and private industry in all the landscapes have been identified as problematic. Governments have accepted NGO and industry help in performing responsibilities supposedly belonging to the state (Netra and Craig, 2009, describe the syndrome plainly, in Laos). Here we confront a Catch-22 situation: governments need such help to strengthen their capacities to govern collaboratively – assuming that is a good thing – but such help often functions to increase dependency and reduce accountability, with adverse effects on governmental capacity building, potentially also affecting local people and forests.

Here we switch from looking at the legalities directly related to forests, trees and land, to three other policies with wide-ranging effects on landscapes and peoples.

Policies that Address the State–Hinterlands Interface

Policies on swidden agriculture, resettlement and participation relate to local peoples' comparative impotence in interaction with states whose power and intrusiveness are growing as communities become more accessible.

Again, we draw on Scott's (2009) interpretations. He juxtaposes the state-based paddy rice cultivation of Zomia with swidden-based systems in the hinterlands. The former are settled, organized, almost captive by nature, and can sustain the large population bases needed for defence and taxation. The latter involve movable fields and habitation, and crop diversity that defies accounting or reliable taxation.

Tropical forest peoples (including those represented in this book) typically practise swidden agriculture – a mode of production with an often undeserved bad name,[14] yet it has been dominant among virtually all tropical forest peoples. The lush tropical forests we struggle to sustain are typically linked to human systems that evolved with swiddening, and swiddens are also an integral part of most tropical forest peoples' cultural systems. Decisions about how communities allocate natural resources among their members, regulations about harvests and sales, norms relating to labour allocation – all part of local governance – are typically inextricably intertwined with swidden agriculture in tropical forests. The persistent and ubiquitous efforts by states to eliminate this system, insofar as the goal is met, reduce the effectiveness of traditional governance locally, typically failing to replace it with effective or benign state involvement (see Peluso 1994). Such sustained state effort, with increasingly effective (or intrusive) interaction between hinterlands and state, can whittle away at cultural diversity, local social capital, individual self-confidence and pride in one's culture. The landscape mosaics created by swidden systems are currently being transformed – often into less sustainable forms – by drivers such as population pressure, better access to markets and changing government policies (see *Human Ecology*, 2009, vol 37, for thorough analyses of such processes in Southeast Asia).

Powerful political decision-makers, in most cases, see the world differently from swiddeners in remote forested areas, often misunderstanding how local agroforestry systems work.[15] Such different perceptions, particularly given the typical political dominance of state actors, have had adverse effects on the lives of swiddeners in many countries. Resettlement policies – the removal of people (forced or voluntary) from their homelands to another location (see Chapter 4) – are one reflection of these misunderstandings.

Scott (2009) notes state efforts to render hinterland peoples and their systems more legible in the search for greater control. The efforts to resettle hinterland populations are an example: by settling them, encouraging eminently legible and taxable cash crop monocultures and defining their agricultural and living spaces, the state can make the people themselves more accessible. A sense of security about one's future is a 'good' to which people normally aspire; there is ample evidence of the dangers externally motivated resettlement poses to such security, for both migrating and host populations. Resettlement has been an issue, with variable ramifications, in all five countries.

Participation is an element in an imagined collaborative governance. People's right to have a voice is a widely accepted democratic principle within the field of governance, and their potential contribution to effective resource management has increasingly been recognized. Yet peoples living in tropical forests – unlikely to be part of their country's elites – typically have little voice. Governments in all five countries have formal laws encouraging citizen participation. Yet there are significant differences in the degree to which such laws translate into genuine or widespread citizen involvement. Policies on participation can be an opening for better dialogue between government actors and forest peoples. But we recognize that communication is a double-edged sword. It can lead to cooperation and mutual understanding, or it can lead to further oppression when the playing field is too tilted (cf Wollenberg et al, 2001) – a potential danger in efforts to govern collaboratively.

Below, we examine each country's policies on swidden agriculture, resettlement and participatory governance.

Cameroon

The illegality of swiddening in Cameroon is implicit. We found no explicit law, but deforestation in the national forest estate, where most swiddeners live and work, is illegal. Temporary deforestation is an integral part of a swidden system, and in Takamanda, people depend on swiddens for their livelihoods. Such legal ambiguity raises uncertainty about the future, particularly in areas zoned as conservation areas. Van Vliet et al (2009) note the absence of a formal agricultural policy as a major factor in land tenure conflicts. Farmers – in the absence of secure tenure – have difficulties getting credit and inputs, thereby also reducing Cameroon's competitiveness on the international stage.

Resettlement has a long history in Cameroon, beginning with German colonialism (Oyono et al, 2008); post-independence governments have removed some populations from protected areas and 'sedentarized' pygmies (Oyono 2005a). Forty-three villages are located within and around the Takamanda landscape, with three enclaved in the Takamanda National Park (Schmidt-Soltau et al, 2001). One village is enclaved inside the Mone Forest Reserve, and some of its agricultural fields extend into the reserve (Asaha and Fru, 2005). The resettlement option to address habitation inside the national park has been much discussed, although never implemented (Curran et al, 2009), despite claims to the contrary by Cernea and Schmidt-Soltau (2006) and Schmidt-Soltau (2003). Resettlement of these enclaves was rejected during the early 2000s, but was proposed as a possible future option in the recent draft management plan for the park (Van Vliet et al, 2009). In this area, it is a fear, rather than a reality, especially with the considerable out-migration from these enclaved communities.

The Cameroonian constitution provides several basic rights related to participation, such as a fair hearing and freedom of expression and association. Still, Njamnshi et al (2008) rated participation in natural resource policies as weak. A regulatory framework calling for civic and environmental education in schools is not reflected in practice (see Etoungou, 2002).

Community involvement in the conservation and management process for those living adjacent to forests has been part of Cameroon's forestry policy since 1994 (Assembe, 2009b). Donor- and government-led initiatives have been designed to facilitate communication at various levels (Ingram and Baan, 2006), but many problems remain (eg, Oyono 2005b; Oyono et al, 2008). In the anglophone regions of Cameroon (including Takamanda), the population has a longstanding discourse on their lack of political voice, entitlements and individual and group rights (Guedje et al, 2007; Konings and Nyamnjoh, 2003; Omosa and Walubengo, 1998; see also Sharpe, 1998). Similar problems for Cameroonian women's participation in natural resource management are identified by Brown and Lapuyade (2001) and Tiani et al (2005). Not surprisingly, trust and confidence remain issues in the relationship between government and its citizens.

Indonesia

Until recently, Bungo District was dominated by rice swiddens and rubber agroforests. Forest cutting and burning in the Indonesian national forest estate are illegal, and thus swiddening is also illegal, if ubiquitous. However, as noted in Chapter 3, the situation is changing rapidly, as both swiddens and rubber agroforests are converted into monocultures of oil palm or rubber (despite the dramatic, if short-lived, 2008 fall in international prices during the economic crisis).

Indonesian resettlement also has a long, uninterrupted history, since Dutch times. Today, much resettlement comes under the transmigration programme and involves people voluntarily resettling within their own or from adjacent provinces; many also move on their own initiative and with their own financial resources (rather than via government programmes). The landscapes we studied included spontaneous and government transmigrants: Tebing Tinggi villagers hosted a transmigration project on clonal rubber in 2000, providing land for a new transmigrant village that some locals joined. Local children now attend a secondary school built in the transmigration village. Our most accessible village, Desa Danau, is part of a huge oil palm transmigration area and has a more substantial transmigrant population (Bonnart, 2008). Conflicts between people indigenous to the area and in-migrants have been common in the past – over land, cropping patterns, government preference, and cultural and religious differences; however, since 2000 these sources of conflict appear to be diminishing in favour of those focused on land tenure.

Indonesia has passed several laws that encourage citizen participation. The Environmental Law (1997) and the Basic Forestry Law (1999) have strong participation elements (Wollenberg et al, 2009). Law No 25/2004, on local governance, mandated citizen involvement in land-use planning at each governmental level and has played an important role in Bungo District (Syamsuddin et al, 2007). As in Cameroon, however, Indonesian implementation lags behind the legal framework. Legal literacy remains a problem. Effective political participation will require major changes in centuries-old patterns, wherein the Java-dominated government has manifested paternalistic attitudes towards groups outside Java and had little interest in citizen input (similar to the Lao situation below and consistent with Scott's 2009 observations). There is some evidence that this has been changing since Indonesia's formal decentralization began in 2001.

Laos

Swidden agriculture for rice production is the dominant system of the mountainous landscape in Laos. There has been a clear, strong and long-lasting political push to eradicate it, notably and most recently through land-use planning designed to restrict swidden-related land use. This restriction has led to shorter fallow periods, fertility declines and weed proliferation (see Fitriana, 2008; Fujita and Phengsopha, 2008; or Baird and Shoemaker, 2005 on Laos generally). Delang (2007) documents plants known and used by the Khmu in various fallow stages (in northwestern Laos), showing some of the adverse effects of policies designed to reduce swiddening.

Laos saw a great deal of human movement in the 1970s, in connection with 'the American war', with a quarter of Laos affected by American bombardments (Phongoudome and Sirivong, 2007). In our Lao site, the people of Phadeng were resettled from their remote home near the Nam Et-Phou Louey National Protected Area to Phousaly, closer to the main road (Fitriana et al, 2009; Chapter 4). The government's 2009 decision to move the people of Phadeng, before the end of that year, came as a surprise and pre-empted a planned, multistakeholder, participatory workshop to identify the best location (see Watts, 2009) – demonstrating the power of the state in interaction with these hinterland peoples.

Fitriana (2008) shows the war-related human movements into our intermediate site, Bouammi, in contrast to the remote Phadeng, where in earlier times, migration was prompted by familial or economic concerns. Phadeng moved short distances several times in the past century as people sought better opium-growing areas (Fitriana, 2008). Some current Bouammi residents had earlier lived in Namtam (inside the nearby protected area); some lived in what is now the Bouammi sub-village of Vangmat; others were resettled in the current site in the 1970s and 1980s. The road that crosses the most accessible village, Muangmuay, was built in 1978, and people moved there between 1990 and 1994. Viengkham District has provided a rare opportunity to look at successive waves of resettlement. Such mobility fits well with Scott's views on the illegibility of local people.

The Lao government, despite formal policies mandating citizen participation, has long marginalized minority groups (including the Hmong and Khmu of our landscape), favouring the politically dominant lowland Lao (see Drouot, 1999; Ovesen, 2004; Milloy and Payne, 1997; or Singh's 2009 article on the 'performance' aspect of participation in Laos). This was evident in our project sites in the unilateral governmental decisions about resettlement (Chapter 4).

Madagascar

Rice swiddens dominate Manompana's agricultural landscape and, indeed, the country's whole eastern escarpment. Laws to restrict such agriculture have been in place for more than a century, but enforcement has been weak and no viable alternative has been identified. Swiddens are central to the traditional Malagasy diet, with rice eaten three times a day (as in Indonesia and Laos), and in its cultural association with ancestors, who represent spiritual 'guides' for local populations. At the same time, population growth has strengthened swidden's role in deforestation. There is a clear lack of coherence between the needs and interests of farmers and political discourses on this practice (cf Keller, 2008).

In Manompana itself, there have been no attempts to resettle populations, although people do move voluntarily in search of agricultural lands, typically from old villages to recently deforested valleys. Our remote village of Maromitety, for instance, is only ten years old. Nearby protected areas include enclaves, for which villagers and conservation officers have negotiated specific rules for resource access. This process has involved some resettlement and considerable conflict in the adjacent Masoala National Park (see Keller, 2008, 2009; Sodikoff, 2009) – events that raise concerns in nearby Manompana.

There is implicit support for citizen participation in Madagascar's decentralization laws, specifically in the management of public affairs and economic growth poles (Randrianatoandro, 2009; Ranjatson, 2009). The GELOSE law calls for 'environmental mediators' for which formal training now exists (see Madagascar's Poverty Reduction Strategy Papers, IMF, 2006). Development projects and officers sometimes prefer participatory methods as a means to strengthen local government actors' abilities in facilitating such processes. But there remain many problems with efforts to enhance citizen participation, especially with inadequate resources, in this rugged and remote landscape (Randrianatoandro, 2009).

The KoloAla programme would seem a good mechanism for community participation in forest governance, although the process is in the early stages. As in other sites, issues of legal literacy[16] remain a constraint to effective citizen participation, along with the governmental tendency to appoint rather than elect officials, regularly silencing them and inhibiting real dialogue (Randrianatoandro, 2009).

Tanzania

Tanzania appears to be further along in a trajectory towards settled agroforestry (see Chapter 5), as recently described for many parts of Southeast Asia (Cramb et al, 2009). In two research sites, cardamom is a central crop, grown in a swidden complex that includes both commercial and subsistence crops (Rantala and Lyimo, 2009b, 2009c). In Kwatango, the most remote site, food crops are central, with recent interest in incorporating oranges, palms and teak (Rantala and Lyimo, 2009a). State regulations and norms on land tenure and management are more accepted within these Landscape Mosaics communities than in the other four countries; the Tanzanian state also appears to provide more avenues for citizen input into decision-making.

Nyerere's Ujama'a policy, followed by 'villagization' in the early 1970s, resulted in dramatic ethnic mixing in the country. There were both spontaneous and government-directed moves for the people in our landscapes, involving power plays and negotiations among colonial and independent Tanzanian government actors, tea planters, forest managers and local people (see Vihemäki, 2009). In the mid-2000s, the establishment of the Derema Corridor (a conservation area designed to link the strictly protected and nearby Amani Nature Reserve to large forest blocks to the north) involved displacing members of five communities and cutting off their access to productive resources; more than 1128 farmers requested compensation. The resulting conflicts have been well described in Chapter 4. The resettlement experience in our sites is said to have been less painful than in other parts of Tanzania because people already lived in compact village settlements and thus mass relocations were avoided; in any event, the result has ultimately been a functioning, ethnically diverse citizenry.

Participation in Tanzania, with its left-wing history and implementation of Ujama'a, differs markedly from participation in the equally leftist Laos. Vihemäki (2009) documents the evolving discourses among conservation actors in Tanzania, moving over time from an instrumental and consultative approach to a much more fundamental co-management and benefit-sharing involvement of local people in conservation. The meshing of formal and various forms of customary governance demonstrates the participatory process at work (Chapter 5).

Hints from the Hinterlands

Our project tried, within the contexts described in this book, to find ways that local stakeholders could govern their areas together more harmoniously. This has meant not only looking at governance from the policy perspectives outlined above but also getting into the nitty-gritty of day-to-day management and governance. It has meant rendering local realities somewhat more legible, at least to the field teams.

Scott (2009) makes clear the diversity that characterizes the cultures and practices of hinterlands peoples – their agriculture, religious beliefs, politics, language, (il)literacy, culture – and the mutability of these features, even within a particular geographical area. Given that we also found such diversity and change, both within and among sites, we clearly cannot address site-level realities here in any comprehensive way. Instead, we have tried to pique readers' interests in exploring the greater diversity that exists in every tropical forest and recognizing the significance of that diversity for any genuinely collaborative governance. Here, we briefly recount the gender and ethnic differences we have observed, simply as samples of such complexity.

Gender equity

There are gender differences in knowledge, use, interests and benefits from forest resources for men and women everywhere, and there is widespread agreement that inequities exist. Southeast Asia is well known for its comparatively egalitarian relations between men and women, although the Hmong have been singled out as a more strongly patrilineal group – and thus potentially less egalitarian – than other Laotian groups (Lastarria-Cornhiel, 2007; see also Hakangard, 1990).

Jambi, Indonesia, represents a mélange of traditions: matrilineal Minangkabau, some stronger patrilineal elements in the indigenous Jambi Melayu (supported by the broader Islamic 'Great Tradition'; Redfield, 1971), bilateral in-migrating Javanese and every combination in between. Matrilineal and bilateral societies typically involve higher female status than patrilineal ones. Regardless of kinship structure, though, the women of Jambi are agriculturally productive (cf Colfer 1991; Sari 2007), a necessary if not sufficient requirement for high female status (Sanday, 1974).

In general, African societies have been less gender equitable. In Takamanda, for instance, many Anyang villagers belong to the two most powerful and respected sacred, secret male societies, the Ekpe and the Makwo (also noted by Egbo, 2009; Schmidt-Soltau, 2001). Although both sons and daughters reportedly inherit land, daughters may be driven away when their father dies (Van Vliet et al, 2009). Mdaihli et al (2002) found the minority Becheve to be particularly antagonistic to female education, and they document women's lower income levels (roughly half that of men), with each family member keeping the cash he or she earns. The changes currently underway, away from NTFP collection by the whole family and towards primarily male cocoa production, could well lower women's status further, as their economic base dwindles.

The ethnic diversity of the East Usambaras has involved associated gender differences. There are legal requirements for women to be involved in formal governance-related groups, but women have had difficulty speaking up in large, mixed groups. Their voices were stronger in smaller, more intimate settings (Rantala et al, 2009; Woodcock, 1995); similar patterns have been found elsewhere (see examples in Colfer's 2005 collection or Arora-Jonsson, 2008). In Madagascar, although women's involvement in agricultural production and household decision-making was acknowledged, their political voices were, as elsewhere, often muted vis-à-vis men's voices.

Although our experience has shown women's customary involvement in managing certain natural resources and their capacity for further involvement in new governance arrangements, there are strong barriers to making such arrangements equitable. One of the highest hurdles is the common governmental assumption – from the state perspective, unfamiliar with daily life in remote forests – that forestry is a man's domain (also common in formal forestry circles; see Chapters 6 and 7). Equitable collaborative governance will require extra effort to acknowledge and build on local women's understandings of NTFPs, the value of clean and abundant water and how to ensure it, the importance of maintaining access to resources for the coming generations, as well as the capacity to pass pertinent knowledge along to them. The invisibility of women in various conservation and development contexts has many parallels with the illegibility of remote peoples to the state.

Inequities among social groups

All our sites include at least two social groupings, with different levels of prestige in the wider society. The Lao situation has already been described, with the Lao Loum considered the most 'civilized', followed by the Khmu, with the Hmong at the bottom. In Indonesia, the situation is not quite so clear, but the in-migrating Javanese have had central government backing and associated power and prestige for centuries; only since 2001 have local ethnic groups (Minang, Jambi Melayu) gained formal, local political power. The Jambi Melayu, and to a lesser extent, the Minang (from neighbouring West Sumatra), have had the power that comes from first or early settlement, 'possession being nine-tenths of the law'.

The Cameroon sites have been inhabited by the Anyang almost exclusively, but there have been ongoing interactions with Nigerians across the border, as well as links to Cameroon's politically dominant francophone majority. The Anyang, like the Minang and Jambi Melayu, have had strength deriving from their association with the land; the Nigerians have had profitable knowledge of the wider world and prices; Cameroonian compatriots have shone in terms of displaying local political knowledge.

In Manompana, the Betsimisaraka have been the most common group in the forests, occasionally interacting with Madagascar's more dominant and commercially savvy ethnic groups, the Betseleo and Merina. Wider perceptions are parallel to those in Indonesia, with the in-migrating Betseleo and Merina taking on the self-defined, more 'civilized' roles. The Tanzanian situation is unique in that the area is a mélange of ethnic groups, so diverse that citizenship has begun to take the place of tribes – a situation that brings another of Scott's (2009) observations to mind:

> *The 'just-so' story of civilization always requires a wild untamed antagonist, usually just out of reach, to eventually be subdued and incorporated. The hypothetical civilization in question – whether French, Han, Burman, Kinh, British or Siamese – is defined by this negation. This is largely why tribes and ethnicity begin, in practice, where sovereignty and taxes stop.* (Scott, 2009, p335)

Ultimately, real collaborative governance will require that these traditional, state-derived perceptions change and that the views, preferences and capabilities of the 'tribes' and 'ethnic groups' be taken seriously—a process that appears in fact to be underway in Tanzania.[17]

Conclusions

Here we have focused on the policy contexts that impinge on peoples living in hinterlands of tropical forests. We have given some exposure to the kinds of incompatibilities and injustices that mark the typical interactions between states and hinterland people, and we have raised some red flags about the dangers we see in improved collaboration. Yet we remain ambivalent. We still see many pragmatic arguments for strengthening links between peoples and their governments, building on customary governance systems in the development of collaborative arrangements:

- Local people understand how customary governance works and usually find it acceptable (perhaps with a little tinkering).
- In situations where states have so far failed to govern effectively, customary governance provides a familiar framework for allocating rights and responsibilities, mediating disputes, and in many cases, imposing sanctions on scofflaws.
- Customary governance takes into account indigenous knowledge about the environment.
- Recognition of customary governance implies respect for local systems, which can facilitate broader collaboration and enhance local self-confidence, which in turn can lead to a more proactive approach to resource management.

The ethical argument for building on such systems has to do with the longstanding disregard for the desires, interests and rights of the peoples living in tropical hinterlands – a disregard acknowledged by governance scholars, whether those looking from the 'top' or the 'bottom', those accepting the legitimacy of states or those questioning it. Granting a greater degree of self-government, in collaboration with large-scale actors, seems the most pragmatic and potentially benign way forward in trying to increase resource sustainability, improve human well-being, and enhance justice. Keeping on track will require that we be mindful of the historical antagonism between hinterlands and states, the dismal track record of state policies and the variability that exists in the hinterlands.

Our attempt, in collaboration with partners and governments, to tinker with governance at the landscape level has been a frustrating, informative and interesting journey. We have documented the formal political systems of our sites in 'governance reports'; assembled a systematic, cross-site, multidisciplinary dataset covering issues of livelihoods, governance and the biophysical environment at various scales; pulled together this book's in-depth examinations of specific issues we found problematic, insightful and/or potentially useful; and continued to work with local communities and districts to fashion more equitable and sustainable elements in local systems on each site.

However, we do continue to wonder whether our efforts to foment collaboration are simply playing into the hands of states, many of whose officials primarily seek the resources (human, natural and economic) of the hinterlands, with little concern for the people or the environments in which they live.

We have imagined a significant role for local people in collaborative governance: as monitors of the behaviour of government and other powerful actors (industry, conservation, carbon traders) in the region. We have seen the people, with help from the many genuinely committed government actors, as potentially balancing the impersonal power of the state. But governance functions within systems of interacting forces. Making good governance a reality is unlikely until the people have the will and the power to have a significant voice in determining their destinies and to curb corruption at all levels.

Yet there are obvious conflicts. To what extent will local officials who are corrupt, for example, help local communities (or their NGO or research partners) become good monitors when good monitoring is likely to reduce illicit benefits, endanger bureaucratic prestige and probably complicate governmental planning?

One common solution, of which all the Landscape Mosaics countries have availed themselves, has been the use of NGOs and other outsiders to perform the duties that formal governments would perform in the West or North of the globe. Higher-level governments have routinely avoided devolving authority or providing resources needed to govern, sometimes using the presence of NGOs as an excuse to deny local governments the hands-on experience they need to govern effectively (Arnst, 1997; Baviskar, 2005; Netra and Craig, 2009).

Special areas of local concern, areas where collaboration with outside actors could play a positive role, include ensuring equity (between men and women, among social groups); gaining higher-level support for efforts to enforce locally determined rules and laws; and social learning that builds on improved analysis of local conditions, planning, administration, mediation, facilitation and conflict management.

Yet, if Scott (2009) is right that the state and the hinterlands have in some sense been defined by their opposition to one another, is it possible to bring these two together – under the new, more tightly linked global conditions in which we find ourselves today – into a lasting and productive collaboration? We readily acknowledge the difficulties we have had trying to do that, but the alternatives to such collaboration seem truly dismal for tropical forest peoples. The world has changed, and collaborative governance – no matter how difficult – appears to be the only potentially just way forward. Success will require proactively learning more about local conditions, working more closely over longer periods of time with local stakeholders and striving to eliminate these policy minefields.

Notes

1. Zomia is 'virtually all the lands at altitudes above roughly 300m all the way from the Central Highlands of Vietnam to northeastern India and traversing five Southeast Asian nations (Vietnam, Cambodia, Laos, Thailand, and Burma) and four provinces of China (Yunnan, Guizhou, Guangxi, and parts of Sichuan)' (Scott, 2009, pix).
2. Whereas in Zomia, the hinterlands (areas similar to our Landscape Mosaics landscapes) are in higher altitudes, in other contexts, remoteness rather than altitude may be the defining feature.
3. Scott (2009) notes the inevitability of state efforts to render hinterland peoples and their systems more 'legible', building on his earlier analysis, *Seeing Like a State*. The efforts to resettle hinterland populations are one example of this effort to make people more legible – by settling them, encouraging eminently legible and taxable cash crop monocultures and defining their agricultural and living spaces, thereby rendering them more accessible.
4. The scope of REDD has broadened with the adoption of a new acronym, REDD+ (UNFCCC, 2009). Whereas REDD was originally intended to incentivize reduced carbon emissions from deforestation and forest degradation, REDD+ also provides incentives for increases in carbon stocks and allows for emission reduction credits from a wider array of forest management practices (Blom et al, 2010, p2).
5. Those who characterize decentralization efforts as failures must ask, Are we comparing the results with our hopes, or with what would have happened with continued centralized rule? 'Developed', 'democratic' nations have oscillated over time between centralization and decentralization (cf Rose et al, 2005 on the United States; Küchli and Blaser, 2005 on Switzerland). Scott (2009), using different terminology, describes a similar historical process in Zomia, as the power of states vis-à-vis the hinterlands has waxed and waned.
6. See Assembe (2009a) on council forests and forest royalties from timber concessions; Zoa (2009) on community Forests; Sandker et al (2009) on community-managed hunting zones.
7. A 2003 World Bank study reported civil servants' perceptions about their forestry colleagues: more than half were believed to be taking bribes.
8. The Landscape Mosaics team did not report on corruption, focusing instead on livelihoods and formal government policies. This may have been related to the political risks of such attention (discussed in Chapter 11).
9. This law was subsequently revised (Law No 26/2007), to strengthen its incentives and disincentives and sanctions against lawbreakers.
10. We thank Heru Komarudin for clarifying these options.
11. Oberndorf reports that LFA has been replaced by a 2009 Participatory Land Use Planning and Allocation Policy (personal communication, Oberndorf, November 2009).
12. Sadly, these are precisely the effects that Colfer witnessed in 1979–80, as a result of similar efforts in the mid-1970s (and later) in Indonesian Borneo. The government resettled swiddeners and hunter-gatherers living in remote areas for the same reasons given by the Laotian government today (Colfer, 2008).
13. The Ujama'a policy of the 1960s involved forced relocation of vast numbers of citizens from various tribal groups into collective farming villages. See Wily (1997).

14. The poor reputation of swiddens contributes to swiddeners' lack of voice and to their likelihood of being resettled.
15. See Colfer and Dudley (1993) on Indonesian myths about swidden systems; Diaw (2005) on the source of such misunderstandings, particularly in Africa; the collection by Duncan (2004) or the more recent special issue of *Human Ecology* (vol 37) on swiddeners in Southeast Asia. It is worth noting that anthropologists and conservation biologists are also likely to have different assessments.
16. Scott (2009, p218) offers interesting perspectives on the advantages of illiteracy: 'If swiddening and dispersal are subsistence strategies that impede appropriation; if social fragmentation and acephaly hinder state incorporation; then, by the same token, the absence of writing and texts provides a freedom of maneuver in history, genealogy, and legibility that frustrates state routines'.
17. One wonders, however, about the terms of this trade. Are there important values the world is losing with the homogenization of cultures?

References

Adnan, H., Tadjudin, D., Yuliani, L., Komarudin, H., Lopulalan, D., Siagian, Y. L. and Munggoro, D. W. (eds) (2008) *Belajar Dari Bungo: Mengelola Sumberdaya Alam di Era Desentralisasi [Learning from Bungo: Managing Natural Resources in an Era of Decentralization]*, CIFOR, Bogor

Arnold, L. L. (2008) 'Deforestation in Decentralised Indonesia: What's Law Got to Do with it?' *Law, Environment and Development Journal*, vol 4, no 2, pp77–101

Arnst, R. (1997) 'International Development Versus the Participation of Indigenous Peoples', in D. McCaskill and K. Kampe (eds) *Development or Domestication? Indigenous Peoples of Southeast Asia*. Silkworm Books, Bangkok, pp441–454

Arora-Jonsson, S. (2008) 'A Different Vantage Point: Decentralization, Women's Organizing and Local Forest Management', in C. J. P. Colfer, G. R. Dahal and D. Capistrano (eds) *Lessons from Forest Decentralization: Money, Justice and the Quest for Good Governance in Asia-Pacific*, Earthscan/CIFOR, London, pp49–66

Asaha, S. and Fru, M. (2005) 'Socio-economic Survey of Livelihood Activities of the Communities in and around the Mone Forest Reserve and the Mbulu Forest Area, South West Province of Cameroon', Report for WCS, Limbe

Assembe, S. (2009a) 'Council Forests: The Case of Dimako', in M. C. Diaw, T. Aseh and R. Prabhu (eds) *In Search of Common Ground: Adaptive Collaborative Management in Cameroon*, CIFOR, Bogor, pp95–115

Assembe, S. (2009b) 'State Failure and Governance in Vulnerable States: An Assessment of Forest Law Compliance and Enforcement in Cameroon', *Africa Today*, vol 55, no 3, pp84–102

Auzel, P., Feteke, F., Fomete, T., Nguiffo, S. and Djeukam, R. (2004) 'Social and Environmental Costs of Illegal Logging in a Forest Management Unit in Eastern Cameroon', in R. M. Ravenel, I. M. E. Granoff and C. A. Magee (eds) *Illegal Logging in the Tropics: Strategies for Cutting Crime*, Haworth Press, New York, pp153–180

Baird, I. G. and Shoemaker, B. (2005) 'Aiding or Abetting? Internal Resettlement and International Aid Agencies in the Lao PDR', Report for Probe International, Toronto

Barr, C., Resosudarmo, I. A. P., Dermawan, A., McCarthy, J., with Moeliono, M. and Setiono, B. (eds) (2006) *Decentralization of Forest Administration in Indonesia:*

Implications for Forest Sustainability, Economic Development and Community Livelihoods, CIFOR, Bogor

Baviskar, A. (2005) 'Between Micro-politics and Administrative Imperatives: Decentralisation and the Watershed Mission in Madhya Pradesh, India', in J. C. Ribot and A. M. Larson (eds) *Democratic Decentralisation through a Natural Resource Lens,* Routledge, London, pp26–40

Blom, B., Sunderland, T. and Murdiyarso, D. (2010) 'Getting REDD to Work Locally: Lessons Learned from Integrated Conservation and Development Projects', *Environmental Science and Policy,* vol 13, no 2, pp164–172

Blomley, T., Ramadhani, H., Mkwizu, Y. and Böhringer, A. (2009) 'Hidden Harvest: Unlocking the Economic Potential of Community-Based Forest Management in Tanzania', in L. German, A. Karsenty and A. M. Tiani (eds) *Governing Africa's Forests in a Globalized World,* Earthscan/CIFOR, London

Bonnart, X. (2008) *How Can the Improved Livelihoods of Rural Community and the Biodiversity Conservation be Integrated in the Landscape Mosaics in Bungo District?* University of Tropical Region, Institut des Régions Chaudes-SupAgro, Montpellier

Brown, K. and Lapuyade, S. (2001) 'Changing Gender Relationships and Forest Use: A Case Study from Komassi, Cameroon', in C. J. P. Colfer and Y. Byron (eds) *People Managing Forests: The Links between Human Well Being and Sustainability,* Resources for the Future/CIFOR, Washington, DC, pp90–110

Cernea, M. M. and Schmidt-Soltau, K. (2006) 'Poverty Risks and National Parks: Policy Issues in Conservation and Resettlement', *World Development,* vol 34, no 10, pp1808–1830

Cerutti, P. and Fomété, T. (2008) 'The Forest Verification System in Cameroon', in D. Brown, K. Schreckenberg, N. Bird, P. Cerutti, F. D. Gatto, C. Diaw, T. Fomété, C. Luttrell, G. Navarro, R. Oberndorf, H. Thiel and A. Wells (eds) *Legal Timber: Verification and Governance in the Forest Sector,* CATIE/RECOFTC/CIFOR/ODI, London, pp135–145

Colfer, C. J. P. (1991) *Toward Sustainable Agriculture in the Humid Tropics: Building on the Tropsoils Experience in Indonesia,* North Carolina State University, Raleigh

Colfer, C. J. P. (ed) (2005) *The Equitable Forest: Diversity, Community and Natural Resources,* Resources for the Future/CIFOR, Washington, DC

Colfer, C. J. P. (2008) *The Longhouse of the Tarsier: Changing Landscapes, Gender and Well-Being in Borneo,* Phillips, Maine, Borneo Research Council, in cooperation with CIFOR and UNESCO

Colfer, C. J. P. and Capistrano, D. (eds) (2005) *The Politics of Decentralization: Forests, Power and People,* Earthscan, London

Colfer, C. J. P., Dahal, G. R. and Capistrano, D. (eds) (2008) *Lessons from Forest Decentralization: Money, Justice and the Quest for Good Governance in Asia-Pacific,* Earthscan/CIFOR, London

Colfer, C. J. P. and Dudley, R. G. (1993) *Shifting Cultivators of Indonesia: Managers or Marauders of the Forest? Rice Production and Forest Use among the Uma'Jalan of East Kalimantan,* FAO, Rome

Comiskey, J. A. and Dallmeier, F. (2003) 'Adaptive Management: A Framework for Biodiversity Conservation in Takamanda Forest Reserve', Cameroon, in J. A. Comiskey, T. C. H. Sunderland and J. L. Sunderland-Groves (eds) *Takamanda: The Biodiversity of an African Rainforest,* Smithsonian Institution, Washington, DC, pp9–17

Contreras, A. P. (ed) (2003) *Creating Space for Local Forest Management in the Philippines,* LaSalle Institute of Governance and Antonio Contreras, Manila

Cramb, R. A., Colfer, C. J. P., Dressler, W., Laungaramsri, P., Le, Q. T., Mulyoutami, E., Peluso, N. L. and Wadley, R. L. (2009) 'Swidden Transformations and Rural Livelihoods in Southeast Asia', *Human Ecology,* vol 37, pp323–346

Curran, B., Sunderland, T. C. H., Maisels, S., Oates, J., Asaha, S., Balinga, M., Defo, L., Dunn, A., Telfer, P., Usongo, L., Von Loebenstein, K. and Roth, P. (2009) 'Are Central Africa's Protected Areas Displacing Hundreds of Thousands of Rural Poor?' *Conservation and Society,* vol 7, no 1, pp30–45

Delang, C. O. (2007) 'Ecological Succession of Usable Plants in an Eleven-year Fallow Cycle in Northern Lao P.D.R.', *Ethnobotany Research and Applications,* vol 5, pp331–350

Diaw, M. C. (2005) 'Modern Economic Theory and the Challenge of Embedded Tenure Institutions: African Attempts to Reform Local Forest Policies', in S. Kant and R. A. Berry (eds) *Sustainability, Institutions, and Natural Resources: Institutions for Sustainable Forest Management,* Springer Publishers, The Netherlands, pp44–82

Drouot, G. (1999) 'Pouvoir et Minorités Ethniques au Laos. De la Reconnaissance Constitutionnelle à la Participation Effective à l'Exercice du Pouvoir [Power and Ethnic Minorities in Laos: From Constitutional Recognition to Effective Participation to the Exercise of Power]', *Mousson: Social Science Research on Southeast Asia,* vol 99, pp53–74

Dudley, R. G. (2000) 'The Rotten Mango: The Effect of Corruption on International Development Projects, Part 1: Building a System Dynamics Basis for Examining Corruption', Paper presented at Sustainability in the Third Millennium - Eighteenth International Conference of the System Dynamics Society, Bergen, Norway

Dudley, R. G. (2002) 'Dynamics of Illegal Logging in Indonesia', in C. J. P. Colfer and I. A. P. Resosudarmo (eds) *Which Way Forward? People, Forests and Policymaking in Indonesia,* Resources for the Future/CIFOR, Washington, DC, pp358–382

Dudley, R. G. (2004) 'A System Dynamics Examination of the Willingness of Villagers to Engage in Illegal Logging', *Journal of Sustainable Forestry,* vol 19, no 1, pp31–53

Duncan, C. R. (ed) (2004) *Civilizing the Margins: Southeast Asian Government Policies for the Development of Minorities,* Cornell University Press, Ithaca, New York

Edmunds, D. and Wollenberg, E. (eds) (2003) *Local Forest Management,* Earthscan, London

Egbe, S. (1997) 'Forest Tenure and Access to Forest Resources in Cameroon: An Overview', Forest Participation Series-Forestry and Land Use Programme, International Institute for Environment and Development (IIED), London

Egbo (2009) www.en.wikipedia.org/wiki/Ekpe (accessed 11 November 2009)

Elias, S. and Wittman, H. (2005) 'State, Forest and Community: Decentralization of Forest Administration in Guatemala', in C. J. P. Colfer and D. Capistrano (eds) *The Politics of Decentralization: Forests, Power and People,* Earthscan/CIFOR, London, pp282–295

Etoungou, P. (2002) *Decentralization Viewed from Inside: The Implementation of Community Forests in East Cameroon.* Report for World Resources Institute, Yaoundé, Cameroon

Fay, C. and Sirait, M. (2001) 'Reforming the Reformists: Challenges to Government Forestry Reform in Post-Soeharto Indonesia', in C. J. P. Colfer and I. A. P. Resosudarmo (eds) *Which Way Forward? Forest People and Policymaking in Indonesia,* Resources for the Future, Washington, DC, pp126–143

Feintrenie, L. and Levang P. (2009) 'Sumatra's rubber agroforests: advent, rise and fall of a sustainable cropping system', *Small Scale Forestry*, vol 8, no 3, pp323–335

Feintrenie, L. and Levang, P. (2010) 'The Alarming Fate of Jambi's Forests: A General Consensus on their Conversion', *Forests, Trees and Livelihoods*

Ferguson, J. (1994) *The Anti-Politics Machine: Development, Depoliticization, and Bureaucratic Power in Lesotho*, University of Minnesota Press, Minneapolis

Fitriana, J. R. (2008) *Landscape and Farming System in Transition: Case Study in Viengkham District, Luang Prabang Province, Lao PDR*. Institut des Régions Chaudes-SupAgro, Montpellier

Fitriana, J. R., Boucard, A., Vongkhamsao, V., Langford, K. and Watts, J. (2009) 'Governance Report for the Landscape Mosaics Project: Laos', Report for CIFOR, Bogor

Foppes, J. (2008) 'Knowledge Capitalization: Agriculture and Forestry Development at "Kum Ban" Village Cluster Level in Lao PDR', Report for LEAP (Laos Extension for Agriculture Project) and NAFES, Vientiane

Freudenberger, K. (2010) Paradise Lost? Lessons from 25 years of environmental programs in Madagascar, USAID/International Resources Group, Washington, DC

Fujita, Y. and Phengsopha, K. (2008) 'The Gap Between Policy and Practice in Lao PDR', in C. J. P. Colfer, G. R. Dahal and D. Capistrano (eds) *Lessons from Forest Decentralization: Money, Justice and the Quest for Good Governance in Asia-Pacific*, Earthscan/CIFOR, London, pp117–132

Gasana, J. K. and Samyn, J.-M. (2009) 'The Africa Forest Law Enforcement and Governance (AFLEG) and Forest Law Enforcement, Governance and Trade (FLEGT) Processes and the Challenges of Forest Governance in African Tropical Timber-exporting Countries', in L. German, A. Karsenty and A. M. Tiani (eds) *Governing Africa's Forests in a Globalized World*, Earthscan/CIFOR, London, pp299–316

German, L., Karsenty, A. and Tiani, A.-M. (eds) (2009) *Governing Africa's Forests in a Globalized World*, Earthscan/CIFOR, London

Guedje, N. M., Zuidema, P. A., During, H., Foahom, B. and Lejoly, J. (2007) 'Tree Bark as a Non-timber Forest Product: The Effect of Bark Collection on Population Structure and Dynamics of *Garcinia lucida Vesque*', *Forest Ecology and Management*, vol 240, pp1–12

Hakangard, A. (1990) 'Women in Shifting Cultivation: Luang Prabang Province Lao P.D.R.', Report for Stockholm University, Stockholm

Hasan, U., Irawan, D. and Komarudin, H. (2008) 'Rio: Modal Sosial Sistem Pemerintah Desa [Rio: Social Capital of the Village Governance System]', in H. Adnan, D. Tadjudin, E. L. Yuliani, H. Komarudin, D. Lopulalan, Y. L. Siagian and D. W. Munggoro (eds) *Belajar dari Bungo: Mengelola Sumberdaya Alam di Era Desentralisasi*, CIFOR, Bogor, pp136–156

Horning, N. R. (2009) 'Bridging the Gap Between Environmental Decision-makers in Madagascar', in L. German, A. Karsenty and A. M. Tiani (eds) *Governing Africa's Forests in a Globalized World*, Earthscan/CIFOR, London

IMF (2006) 'Republic of Madagascar Poverty Reduction Strategy Paper Annual Implementation Report 2005', Report for International Monetary Fund, Washington, DC

Ingram, V. and Baan, P. D. (2006) 'Giving Birth to Good Forestry Governance', SNV (occasional paper), Yaoundé, Cameroon

Irawan, D., Hasan, U. and Komarudin, H. (2008) 'Penataan Ruang untuk Memperkuat Hak Properti Masyarakat [Land-use Planning for Strengthening

Community Property Rights]', in H. Adnan, D. Tadjudin, L. Yuliani, H. Komarudin, D. Lopulalan, Y. Siagian and D. W. Munggoro (eds) *Belajar dari Bungo: Mengelola Sumberdaya Alam di Era Desentralisasi*, CIFOR, Bogor, pp388–403

Kadjatmiko (2008) 'Implementing Decentralization: Lessons from Experiences in Indonesia', in C. J. P. Colfer, G. R. Dahal and D. Capistrano (eds) *Lessons from Forest Decentralization: Money, Justice and the Quest for Good Governance in Asia-Pacific*, Earthscan/CIFOR, London, pp149–162

Karsenty, A. (2009) 'The New Economic "Great Game" in Africa and the Future of Governance Reforms in the Forestry Sector', in L. German, A. Karsenty and A. M. Tiani (eds) *Governing Africa's Forests in a Globalized World*, Earthscan/CIFOR, London, pp79–100

Keller, E. (2008) 'The Banana Plant and the Moon: Conservation and the Malagasy Ethos of Life in Masoala, Madagascar', *American Ethnologist*, vol 35, no 4, pp650–664

Keller, E. (2009) 'Who are "They"? Local Understandings of NGO and State Power in Masoala, Madagascar', *Tsantsa*, vol 4, pp76–85

Komarudin, H., Siagian, Y. and Colfer, C. J. P. (forthcoming) 'The Role of Collective Action in Securing Property Rights for the Poor: A Case Study in Jambi Province, Indonesia', in E. Mwangi, H. Markelova, and R. Meinzen-Dick (eds) *Collective Action and Property Rights for Poverty Reduction*, University of Pennsylvania Press, Philadelphia

Komarudin, H., Siagian, Y. L. and Colfer, C. J. P., with Neldysavrino, Yentirizal, Syamsuddin and Irawan, D. (2008) *Collective Action to Secure Property Rights for the Poor: A Case Study in Jambi Province, Indonesia*, Washington, DC, Collective Action and Property Rights System-Wide Initiative

Konings, P. and Nyamnjoh, F. B. (2003) *Negotiating an Anglophone Identity: A Study of the Politics of Recognition and Representation in Cameroon*, Brill, Leiden

Küchli, C. and Blaser, J. (2005) 'Forests and Decentralization in Switzerland: A Sampling', in C. J. P. Colfer and D. Capistrano (eds) *The Politics of Decentralization: Forests, Power and People*, Earthscan/CIFOR, London, pp152–165

Kusumanto, Y. (2001) 'Context Study Report – Adaptive Collaborative Management (ACM) Research in Jambi, Indonesia (Policies)', Report for CIFOR, Bogor

Laird, S. A., Ingram, V., Awono, A., Ndoye, O., Sunderland, T., Lisinge, E. and Nkuinkeu, R. (2010) 'Bringing Together Customary and Statutory Systems: The Struggle to Develop a Legal and Policy Framework for NTFPs in Cameroon', in S. A. Laird, R. J. Mclain and R. P. Wynberg (eds) *Wild Product Governance*, Earthscan, London, pp53–70

Larson, A., Barry, D., Dahal, G. R. and Colfer, C. J. P. (eds) (2010) *Forest for People: Community Rights and Forest Tenure Reform*, Earthscan/CIFOR, London

Lastarria-Cornhiel, S. (2007) 'Who Benefits From Land Titling? Lessons from Bolivia and Laos', IIED Gatekeeper Series, vol 132, p24

Lehébel-Péron, A. (2008) *Evaluation of the Production Potential of Complex Agroforests: The Example of Rubber Agroforests in Lubuk Beringin (Indonesia)*, University of Technology and Sciences, Montpellier

Logo, P. B. (2010) 'Governance of Decentralized Forest Revenue in Central Africa: For Better or For Worse?', in L. German, A. Karsenty and A. M. Tiani (eds) *Governing Africa's Forests in a Globalized World*, Earthscan/CIFOR, London, pp174–190

Markelova, H. and Swallow, B. (2008) 'Bylaws and their Critical Role in Natural Resource Management: Insights from African Experience', Paper Presented at 12th Biennial Conference of the International Association for the Study of Commons, Cheltenham, UK, 14–18 July 2008

Mdaihli, M., Feu, T. D. and Ayeni, J. S. O. (2003) 'Fisheries in the Southern Border Zone of Takamanda Forest Reserve, Cameroon', in J. A. Comiskey, T. C. H. Sunderland and J. L. Sunderland-Groves (eds) *Takamanda: The Biodiversity of an African Rainforest.* Smithsonian Institution, Washington, DC, pp141–154

Mdaihli, M., Schmidt-Soltau, K. and Ayeni, J. S. O. (2002) 'Socio-economic Baseline Survey of the Villages in and Around the Takamanda Forest Reserve', Report for MINEF/GTZ Project for the Protection of Forests around Akwaya (PROFA), Yaoundé

Milledge, S. (2009) 'Forestry Governance and Trade Transformations: Experiences from Tanzania and Implications for Sustainable Development', in L. German, A. Karsenty and A. M. Tiani (eds) *Governing Forests in a Globalizing World,* Earthscan/CIFOR, London, pp282–298

Milloy, M. J. and Payne, M. (1997) 'My Way and the Highway: Ethnic People and Development in the Lao PDR', in D. McCaskill and K. Kampe (eds) *Development or Domestication? Indigenous Peoples of Southeast Asia,* Silkworm Books, Bangkok, pp398–440

Moeliono, M., Wollenberg, E. and Limberg, G. (eds) (2009) *The Decentralization of Forest Governance: Politics, Economics, and the Fight for Control of Forests in Indonesian Borneo,* Earthscan/CIFOR, London

Montagne, P., Razanamaharo, Z. and Cooke, A. (2007) 'Tanteza – Le Transfert de Gestion à Madagascar, 10 ans d'Efforts' [Tanteza – Management Transfer in Madagascar, 10 Years of Effort], Report for CIRAD, Montpellier

Morris, J., Hicks, E., Ingles, A. and Ketphanh, S. (2004) *Linking Poverty Reduction with Forest Conservation: Case Studies from Lao PDR,* IUCN, Bangkok

Netra, E. and Craig, D. (2009) 'Accountability and Human Resource Management in Decentralised Cambodia', Report for Cambodia Development Research Institute, Phnom Penh

Njamnshi, A., Nchunu, J. S., Galega, P. T. and Chili, P. C. (2008) 'Environmental Democracy in Cameroon: An Assessment of Access to Information, Participation in Decision-making, and Access to Justice in Environmental Matters', Report for The Access Initiative Cameroon (TAI), Yaoundé

Oberndorf, R. (2009) 'Village Forestry in the Lao PDR: Protecting Forest Resources and Supporting Local Livelihoods', CIFOR/RECOFTC

Ojha, H. R. (2008) *Reframing Governance: Understanding Deliberative Politics in Nepal's Terai Forestry,* Adroit Publishers, New Delhi

Omosa, E. and Walubengo, D. (1998) 'FTPP in Anglophone Africa', Report for University of Makerere, Kampala

Ovesen, J. (2004) 'All Lao? Minorities in the Lao People's Democratic Republic', in C. R. Duncan (ed) *Civilizing the Margins: Southeast Asian Government Policies for the Development of Minorities,* Cornell University Press, Ithaca, New York, pp214–240

Oyono, P. R. (2004) 'One Step Forward, Two Steps Back? Paradoxes of Natural Resources Management Decentralisation in Cameroon', *Journal of Modern African Studies,* vol 421, pp91–111

Oyono, P. R. (2005a) 'From Diversity to Exclusion for Forest Minorities in Cameroon', in C. J. P. Colfer (ed) *The Equitable Forest: Diversity, Community, and Resource Management,* RFF/CIFOR, Washington, DC, pp113–130

Oyono, P. R. (2005b) 'Profiling Local-level Outcomes of Environmental Decentralizations: The Case of Cameroon's Forests in the Congo Basin', *Journal of Environment and Development,* vol 14, no 2, pp1–21

Oyono, P. R., Kombo, S. S. and Biyong, M. B. (2008) 'New Niches of Community Rights to Forests in Cameroon: Cumulative Effects on Livelihoods and Local Forms of Vulnerability', Report for CIFOR (RRI), Bogor

Peluso, N. L. (1994) *The Impact of Social and Environmental Change on Forest Management: A Case Study from West Kalimantan, Indonesia,* FAO, Rome

Phongoudome, C. and Sirivong, K. (2007) 'Forest Restoration and Rehabilitation in Lao PDR', in D. K. Lee (ed) *Keep Asia Green,* vol 1, Southeast Asia, AKECOP, Yuhan-Kimberly, IUFRO, Vienna, pp57–84

Rabenandrasana, J. C. (2007) 'Actualisation des Données de Base sur la Production et la Consommation en Produit Forestier Ligneux à Madagascar' [Realization of the Baseline Data on Production and Consumption of Ligneous Forest Products in Madagascar], Report for IRG, USAID, Antananarivo, Madagascar

Randrianatoandro, E. A. (2009) *Decentralisation – Information et Communication sur les Questions Aménagements du Territoire et Gestion des Ressources Naturelles: Etudes de Cas: La Gestion Décentralisée et Adaptative du Corridor Forestier de Manompana Suivant le Processus "Kolala"* [Decentralization – Information and Communication on Administrative Questions Related to Lands and Management of Natural Resources: Case Studies: Decentralized and Adaptive Management of the Manompana Forest Corridor According to the KoloAla Process], Memoire de fin d'Etude en Vue d'Obtention d'un Diplome d'Etude Approfondie en Foresterie, Développement et Environnement, CIFOR

Ranjatson, J. P. (2009) 'La Gouvernance Nationale du Processus KoloAla et Quelques Implications pour le Projet KAM' [National Governance of the KoloAla Process and Some Implications for the KAM Project], Report for ESSA-Forêts, Université d'Antananarivo, Antananarivo, Madagascar

Ranjatson, P. (2008) 'Prestation de Service pour l'Analyse des Politiques Publiques dans le Cadre du Projet "Landscape Mosaics"' [Performance of service for the Analysis of Public Policy in the Framework of the Landscape Mosaics Project], Report for CIFOR, Bogor

Rantala, S. and Lyimo, E. (2009a) 'Summary on Livelihoods and Conservation – Kwatango', Report for CIFOR, Bogor

Rantala, S. and Lyimo, E. (2009b) 'Village Feedback Workshop Report – Misalai Village', Report for CIFOR, Bogor

Rantala, S. and Lyimo, E. (2009c) 'Village Feedback Workshop Report – Shambangeda Village', Report for CIFOR, Bogor

Rantala, S., Lyimo, E., Powell, B., Kitalyi, A. and Vihemäki, H. (2009) 'Natural Resource Governance and Stakeholders in the East Usambara Mountains: Governance Report', Report for CIFOR and ICRAF, Bogor

Redfield, R. (1971) *The Little Community and Peasant Society and Culture,* University of Chicago Press, Chicago

Resosudarmo, I. A. P. and Dermawan, A. (2002) 'Forests and Regional Autonomy: The Challenge of Sharing the Profits and Pains', in C. J. P. Colfer and I. A. P. Resosudarmo (eds) *Which Way Forward? People, Forests and Policymaking in Indonesia,* Resources for the Future/CIFOR, Washington, DC, pp325–357

Ribot, J. (2008) *Building Local Democracy Through Natural Resource Interventions: An Environmentalist's Responsibility,* World Resources Institute,

Washington, DC
Ribot, J. C. and Larson, A. M. (eds) (2005) *Democratic Decentralisation through a Natural Resource Lens*, Routledge, London
Rights and Resources Initiative (2010) *The End of the Hinterland: Forests, Conflict and Climate Change*, RRI, Washington, DC
Roda, J.-M. (2009) 'On the Nature of Intergenerational and Social Networks in the African Forest Sector: The Case of Chinese, Lebanese, Indian and Italian Business Networks', in L. German, A. Karsenty and A. M. Tiani (eds) *Governing Africa's Forests in a Globalized World*, Earthscan/CIFOR, London
Rose, G. A., with MacCleery, D. W., Lorensen, T. L., Lettman, G., Zumeta, D. C., Carroll, M., Boyce, T. C. and Springer, B. (2005) 'Forest Resources Decision-making in the US', in C. J. P. Colfer and D. Capistrano (eds) *The Politics of Decentralization: Forests, Power and People*, Earthscan/CIFOR, London, pp238–252
Sanday, P. (1974) 'Female Status in the Public Domain', in M. Z. Rosaldo and L. Lamphere (eds) *Woman, Culture and Society*, Stanford University Press, Stanford, California, pp189–206
Sandker, M., Campbell, B. M., Nzooh, Z., Sunderland, T., Amougou, V., Defo, L. and Sayer, J. (2009) 'Exploring the Effectiveness of Integrated Conservation and Development Interventions in a Central African Forest Landscape', *Biodiversity Conservation*, vol 18, pp2875–2892
Sari, E. P. (2007) 'Kembali ke Ladang, Menggapai Asa' [Return to the Upland Ricefield, Reach for your Dream], in Y. Indriatmoko, E. L. Yuliani, Y. Tarigan, F. Gaban, F. Maulana, D. W. Munggoro, D. Lopulalan and H. Adnan (eds) *Dari Desa ke Desa: Dinamika Gender dan Pengelolaan Kekayaan Alam*, CIFOR, Bogor, pp71–82
Schmidt-Soltau, K. (2001) 'Human Activities in and Around the Takamanda Forest Reserve: Socio-economic Baseline Survey [Draft Report 27/08/01]', Report for PROFA (Cameroonian-German Project for the Protection of Forests around Akwaya), Bogor
Schmidt-Soltau, K. (2003) 'Conservation-related Resettlement in Central Africa: Environmental and Social Risks', *Development and Change*, vol 34, no 3, pp1–27
Scott, J. C. (2009) *The Art of Not Being Governed: An Anarchist History of Upland Southeast Asia*, Yale University Press, New Haven, Connecticut
Sharpe, B. (1998) 'First the Forest: Conservation, "Community" and "Participation" in South-West Cameroon', *Africa*, vol 68, no 1, pp25–45
Singh, S. (2009) 'World Bank-directed Development? Negotiating Participation in the Nam Theun 2 Hydropower Project in Laos', *Development and Change*, vol 40, no 3, pp487–507
Sodikoff, G. (2009) 'The Low-wage Conservationist: Biodiversity and Perversities of Value in Madagascar', *American Anthropologist*, vol 111, no 4, pp443–455
Stuart-Fox, M. (2006) 'Historical and Cultural Constraints on Development in the Mekong Region', paper presented at Accelerating Development in the Mekong Region: The Role of Economic Integration, Siem Reap, Cambodia, 26–27 June 2006
Sunderland, T., Asaha, S., Balinga, M. and Isoni, O. (2010) 'Regulatory Issues for Bush Mango (*Irvingia* spp.) Trade in Southwest Cameroon and Southeast Nigeria', in S. A. Laird, R. J. Mclain and R. P. Wynberg (eds) *Wild Product Governance*, Earthscan, London, pp77–84

Sunderland-Groves, J. L. and Maisels, F. (2003) 'Large Mammals of Takamanda Forest Reserve, Cameroon', in J. A. Comiskey, T. C. H. Sunderland and J. L. Sunderland-Groves (eds) *Takamanda: The Biodiversity of an African Rainforest,* Smithsonian Institution, Washington, DC, pp111–127

Sunderlin, W. D. (2006) 'Poverty Alleviation Through Community Forestry in Cambodia, Laos, and Vietnam: An Assessment of the Potential', *Forest Policy and Economics,* vol 8, pp386–396

Sunderlin, W. D. (2008) 'From Exclusion to Ownership? Challenges and Opportunities in Advancing Forest Tenure Reform', Report for Rights and Resources Initiative, Washington, DC

Syamsuddin, Neldysavrino, Komarudin, H. and Siagian, Y. L. (2007) 'Are Community Aspirations Being Accommodated in Development Plans? A Lesson from Collective Action in Jambi', CIFOR Governance Briefs, vol 34, p12

Tiani, A. M., Akwah, G. and Nguiébouri, J. (2005) 'Women in Campo Ma'an National Park: Uncertainties and Adaptations', in C. J. P. Colfer (ed) *The Equitable Forest: Diversity, Community and Resource Management,* RFF/CIFOR, Washington, DC, pp131–149

Toillier, A., Lardon, S. and Hervé, D. (2008) 'An Environmental Governance Support Tool: Community-based Forest Management Contracts (Madagascar)', *International Journal of Sustainable Development,* vol 11, no 2/3/4

UNFCCC (United Nations Framework Convention on Climate Change) (2009) Proceedings of the 5th session of the ad hoc working group on long-term cooperative action under the convention, United Nations Framework Convention on Climate Change, Bonn

USAID (2009) 'USAID-funded Transparency Initiative Improves Good Governance Across all Sectors: Fighting Corruption in Madagascar', *USAID Telling Our Story,* available at www.usaid.gov/stories/madagascar/cs_madagascar_corruption.html (accessed 16 November 2009)

Van Vliet, N., Asaha, S., Ingram, V. and Sunderland, T. (2009) 'Governance Report, Cameroon Site: Takamanda Mone Technical Operations Unit', Report for CIFOR, Bogor

Vihemäki, H. (2009) 'Participation or Further Exclusion? Contestations Over Forest Conservation and Control in the East Usambara Mountains, Tanzania', PhD dissertation, University of Helsinki

Watts, J. (2009) 'Phadeng Field Trip Report: 3–9 June 2009', Report for CIFOR, Bogor

Wells, A. (2008) 'Verification of Legal Compliance in Indonesia', in D. Brown, K. Schreckenberg, N. Bird, P. Cerutti, F. D. Gatto, C. Diaw, T. Fomété, C. Luttrell, G. Navarro, R. Oberndorf, H. Thiel and A. Wells (eds) *Legal Timber: Verification and Governance in the Forest Sector,* CATIE, RECOFTC, CIFOR, ODI, London, pp173–186

Wescott, C. (2003) 'Combating Corruption in Southeast Asia', in J. B. Kidd and F. J. Richter (eds) *Fighting Corruption in Asia: Causes, Effects and Remedies,* World Scientific Press, Singapore

Wily, L. A. (1997) *Finding the Right Institutional and Legal Framework for Community-Based Natural Forest Management: The Tanzanian Case,* CIFOR, Bogor

Wily, L. A. and Dewees, P. (2001) 'From Users to Custodians: Changing Relations Between People and the State in Forest Management in Tanzania', World Bank Policy Research Working Paper, vol 2569, pp1–31

Wollenberg, E., Anderson, J. and Edmunds, D. (2001) 'Pluralism and the Less Powerful: Accommodating Multiple Interests in Local Forest Management', *International Journal of Agricultural Resources, Governance and Ecology*, vol 1, no 3/4, pp199–222

Wollenberg, E., Moeliono, M. and Limberg, G. (2009) 'Between State and Society: Decentralization in Indonesia', in M. Moeliono, E. Wollenberg and G. Limberg (eds) *The Decentralization of Forest Governance: Politics, Economics and the Fight for Control of Forests in Indonesian Borneo*, Earthscan/CIFOR, London, pp3–23

Woodcock, K. A. (1995) 'Indigenous Knowledge and Forest Use: Two Case Studies from the East Usambaras, Tanzania', Tanga, Ministry of Natural Resources and Tourism, Tanzania; Forestry and Beekeeping Division, Department of International Development Cooperation, Finland, Finnish Forest and Park Service

World Bank (2003) 'Combating Corruption in Indonesia: Enhancing Accountability for Development', Report for World Bank – East Asia Poverty Reduction and Economic Management Unit, Jakarta

Yen, M. H., Preece, L. D. and Lan, N. N. (in preparation) 'A Review of Conservation Area Governance in Cambodia, Laos and Vietnam', in T. C. H. Sunderland (ed) *Conservation Area Governance in the Lower Mekong*, CIFOR, Bogor

Zoa, M. (2009) 'Community Forests: Reconciling Customary and Legal Concepts', in M. C. Diaw, T. Aseh and R. Prabhu (eds) *In Search of Common Ground: Adaptive Collaborative Management in Cameroon*, CIFOR, Bogor, pp117–138

11

The Essential Task of 'Muddling Through' to Better Landscape Governance

Carol J. Pierce Colfer, Jean-Laurent Pfund and Terry Sunderland

In 2007, we planned to facilitate negotiations and support agreements between communities and governments relating to selected policy processes that would include attention to both local people's lives and environments. We hoped to have strengthened the abilities of communities within each landscape and their subgroups to deal effectively with powerful external actors (such as government,[1] industry and conservation officers) and with their own elites. Land-use planning was the policy terrain in which we worked in Laos, Indonesia and Tanzania; forest management and conservation were the issues in Cameroon and Madagascar. We expected that our facilitation could enhance social learning and collective action processes for empowerment at both community and district levels. Although we have made some strides in this endeavour, with rates differing from site to site, we have faced a series of problems and surprises, ranging from administrative difficulties in launching activities to relationships among partners, to totally unexpected events like political instability in Madagascar, gold mining in Laos and rivalries linked to the location of park headquarters in Cameroon. Our experience has demonstrated clearly that more time, resources and a long-term focus on the adaptive capacity at all levels are needed, as is follow-up, in terms of action and research, on these early findings.

Table 11.1 *Significant issues in collaborative landscape governance*

The powerful duo of government and industry

- Oil palm expansion, Sumatra (Chapter 3)
- Officials' intervention and collusion in governance issues related to exploitation of NTFPs, protected species and timber (Chapters 8, 10)

Risks linked to some national policies

- Differing resettlement and displacement experiences, Laos and Tanzania (Chapter 4), and wider concerns about such policies (Chapter 10)
- Forest management focus on men and timber, without complementary income-generating and gender-balanced activities, Madagascar (Chapter 6)
- Reductions in future landscape options with moves towards monocultures, Indonesia (Chapter 3)
- Illegality of swiddens (predominant, diversified source of local foods), all sites (Chapter 10)

Complexities of pluralistic governance

- Shared development of local governance norms in multi-ethnic contexts, Tanzania (Chapter 5)
- Agricultural, rather than forestry, orientation of some forest-dwelling ethnic groups, Madagascar (Chapter 6)
- Adaptability of villagers to changing contexts (Chapters 5, 6, 7, 8)
- Differing relations between hinterland groups and governments (Chapters 5, 6, 10)

Differences in cultural significance and governance of NTFPs, including differentiation in roles between sexes and among social groups

- Varying labour allocation, values, uses, regulation and benefit distribution in NTFP management, among sites (Chapters 7, 8)
- Timber vs *Pandanus* management, Madagascar (Chapter 6)

Discontinuity between national laws and swidden agroforestry systems

- Legal blindness to existence of swidden systems and potential rationality and utility in governance (Chapters 1, 10)

New potential dangers for hinterland peoples from international sources

- Risks of exclusion linked to international encouragement of proliferation of official protected areas, all sites (Chapters 4, 8, 10)
- Possibility of significant carbon payments to unaccountable governments, with potential to expand protected areas, compel resettlement and restrict swiddens, all sites (Chapter 10)

The process of trying to accomplish our ambitious goals has, however, led to useful understandings and insights. We have identified recurring issues that are likely to impinge on any further efforts to work collaboratively with tropical forest communities and landscapes – governmental policies with complex, diverse and often unpredictable effects, varying interfaces between customary and formal legal systems, intriguing differences in the use and governance of non-timber forest products (NTFPs), and the potential even within collaborative governance for harm (win-win solutions are unlikely always to be an option, and many argue that trade-offs are the norm).

Table 11.1 summarizes six major issues and refers to the relevant field locations and the chapters that describe them.

Most of those issues demonstrate the global variation geographically and over time in contexts, peoples and regimes governing natural resources. Such diversity and dynamism – about which local residents are among the most knowledgeable experts – reinforce the desirability of (1) strengthening and supporting their involvement in their own governance; and (2) tailoring any interventions to the specificities of any locale. Indeed, implementation of the latter probably requires the implementation of the former. Such local involvement becomes even more central if we take seriously the historical antagonisms between governments and many hinterland peoples, discussed in Chapter 10. Formal governmental shortcomings strengthen the argument for stronger citizen involvement – to monitor and ultimately constrain such power.

In this concluding chapter, we first give 'the bad news' from our experience in trying to facilitate collaborative landscape governance. We then provide some more encouraging perspectives, concluding with a call to action, based on our perception that there really is no viable, known, alternative approach for governing tropical forests and their people.

The Bad News

We find ourselves squarely on the horns of what some might consider a predictable dilemma. Clearly, improved governance will mean changing the behaviour of actors at various scales (community, landscape or district, national and international). But the higher up one goes in this hierarchy of scales, the less leverage ordinary citizens (and researchers) have to foment or require change. In our Landscape Mosaics project and in other action research efforts, such power discrepancies have to be considered when setting goals, as they can reduce the efficacy of our efforts.

One often-proposed response to this reality is to empower local communities and/or individuals through formal democracy (see Ribot, 2004) as the means to balance such power – a process that has proceeded slowly, globally. Another, which we have tried to implement, is to use social learning, networking and collective action to contribute to a more level playing field. We know this approach can work. But perhaps it has worked best when actions were so local that they occurred in a first stage, below governmental 'radar'.

On the one hand, our actions, both as research teams and with communities, are subject to the constraints of state power, notably through often-burdensome procedures and complex bureaucracies. These can serve as a buffer when decision-makers want to ensure that empowered communities or researchers' findings will not create a future constraint for them. Our own actions can also precipitate changes of attitude and in some cases exacerbate power tensions. But all in all, governmental power, vis-à-vis national citizens, is obviously greater. Such 'rules of the game' have to be taken seriously in governance research and in the planning of action research activities, especially in terms of time. Whereas the risks for researchers are, finally, only temporary, risks for local communities are much more significant.[2] Members of hinterland communities are typically more aware of such dangers than are outsiders. If government officials do not like the decisions or choices that communities make in the course of an action research process, for instance, in many countries they can simply render local decisions illegal or impose their will, by fiat. We cannot expect rapid and important changes in this power relationship, but such behaviour highlights a central dilemma: to express their voices and follow collaborative decisions, local communities will have to be confident that joint decisions will remain valid for a sufficient length of time. Changes in behaviour are needed at all scales, including among the powerful people, who are least subject to persuasion and may also be inclined to modify decisions according to new constraints or opportunities emerging at higher levels. The process of making the changes needed to govern in ways that will protect the environment while enhancing human well-being is inevitably a messy one, which Wollenberg et al (2004) and Sayer et al (2008) have called 'muddling through', notably towards cooperation.

A second challenge relates to the actual conduct of research. To be effective, action research teams need to work at various governance levels simultaneously; they need to have skills from a variety of disciplines but also diplomacy specific to each level. They need long time periods to develop rapport with various groups of decision-makers and citizens.

We have found these requirements difficult to meet. Within the realm of social sciences alone, numerous skills are required. Understanding social patterns, norms and micro-politics is likely to demand ethnographic skills; dealing with government officials, understanding policy development processes and legal constraints and niceties are the skills of a political scientist or a policy analyst; facilitating equitable and effective community and district-level processes entails abilities in community organizing. The classical training of foresters, ecologists, anthropologists and political scientists – each of whose disciplinary skills are essential – rarely prepares researchers to recognize sufficiently the value and complexity of transdisciplinarity, especially of the social skills needed to understand and build confidence among a large variety of partners.

Hinterland areas are, by definition, difficult to reach, and working at the landscape scale introduced another significant problem. Remoteness yields inevitable logistical nightmares, often including exorbitant transport and time costs to get the work done. Such difficulties are magnified when long-term collaborative governance efforts are attempted. On the one hand, people with the right skills need to remain for long periods of time in communities, and on the other, someone with different skills needs to be in routine contact with higher-level policy-makers. Finally, the researcher(s) must link the information flowing from these two levels into a comprehensive dataset and coherent understanding, in one form or another.

Within the developing world, skilled researchers are often in short supply. They may be reluctant to live in or regularly visit communities (sometimes despite agreement to do so at the outset); or their skills may be so much in demand that they simply cannot practically spend long periods in the field. Whatever the reasons, the maintenance of qualified field personnel in communities for a sufficient time has been an often-unmet challenge.

Knowing that the typical funding cycle, two to four years, was insufficient to understand the local context, establish rapport with relevant stakeholders and facilitate a participatory approach from start to finish, we tried in all cases to identify partners who could follow up on the process. But most projects as well as official partners follow fixed agendas and lack the flexibility and time to reach the objectives generally proposed for these complex landscapes, so valuable for biodiversity conservation (cf Pfund, 2010).

The Somewhat Good News

Although the difficulties outlined above are real, a number of impressive changes have occurred with collaborative approaches. With sufficient time and facilitation, small groups of villagers have learned to analyse their own situations, develop plans to address their problems, implement and monitor those plans, and make course corrections. Women have gained the confidence to speak up in public meetings and share their goals, interests and knowledge with others, influencing important decisions. Ethnic groups whose voices were muted have been able to express their concerns and join more effectively in resource management. Communities have successfully held off powerful industries that offered inequitable terms of engagement, and they have learned ecological lessons that help them to manage their resources better. Such examples are many: they include Komarudin et al (2008) and collections by Colfer (2005), Vernooy (2006), Fisher et al (2007) and Mandondo et al (2008). We have seen sufficient growth in people's skills and confidence to lead us to believe that such empowerment, including facilitated dialogue between communities and landscape- or district-level actors can lead to real improvements in local governance. But, although our Landscape Mosaics work has contributed to such conclusions, it has not proved this belief.

We have seen the concerns of villagers (e.g. land tenure or protection of water resources) align with those of district or conservation officials (e.g. land-use or conservation planning) and in some cases with industry (e.g. contract farming agreements). Facilitation has often enhanced people's abilities to express their concerns about national or regional policies or procedures, sometimes adapting the rules to fit better within the local context and thereby benefit the community. But in all cases, the process has involved 'muddling through', never yet a 'straight shot' at success (however that may be defined). Indeed, the lack of agreement about what constitutes success in any context is a central argument for democracy and the muddling process itself, as well as for adaptive management.

In all our sites, there have been honest and committed actors at all scales (government, industry, conservation projects and NGOs), with whom we have worked. We see such actors, like empowered citizens and communities, as equally important keys to success at collaborative governance. Such actors can provide leadership, knowledge and networking that can open doors, maintain momentum and access to resources, and help bring processes to conclusion. These include agreements, land-use plans, bylaws, sanctions and mechanisms to enforce them, as well as negotiated settlements.[3] Such actors also represent potentially powerful forces for change within their own institutions. There is real power in setting a good example; a respected colleague's refusal to 'go along' can change norms and social dynamics among the powerful.

The Challenge

Our own conclusion is that this messy process – working and negotiating with communities and government officials to manage landscapes and govern collaboratively – is in fact our only choice in the search for healthy and productive landscapes and people. The task of finding consensus about ways to protect the environment and people's livelihoods must fall primarily to those who inhabit that environment and derive their livelihoods from it. Their motivation to perform this function will be strongly influenced by the degree to which they are satisfied that their formal governments are behind them and that governance is generally just and reasonably equitable.

Looming climatic changes lend greater urgency to the work of collaborative governance. Many people have noted that the expected consequences from climate change may disproportionately affect 'the poor', a category that encompasses most tropical forest dwellers. Real adaptation by forest peoples to such changes will require their genuine and meaningful contributions in developing and implementing coping strategies.

We know that wealth in places where the common people have little power can exacerbate corruption and inequity (Dove, 1993). The urgency of addressing this pattern has risen exponentially as global interest has grown in forest protection as a mechanism to tackle climate change. Reducing Emissions from Deforestation and Forest Degradation (REDD) and REDD+[4] include significant probable financial incentives, most likely in the hands of national governments. Despite the potentially positive results for forests and forest peoples, these initiatives also carry risks. The high level of corruption common in tropical forests, combined with the historic marginalization of forest peoples, suggests a need for both serious reflection and action. Brown et al (2008) agree:

> *By conferring new value on forest lands, REDD could create incentives for government and commercial interests to actively deny or passively ignore the rights of indigenous and other forest-dependent communities to access and control forest resources.* (p113)

Given the common attitudes towards 'uncivilized' forest peoples and 'destructive' swidden agriculture, particularly combined with opportunities for illicit financial gain at the national level, this outcome is a real danger.

Assessment, monitoring and verification can go part of the way towards avoiding such outcomes (Brown et al, 2008). Lessons from integrated conservation and development projects may contribute to better REDD efforts (see Blom et al 2010; Springate-Baginski and Wollenberg 2010). The recent overview of climate change and its links to rural peoples by Robledo et al (2010) provides a helpful context for this work. The governance assessment tool provided in Chapter 9 represents one useful approach.

However, real solutions will require the more substantial and proactive involvement of rural communities and district governments or landscape-level actors – as proposed in Locatelli et al (2008). The constraints identified above will have to be addressed fully. Longer-term and more substantial funding is needed. More genuinely interdisciplinary teams that represent social, political, ethnographic and facilitation skills must be convened to conduct or coordinate such efforts. Some personnel must remain in or regularly visit field sites and local governments. Governmental actors at all levels need to develop respectful and collaborative attitudes towards their own citizens and improve their record on downward accountability. Like donors, they must reduce the rigidity of their bureaucracies, enable their officials to take more risks and begin to learn from failures – integral parts of any adaptation process.

These are tall orders that require significant changes from all, including the powerful. We do not know whether such change is possible, given the long history of governance issues in the world and the yawning gulf between governments and hinterland peoples. But we have observed the creativity, energy and adaptability of rural peoples, which has given us hope. We believe that changes such as those listed above – difficult as they are – provide the keys to managing natural resources successfully while protecting people's livelihoods and cultural preferences.

Notes

1. The glossary of terms is available in Appendix 1.1.
2. In another CIFOR action project in Zimbabwe and Nepal, such risks included death threats.
3. This discussion implies a differentiation between communities and the other sectors mentioned that may mislead. Certainly members of communities can also perform these leadership, networking and knowledge-related functions, but usually on a smaller scale.

References

Blom, B., Sunderland, T. and Murdiyarso, D. (2010) 'Getting REDD to Work Locally: Lessons Learned From Integrated Conservation and Development Projects',

Environmental Science and Policy, available at http://www.sciencedirect.com/science?_ob=MImg&_imagekey=B6VP6-4YDTS1G-1-1&_cdi=6198&_user=492137&_pii=S1462901110000043&_orig=search&_coverDate=02%2F18%2F2010&_sk=999999999&view=c&wchp=dGLbVlb-zSkWz&md5=f2a3b1628bbb0b1e34c5aea3ea4e3833&ie=/sdarticle.pdf

Brown, D., Seymour, F. and Peskett, L. (2008) 'How do we Achieve REDD Co-benefits and Avoid Doing Harm?, in A. Angelsen (ed) *Moving Ahead with REDD: Issues, Options and Implications,* CIFOR, Bogor, pp107–118

Colfer, C. J. P. (ed) (2005) *The Equitable Forest: Diversity, Community and Natural Resources,* RFF/CIFOR, Washington, DC

Dove, M. (1993) 'A Revisionist View of Tropical Deforestation and Development', *Environmental Conservation,* vol 20, no 1, pp17–55

Fisher, R. J., Prabhu, R. and McDougall, C. (eds) (2007) *Adaptive Collaborative Management of Community Forests in Asia: Experience from Nepal, Indonesia and the Philippines,* CIFOR, Bogor

Komarudin, H., Siagian, Y. L., Colfer, C. J. P., with Neldysavrino, Yentirizal, Syamsuddin and Irawan, D. (2008) *Collective Action to Secure Property Rights for the Poor: A Case Study in Jambi Province, Indonesia,* Collective Action and Property Rights System-Wide Initiative, Washington, DC

Locatelli, B., Kanninen, M., Brockhaus, M., Colfer, C. J. P., Murdiyarso, D. and Santoso, H. (2008) *Facing an Uncertain Future: How Forests and People Can Adapt to Climate Change,* CIFOR, Bogor

Mandondo, A., Prabhu, R. and Matose, F. (eds) (2008) *Coping Amidst Chaos: Studies on Adaptive Co-management from Zimbabwe,* CIFOR, Bogor

Pfund, J.-L. (2010) 'Landscape-scale Research for Conservation and Development in the Tropics: Fighting Persisting Challenges', *Current Opinion in Environmental Sustainability,* vol 2, pp117–126

Ribot, J. C. (2004) 'Waiting for Democracy: The Politics of Choice in Natural Resource Decentralization', Report for World Resources Institute, Washington, DC

Robledo, C., Blaser, J. and Byrne, S. (2010) Climate Change: What are its Implications for Forest Governance?, in L. German, A. Karsenty and A. M. Tiani (eds) *Governing Africa's Forests in a Globalized World,* Earthscan/CIFOR, London, pp355–376

Sayer J. L., Bull, G. and Elliott, C. (2008) 'Mediating Forest Transitions: 'Grand Design' or "Muddling Through"', *Conservation and Society,* vol 6, no 4, pp320–32

Springate-Baginski, O. and Wollenberg, E. (2010) *REDD, Forest Governance and Rural Livelihoods: The Emerging Agenda,* CIFOR, Bogor

Vernooy, R. (ed) (2006) *Social and Gender Analysis in Natural Resource Management: Learning Studies and Lessons from Asia,* Sage Publications, London

Wollenberg, E., Iwan, R., Limberg, G., Moeliono, M., Rhee, S. and Sudana, I. M. (2004) 'Muddling Towards Cooperation: A CIFOR Case Study of Shared Learning in Malinau District, Indonesia', *Currents,* vol 33, pp20–24

Index

Adaptive Collaborative Management (ACM) 43, 45–46, 219
African Development Bank, resettlement policy 81
agriculture
  forest impacts 133
  see also swidden agriculture
ala fady (taboos) (Madagascar) 141–144, 151
Allanblackia stuhlmannii see msambu
Amani Nature Reserve (Tanzania) 22
Ambofampana (Madagascar)
  forest use 139–150
  Landscape Mosaics site 20, 137–138
Ambohimarina (Madagascar), Landscape Mosaics site 20
analytic hierarchy process (AHP) 45
Antsahabe (Madagascar)
  forest use 139–150
  Landscape Mosaics site 137–138
Anyang ethnic group (Cameroon) 186, 256, 257
Assam (Cameroon), Landscape Mosaics site 10–11
Association Intercooperation Madagascar (AIM) 5

bamboo, Laos 174–175
Betseleo ethnic group (Madagascar) 257
Betsimisaraka ethnic group (Madagascar) 20, 137–150, 257
Bevalaina (Madagascar)
  forest use 139–150
  Landscape Mosaics site 20, 137–138
biodiversity loss, forests 36
Bouammi (Laos)
  Landscape Mosaics site 17
  resettlement 253
Bungo District (Indonesia)
  district government 55–75
  governance assessment 224–230
  Landscape Mosaics sites 12–15
  state governance 245–246
bush mango (*Irvingia* spp.), Cameroon 164–165, 188–190, 193–210
bush pepper (*Piper guineensis*), Cameroon 195, 198, 202, 207
bush pigs (*Potamochoerus porcus*), Tanzania 159–160

Cameroon
  bush mango (*Irvingia* spp.) 164–165, 188–190, 193–210
  corruption 201, 210n8, 239–243
  customary structures 201–205
  decentralization 236–239
  eru (*Gnetum* spp.) 171–172, 190–192, 193–210
  ethnic groups 257
  forest classifications 199
  Forest Law (1994) 190, 205, 244
  forest management 198–210
  gender differences 256
  gorillas 9, 158–159
  government institutions 198–201
  Landscape Mosaics sites 7–12
  national policy linkages 47
  non-timber forest products (NTFPs) 183–210
  participation 251–252
  resettlement policies 251
  Ricinodendron heudelotii 192–193, 198, 207
  state land control 244–245

swidden agriculture 251
cardamom, Tanzania 22, 23, 96, 98, 119, 254
cat's-eye (*Dimocarpus longan*), Indonesia 173
cattle stick (*Carpolobia* spp.), Cameroon 194, 195, 197–198, 202, 203, 207, 209
Center for International Forestry Research (CIFOR) 26n4, 45, 47, 56, 75–76n3, 82
Centre International de Recherche en Agronomie du Développement (CIRAD) 56
Chama Cha Mapinduzi (CCM) (Party of the Revolution) (Tanzania) 111
citizen participation 220–221
climate change 276–277
Collective Action and Property Rights (CAPRi) 46, 50n1
co-management, governance 43–44
commercialization, non-timber forest products (NTFPs) 162–167, 178
Communauté de Base (COBA) (Madagascar) 134, 151–152
communications, remote communities 5
Community Based Forest Management (CBFM), Tanzania 113–114
community-level governance 39–40
  Cameroon 203–204
  Madagascar 134
  Tanzania 107–129
compensation, Derema Corridor (Tanzania) 95–99
conflicts
  Indonesia 60, 70–73, 74–75, 229
  Madagascar 149
conservation
  landscape approaches 44–45
  Tanzania 118
corruption
  Cameroon 201, 210n8
  governance 239–243
  Indonesia 219
Cross River gorillas (*Gorilla gorilla diehli*), Cameroon 9, 158–159, 161, 186
cultural values, Cameroon 205–206
customary law
  Cameroon 201–205
  Indonesia 64
  Tanzania 107–110, 114–118, 126–128
customary rights, Madagascar 139–146, 148–150

decentralization
  governance 42–43, 236–239
  Indonesia 60–61
  Madagascar 134
  Tanzania 110–114
deforestation
  Madagascar 133–134
  see also forest fragmentation
Department of Planning (BAPPEDA) (Indonesia) 59, 62–63, 66–67, 74
Derema Corridor (Tanzania), resettlement 80, 90–101, 255
Desa Danau (Indonesia)
  Landscape Mosaics site 14–15
  transmigration 252
displacement *see* resettlement
district government, Indonesia 55–75
Drongo bird (Madagascar) 144
durian (*Durio zibethinus*), Indonesia 14, 172–173

East Usambara Catchment Forest Project (EUCFP) (Tanzania) 118
East Usambara Conservation Area Management Programme (EUCAMP) (Tanzania) 92, 94, 96, 118
East Usambara Forest Landscape Restoration Project 122
East Usambara Mountains (Tanzania)
  gender differences 256
  Landscape Mosaics sites 21–24
  local management systems 107–129
  resettlement 90–102
ebony (*Diospyros*), Madagascar 170
ékpe sacred societies, Cameroon 201–202, 205, 256
eru (*Gnetum* spp.), Cameroon 171–172, 190–192, 193–210
ethnic groups
  inequalities 257–258
  Laos 16

fady (taboos) (Madagascar) 140–144, 151
Ferguson, J. 48
firewood collection, Tanzania 126
fishing, Laos 175–176, 177
food security, subsistence species 171–177
forest fragmentation 138–139
   Madagascar 141–142, 146–147, 150–152
forest management
   Cameroon 198–210
   Indonesia 55–75
   Madagascar 133–152
   Tanzania 92
forest rights
   Madagascar 139–142
   Tanzania 122–125

gender differences
   bamboo collection 174
   food collection 177
   inequities 256–257
   Madagascar 145–146
   non-timber forest products (NTFPs) 168, 257
   see also women
global governance 40–41
Gnetum spp. *see* eru
gorillas *see* Cross River gorillas
governance
   assessment 217–230
   co-management 43–44
   community-based 39–40, 107–129
   corruption 239–243
   critical perspectives 48–49
   decentralization 42–43, 236–239
   definition 6, 26n1
   hinterland 234–236, 260n3
   improvement 220–221, 271–277
   multilevel 38–44
   national and global level 40–41
   participation 220–221, 250–251
   problems 233–258
   stakeholder relationships 219
   state 236–243
   structure and function 220
governance assessment tool (GAT) 221–230

government institutions, Cameroon 198–201

hinterlands
   collaborative governance 258–259
   gender differences 256–257
   governance 234–236, 260n3
   resettlement policies 250
   social inequalities 257–258
   see also remote communities
Human Development Index 6–7, 9
hunting
   gorillas 158–159, 161
   wild pigs 159–162
hutan desa (village forests) (Indonesia) 47, 67

Indonesia
   conflicts 60, 70–73, 74–75, 229
   corruption 219, 239–243
   decentralization 236–239
   district government 55–75
   ethnic groups 257
   gender differences 256
   governance assessment 224–230
   improved governance 221
   Landscape Mosaics sites 12–15
   national policy linkages 47
   participation 252–253
   resettlement 252
   state land control 245–246
   swidden agriculture 252
   transmigration programmes 57–58, 59, 69–70, 71–73, 75n1, 252
   wild fruits 172–173
   wild pigs 160–161
inequalities
   gender differences 256–257
   social groups 257–258
Institut de Recherche pour le Développement (IRD) 56
institutions, landscape-level 45–46
integrated approaches, landscape governance 44–45
Irvingia spp. *see* bush mango

jackfruit (*Artocarpus integra*), Indonesia 173

Jambi Province (Indonesia)
  district government 55–75
  governance assessment 224–229
  Landscape Mosaics sites 12–15, 46, 56
jengkol (*Archidendron jiringa*), Indonesia 173

kabau (*Archidendron bubalinum*), Indonesia 173
Kagwene Gorilla Sanctuary (Cameroon) 9, 186
Kilindi leaders (Tanzania) 114–115, 127
KoloAla programme, Manompana (Madagascar) 5, 19, 47, 134–135, 148–149, 150–152, 247–248, 254
Kuamang Kuning (Indonesia) 69–70
kumban (village cluster) (Laos) 16, 47, 238
Kwatango (Tanzania)
  agroforestry 254
  forest rights 122–125
  land rights 119–122
  Landscape Mosaics site 23, 110, 119
  rights to trees on farms 126
Kwezimagati Forest Reserve (Tanzania) 23
Kwezitu village (Tanzania), resettlement 91–92

landownership, legal framework 243–249
land rights
  customary 111–112
  Tanzania 119–122
landscape ecology, definition 36–37
landscape governance
  improving 271–277
  integrated approaches 44–45
  see also governance
Landscape Mosaics
  challenges 276–277
  governance study 6
  policy process links 47
  population statistics 7, 9
  problems 274–275
  project design 4–6, 46–47
  sites 6–24
  size 37, 38

landscapes, multifunctional 35–38
land-use planning 45, 243–249
Laos
  bamboo 174–175
  corruption 239–243
  decentralization 236–239
  ethnic groups 16, 257
  fishing 175–176, 177
  gender differences 256
  Land and Forest Allocation (LFA) 242, 246–247
  Landscape Mosaics sites 15–18
  national policy linkages 47
  participation 253
  peuak meuak (*Boehmeria malabarica*) 163–164, 168, 177, 178
  resettlement 80, 83–90, 99–101, 253
  state land control 246–247
  swidden agriculture 253
Launaea cornuta see mchunga
literacy 254, 261n16
Lowe, Celia 48–49
Lubuk Beringin (Indonesia), Landscape Mosaics site 13–14

Madagascar
  conflicts 149
  corruption 239–243
  customary rights 139–146
  decentralization 236–239
  ethnic groups 257
  forest management 133–152
  gender differences 145–146, 256
  La Gestion Contractualisée des Forêts (GCF) 247
  La Gestion Locale Sécurisée (GELOSE) 134, 247, 254
  Landscape Mosaics sites 18–20
  national policy linkages 47
  Pandanus 139, 143, 145–146, 151–152, 167
  participation 254
  resettlement 254
  state land control 247–248
  swidden agriculture 254
  timber products 144–145, 170
makwo sacred society, Cameroon 201–202, 205, 256
management, definition 26n2

Manompana (Madagascar)
  forest management 133–152
  KoloAla programme 5, 19, 47, 134–135, 148–149, 150–152, 247–248, 254
  Landscape Mosaics sites 18–20
  resettlement 254
Maromitety (Madagascar)
  forest use 139–150
  Landscape Mosaics site 20, 137–138
  resettlement 254
mchunga (*Launaea cornuta*), Tanzania 173–174, 177
megafauna, management strategies 158–162
Merina ethnic group (Madagascar) 18, 257
migration *see* transmigration
Milicia excelsa see mvule
Misalai (Tanzania)
  forest rights 122–125
  land rights 119–122
  Landscape Mosaics site 23, 110, 119
  rights to trees on farms 126
Mpisikidy (Madagascar) 140–141, 143
msambu (*Allanblackia stuhlmannii*), Tanzania 23, 24, 92, 165–166, 168
Msasa IBC (Tanzania), resettlement 91–98
Muangmuay (Laos), Landscape Mosaics site 17–18
Mukonyong (Cameroon), Landscape Mosaics site 11–12
multicriteria analysis 45
multidisciplinary landscape assessments 45
multifunctional landscapes
  definition 36–38
  governance 35, 38–44
mvule (*Milicia excelsa*), Tanzania 125, 169–170

Nam-et Phou Louey National Protected Area (Laos) 16, 17, 83, 253
nanto (*Capurodendron*), Madagascar 170
National Agriculture and Forestry Research Institute (NAFRI) (Laos) 82, 88–89

national-level governance 40–41
Nepal
  collaborative management 46
  good governance 220–221
Nigeria, NTFP trade 191–192, 202–203, 204–205
non-governmental organizations (NGOs)
  governance role 38, 102, 249, 259
  Indonesia 56, 59, 66–67, 74
  Tanzania 118, 127, 165
non-timber forest products (NTFPs)
  Cameroon 183–210
  commercialization 162–169, 178, 186–188
  definition 210n1
  gender roles 168, 257
  governance 196–206, 207–210
  Madagascar 142–144
  socio-economic impacts 193–195
  subsistence foods 171–175
  sustainability 206–207
  trade 183–184, 186–188, 208–209
  value chains 183–210
Northern Agriculture and Forestry Extension Center (NAFReC) (Laos) 82, 89
Novella Africa Initiative 165
Nucleus Estates and Smallholders project (NES) (Indonesia) 69–70

oil palm plantations, Indonesia 57–59, 71–73
Okpambe (Cameroon), Landscape Mosaics site 11
Operation vijiji (Tanzania) 109, 112
opium cultivation, Laos 83, 88, 253

Palissandre (*Dalbergia* sp.), Madagascar 248
panarchy 44
Pandanus, Madagascar 139, 143, 145–146, 151–152, 167
participation
  Cameroon 251–252
  governance 220–221, 250–251
  Indonesia 252–253
  Laos 253
  Madagascar 254
  Tanzania 255

participatory action research (PAR) 2–3, 4, 45–46
payments for environmental services (PES) 217, 235
perceptions of governance, Indonesia 64–65
Perkebunan Inti Rakyat project (PIR) (Indonesia) 69
pests, wild animals 158–162
petai (*Parkia speciosa*), Indonesia 14, 172–173
peuak meuak (*Boehmeria malabarica*), Laos 163–164, 168, 177, 178
Phadeng (Laos)
 Landscape Mosaics site 17
 resettlement 80, 83–90, 253
Phoukhong (Laos), resettlement village 85–90
Phousaly (Laos), resettlement village 86–90, 253
pigs
 Indonesia 160–162
 Tanzania 159–160
plants, commercialization 162–167
population figures 7, 9

rattan, Indonesia 173
Reducing Emissions from Deforestation and Forest Degradation (REDD) 3, 217, 235, 260n4, 276–277
 Cameroon 186, 210
 Indonesia 67
remote communities
 governance 234–235
 study of 4–5, 274–275
 see also hinterlands
resettlement 79–102, 250
 Cameroon 251
 causes 81–82
 Indonesia 252
 Laos 80, 83–90, 99–101, 253
 Madagascar 254
 Tanzania 80, 90–101, 255
 see also transmigration programmes
resource access 219–220
resource management
 community-based 39–40, 107–129

integrated approaches 44–45
non-timber forest products (NTFPs) 209–210
rice production
 Indonesia 68
 Madagascar 133, 146–147
Ricinodendron heudelotii, Cameroon 192–195, 198, 207
Rights and Resources Initiative 3
Roe, E. M. 48
rosewood (*Dalbergia*), Madagascar 170
rubber plantations, Indonesia 57–58, 68

sanam (hamlet) (Laos) 89–90
sandrana (prohibitions) (Madagascar) 142–144, 151
Scott, James C. 48, 234–235, 249–250
Shambaa ethnic group (Tanzania) 21, 91, 109, 114–126, 159–160
Shambangeda (Tanzania), Landscape Mosaics site 24
shifting cultivation
 Laos 83
 Madagascar 133
social groups *see* ethnic groups
social impact assessment (SIA), Tanzania 94–95
stakeholders, relationships 219
state, governance 236–243
subsistence species, food security 171–177
sustainable livelihood approach, Madagascar 136
swidden agriculture 249–250, 261n14
 Cameroon 251
 Indonesia 252
 Laos 253
 Madagascar 254
 Tanzania 254

taboos, Madagascar 140–144
Takamanda-Mone (Cameroon)
 bush mango (*Irvingia* spp.) 164–165, 188–190, 193–210
 Landscape Mosaics sites 7–12
 land-use planning 244–245
 non-timber forest products (NTFPs) 183–210

Takamanda-Mone Technical Operations Unit (TOU) 9, 47, 184–186, 198–201
Takamanda National Park (Cameroon) 186, 200, 205–206, 244–245, 251
Tanzania
　bush pigs (*Potamochoerus porcus*) 159–160
　corruption 239–243
　customary law 107–110, 114–118, 126–128
　decentralization 110–114, 236–239
　ethnic groups 257–258
　gender differences 256
　Landscape Mosaics sites 21–24
　local management systems 107–129
　mchunga (*Launaea cornuta*) 173–174, 177
　msambu (*Allanblackia stuhlmannii*) 23, 24, 92, 165–166, 168
　national policy linkages 47
　participation 255
　precolonial period 114–117
　resettlement 80, 90–101, 255
　state land control 248–249
　swidden agriculture 254
　timber species 169–170
　Ujama policy 249, 255, 260n13
　Village Land Act (1999) 22, 111–112, 120, 239, 248
Tanzania Forest Conservation Group (TFCG) 22, 118, 122–123
tavy (mountain rice) (Madagascar) 133, 146–147
Tebing Tinggi (Indonesia)
　Landscape Mosaics site 14
　transmigration 252
timber products
　Madagascar 144–145, 170
　Tanzania 169–170

trade, non-timber forest products (NTFPs) 183–184, 186–188, 208–209
transmigration programmes
　Indonesia 57–58, 59, 69–70, 71–73, 75n1, 252
　see also resettlement
Transparency International 239–240
trees
　rights 116, 126, 171
　see also timber products
Tsing, Anna Lowenhaupt 48–49

value chains, non-timber forest products (NTFPs) 183–210
Viengkham District (Laos), Landscape Mosaics sites 15–18
village governance 39–40
　Indonesia 62–63
　Tanzania 110–114

Warung Konservasi (WARSI) (Indonesia) 66–67, 75–76n3
water resource management 44
'wicked problems' 3, 26n3
wild fruits, Indonesia 172–173
wild pigs, Indonesia 160–161
wild species, management strategies 157–179
women
　land rights 121–122
　Madagascar 145–146, 151–152, 167
　see also gender differences
World Agroforestry Center (ICRAF) 26n4, 50n1, 56, 75–76n3
World Bank, compensation payments 97–98
World Resources Institute (WRI), governance assessment tool 221
World Wide Fund for Nature (WWF), Tanzania 97–98, 118, 248

Zomia 234–235, 249–250, 260n1